LASER LIGHT SCATTERING

LASER LIGHT SCATTERING

Basic Principles and Practice
Second Edition

Benjamin Chu
Department of Chemistry
Stony Brook University

DOVER PUBLICATIONS, INC.
MINEOLA, NEW YORK

Bibliographical Note

This Dover edition, first published in 2007, is an unabridged republication
of the second edition, published by Academic Press, San Diego, California,
1991. The author has prepared a new preface especially for the Dover edition.

Library of Congress Cataloging-in-Publication Data

Chu, Benjamin.
    Laser light scattering : basic principles and practice / Benjamin Chu. —
2nd ed.
        p. cm.
    "This Dover edition, first published in 2007, is an unabridged republi-
cation of the second edition, published by Academic Press, San Diego,
California, 1991."
    Includes bibliographical references and index.
    ISBN 0-486-45798-2 (pbk.)
        1. Laser beams—Scattering. I. Title.

QC446.2.C49 2007
621.36'6—dc22

2007000975

Manufactured in the United States of America
Dover Publications, Inc., 31 East 2nd Street, Mineola, N.Y. 11501

CONTENTS

PREFACE

More than fifteen years have passed since I wrote the first edition of *Laser Light Scattering* in 1974. The technique has matured, and dynamic light scattering has become one of the routine analytical tools used in particle sizing. There have been diversified and important developments on its applications to many fields in biology, chemistry, engineering, medicine, and physics. However, the basic principles of intensity correlation-function measurements have remained essentially unchanged. The original aims of the second edition were to try to delete portions of the book, such as Chapter 2 in the first edition, which were difficult to follow for the readers without formal training in electrodynamics, and to update the optical arrangements and correlator schemes commonly used in laser light scattering. This exercise turned out to be much more difficult than I originally anticipated. There was always a conflict of approach: whether to present a qualitative but less rigorous answer or a quantitative but more difficult one.

The subject matter concentrates exclusively on the technical aspects of laser light scattering, including the basic principles and practice. Only the application to dilute polymer solution characterizations is included in some detail, leaving the diverse developments on applications to many fields for other texts (see Chapter 1) and selected papers (see Appendix in Chapter 8). In introducing the Laplace inversion in Chapter 7, the level of discussions unavoidably becomes more mathematical. Thus, the reader who is not interested in the details should skip some of the sections.

I am grateful to many of my coworkers and colleagues who have contributed to this second edition, and to M. Adam, S. H. Chen, T. Lodge, H. S. Dhadwal, and G. Patterson for reading the manuscript and making valuable comments. The thoughtful criticisms by Patterson and Dhadwal were particularly appreciated.

Benjamin Chu

PREFACE TO THE DOVER EDITION

The first edition of *Laser Light Scattering* (1974) was intended to convey to the reader the fundamentals of experimental practice. The author went through a tortuous path in learning what the principles of laser light scattering are, and how they could be used in practice. Thus, one of the aims of the book was to lead the reader through an easier pathway in the learning process. Even with kind advice from such great masters as Debye and Zimm, and such contemporary experts as Benedek, many chapters in the first edition, based on the Maxwell equations, were essential to basic understanding of the physics of laser light scattering. However, they masked the essentials for those mainly interested in performing laser light-scattering experiments. Thus, the second edition (1991) emphasized optics, cell design, and detection for light-scattering experiments. The author is pleased to have the 1991 book available again in this new Dover edition.

In the intervening sixteen years since 1991, new developments in digital electronics, diode lasers, and micro-fluidics have changed the field. The reader may be interested to learn where the 1991 book stands in terms of those recent advances. It is necessary to mention first that while improvements have miniaturized the instrumentation, making the laser source more reliable and the detector more sensitive, the approach and basic experimental practices have remained essentially unchanged. The updates can be summarized as follows:

The fundamentals of laser light scattering, using the Maxwell equations as a foundation, remain the same. The principles of coherence consideration, the optical designs, in terms of measurements on the angular dependence of scattered light, of intensity fluctuations, and of depolarization, follow the same pathway. Thus, in this respect, the 1991 book remains current. What are the changes in instrumentation the reader should be aware of? These are 1) digital electronics; 2) diode lasers; and 3) micro-fluidics.

1) Digital electronics: The advances in digital electronics are such that we can simply use software with a personal computer

to calculate the desired time-correlation function. Hard-wired digital computers, designed to carry out specific time-correlation function computations *in situ*, known as digital correlators, can be reduced in size, often to less than that of a matchbox.

2) Diode lasers: Instead of He-Ne or argon-ion lasers, much smaller and more reliable diode lasers can be used as the incident light source.

3) Micro-fluidics: With specific designs in modules for dilution, mixing of solutions, filtration, and cell cleaning by means of micro-fluidic channels and valves, the easier removal of unwanted dust particles and the ability to handle small amounts of solutions could make laser light scattering into a powerful technique, especially for biological and industrial applications.

Finally, although avalanche photodiodes have been used since 1991, recent advances, including better temperature control, smaller packing and improved after-pulsing characteristics, should make such devices worthy alternatives to photomultiplier tubes.

The combination of the above factors makes miniaturization of a light scattering instrument a useful and powerful tool that can help make it more practical to elucidate the size, shape, and molecular conformation of macromolecules in solution, and colloidal particles in suspension, especially for experts in biological, chemical, and medical sciences and engineering.

With synchrotron X-rays and forthcoming X-ray lasers, similar scattering measurements over larger ranges on the magnitude of the momentum transfer vector ($q = (4\pi/\lambda)\cdot\sin(\theta/2)$ with λ and θ being the wavelength of electromagnetic radiation in the scattering medium and the scattering angle, respectively) will be the next logical development. In addition, we can use cross-correlation function techniques to eliminate after-pulsing or to extract single scattering information in the presence of multiple scattering. While some of these new developments were not discussed in the 1991 book, they will be available in forthcoming books on laser light and X-ray scattering.

Benjamin Chu
January 2007

I

INTRODUCTION

Successful applications of scattering techniques, such as light scattering and neutron scattering, often involve a considerable amount of interdisciplinary knowledge. Chemists and physicists have utilized light scattering, small angle x-ray scattering (SAXS) and small-angle neutron scattering (SANS) to study the size and shape of macromolecules in solution as well as a whole range of materials including colloidal suspensions and solid polymers. Meteorologists have used microwaves to observe the scattering by rain, snow, hail, and other objects in the atmosphere, while astrophysicists have been interested in the scattering of starlight by interplanetary and interstellar dust. The same basic scattering principles govern all such phenomena. Wave interference yields information on particle size whenever the wavelength of the electromagnetic radiation (or the de Broglie wavelength of neutrons) is of the same order of magnitude as the size of the scatterer. Before the advent of lasers, most scattering studies, as we have discussed above, were concerned with the time-averaged scattered intensity.

Historically there have been two different approaches to light-scattering theory: scattering theory based on solving the Maxwell equations of a single particle of well-defined geometry, and that based on fluctuation phenomena. In particle scattering, we can consult the classical text by Van de Hulst (1957) and the comprehensive book by Kerker (1969) on "The Scattering of Light and Other Electromagnetic Radiation." Another book, edited by Huglin

(1972), contains extensive descriptions of many aspects of light scattering from polymer solutions. Huglin (1977) has since reviewed the practice of light scattering intensity measurements.

In condensed media or whenever the scatterers are close to one another, a detailed computation of the induced electromagnetic field surrounding a particle becomes very complex because intermolecular interactions have to be taken into account. Einstein (1910) was able to bypass the difficulties inherent in particle-scattering analysis in the presence of interactions. He assumed that local density fluctuations in neighboring volume elements could be independent of one another, and carried out a quantitative calculation of the mean squared amplitude of those fluctuations. Although Einstein's theory was able to explain the scattering from pure liquids and to predict the enormous increase in the scattering as the liquid–gas critical point was approached (the so-called *critical opalescence*), it failed to account for the angular dissymmetry of the strong scattered intensity in critically opalescent systems. Later, Ornstein and Zernike (1914, 1915, 1916, 1926) tried to include the effects of correlation between fluctuations of neighboring volume elements. Again, there have been extensive reviews (e.g. see Chu, 1967) on the fluctuation theory.

A typical light-scattering geometry is shown in Fig. 1.1(a), where the incident beam (I_{INC}) impinges on the scattering medium (dots), with most of the radiation being transmitted (I_t) and a small portion being scattered (I_s). In laser light scattering (Chu, 1968), we study not only the changes in the number (intensity) and the direction (momentum) of each type of photon in the incident and the emerging light beams, but also the corresponding frequency (energy) changes. Whereas the scattered intensity can be related to the structure of the particles, the optical spectra reveal the dynamical motions of particles. The word "laser" is included in the title in order to emphasize its essential role in obtaining effective measurements of optical spectra where wavefront matching of scattered light must be accomplished. Thus, a laser light source is needed in dynamic light scattering. Furthermore, "laser light scattering" is retained as the title so as to remind us of the importance of time-averaged light-scattering intensity measurements.

The scattering process that will be discussed in this book can be described classically and is confined to using lasers as a light source. Quantum-mechanical phenomena, such as the Raman effect, are excluded. There now exist many reviews, books, and proceedings, all of which cover one or more aspects of topics related to laser light scattering. A listing is provided in Appendix 1. A of this chapter.

In view of the extensive range of books available on laser light scattering, we shall give only pertinent references that are intended as guides to the interested reader or that have historical significance. The problem with the available

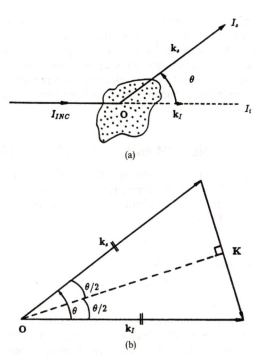

(a)

(b)

FIG. 1.1.   (a) Scattering geometry. I_{INC}, I_s, and I_t are, respectively, incident, scattered, and transmitted intensity. θ is the scattering angle. The magnitude of the scattering vector (**k**) for the incident beam is $k_I = 2\pi/\lambda_I$ and $\lambda_I \simeq \lambda_s = \lambda_0/n$ with λ_0 and n being the wavelength *in vacuo* and the refractive index of the scattering medium, respectively. (b) Conservation of momentum (Eq. (1.1)) in the scattering process. From the law of cosines and $|\mathbf{k}_I| \simeq |\mathbf{k}_s| = (2\pi n/\lambda_0)$, we have $K^2 = |\mathbf{k}_I - \mathbf{k}_s|^2 = k_I^2 + k_s^2 - 2\mathbf{k}_I \cdot \mathbf{k}_s = 2k_I^2 - 2k_I^2 \cos\theta = 4k_I^2 \sin^2(\theta/2)$, or $K = (4\pi/\lambda_I)\sin(\theta/2)$. This diagram represents a simple derivation for the magnitude of **K**.

books and proceedings, many of which are excellent, is that it remains somewhat difficult for uninitiated readers, especially those who are less physics-oriented, to utilize laser light scattering for their applications. Thus, in this book, I hope to cover some aspects of basic light-scattering theory and practice, leaving the advanced discussions to the monographs, proceedings, and research articles. More specifically, this book is directed primarily toward the experimental and technical aspects of laser light scattering, with emphasis on how such experiments can be performed (e.g., see Chapter VI), while the books by Berne and Pecora (1976) and Schmitz (1990) deal with the relations between the measured parameters and the molecular quantities of interest.

In the remaining portion of this introduction, the inter relationship between laser light scattering and other types of scattering techniques which use x-rays and neutrons will be discussed. We want to get a general idea of how to find appropriate scattering techniques for studying structural and dynamical properties of the system of interest. It is clear that each physical technique has its limitations. The important point is to learn when it becomes useful and how it works.

1.1. MOMENTUM AND ENERGY TRANSFERS

In using thermal neutrons or photons, the probing radiation is usually weakly coupled to the system of interest. Interaction of radiation with matter can be expressed in terms of two characteristic parameters, namely, the momentum transfer ($\hbar K$) obeying

$$\hbar \mathbf{K} = \hbar(\mathbf{k}_I - \mathbf{k}_s),\qquad(1.1.1)$$

and the energy transfer ($\hbar \Delta\omega$) obeying

$$\hbar\Delta\omega = \hbar(\omega_I - \omega_s),\qquad(1.1.2)$$

where $\mathbf{k}_I$ ($\mathbf{k}_s$) and ω_I (ω_s) are, respectively, the incident (scattered) wave vectors and angular frequencies as shown in Fig. 1.1. The incident and scattered wave vectors $\mathbf{k}_I$ and $\mathbf{k}_s$ are in the directions of the incident and scattered beams, respectively. Equations (1.1.1) and (1.1.2) are known, respectively, as the momentum and energy conservation equations. They also imply that the scattered light from a well-defined incident beam ($\hbar\mathbf{k}_I, \hbar\omega_I$) has a distribution of wave vectors and that, at each wave vector $\mathbf{k}_s$, there can be a distribution of frequencies depending on the dynamics in the scattering medium. As the energy transfers in terms of frequency changes ($\Delta\omega$) depend on the scattering dynamics, the scattered intensity fluctuates as a function of time.

The transfer of momentum ($\hbar\mathbf{K}$) corresponds to probing the structures of the system with a spatial resolution $R \sim K^{-1}$. Measurements of the time-averaged scattered intensity $\langle I(K)\rangle$ can be related to structural studies and are known as classical light-scattering measurements. Measurements of the time (or frequency) dependence of the scattered intensity $I(\mathbf{K}, \Delta\omega)$ as a function of $\mathbf{K}$ constitute dynamic light-scattering measurements. The subject matter of this book includes both classical and dynamic light scattering, with emphasis on the latter. Appendix 1.B of this chapter lists some typical spatial and temporal correlation ranges based on Eqs. (1.1.1) and (1.1.2).

The ranges covered by different complementary scattering techniques are summarized in Fig. 1.1.1. The figure shows typical regions of momentum- and energy-transfer space covered by different types of scattering techniques. Using visible light, it clearly illustrates that optical scattering (i.e., light scat-

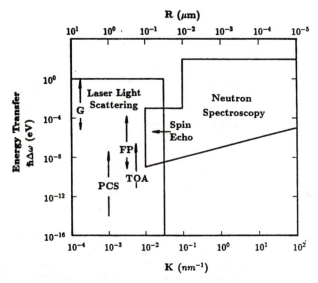

Fɪɢ. 1.1.1. Momentum- and energy-transfer space diagram showing regions covered by scattering techniques. The operating energy domains of optical mixing spectrometers are illustrated more explicitly in Fig. 1.1.2. G: grating instrument; FP; Fabry–Perot interferometry; TOA: time-of-arrival photoelectron statistics; PCS: photon correlation spectroscopy, which is the preferred technique for optical mixing spectroscopy.The ends of the arrows suggest approximate limits of energy transfer by various techniques.

tering) experiments follow the energy-transfer axis and involve only small momentum transfers (i.e., small K-values). Extension of light scattering measurements to cover larger momentum transfers will be limited even if the incident wavelength (λ_i) is extended to the ultraviolet region. On the other hand, optical mixing spectroscopy (see Fig. 1.1.2) enables us to observe energy changes that are many orders of magnitude smaller than those detectable by other spectroscopic methods.

By introducing the time-of-arrival (TOA) photoelectron statistics (Dhadwal et al., 1987), the experimental gap along the energy-transfer axis between photon correlation spectroscopy (PCS) and optical Fabry–Perot interferometry (FP) was filled. Energy changes that should be within reach, in principle, between PCS and FP were difficult to achieve in practice because it was difficult to perform PCS experiments with sample (or delay) time increments < 50 nsec or FP interferometry with a resolution ~ 1 MHz. Here, the uninitiated reader may feel uncomfortable with abbreviations such as TOA, PCS, and FP or expressions such as "time scales of 50 nsec" and frequency resolutions of 1 MHz. All such terms will be explained in

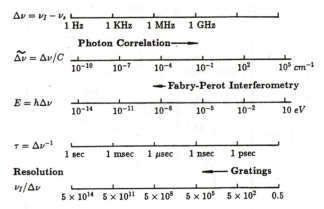

Fig. 1.1.2. Comparison of energy-transfer ranges for photon correlation spectroscopy (PCS), Fabry–Perot interferometry (FP), and grating spectrometry. The ends of the arrows suggest approximate limits of energy transfer. For PCS, which is an extremely high-resolution spectroscopic technique, we are concerned with the shortest delay-time increment achievable (on the order of nanoseconds). Longer delay-time increments test the patience of the experimenter; e.g., G. Patterson has collected good data out to delay-time increments of the order of 100 sec. For FP, the best interferometers can search the megahertz (or even kilohertz) range, while grating spectrometers usually have much lower resolutions (see Section 3.2.1). In comparing spectral frequencies and correlation times, $\tau = \Delta\nu^{-1}$ was used. More precisely, one could use $\tau = \Delta\omega^{-1} = (2\pi\,\Delta\nu)^{-1}$ with $\Delta\omega$ in radians per second, because in a Lorentzian curve for the power spectrum, the characteristic line width is expressed in radians per second. Thus, there is a factor of 2π which needs to be taken into account.

subsequent chapters. It is sufficient to realize at this time that the frequency ranges covered by PCS (low frequency range) and FP (high frequency range) can be connected using TOA. Furthermore, the reader should know that Fabry–Perot interferometers and grating spectrometers are used to measure $\langle I(\mathbf{K}\,\Delta\omega)\rangle$, a long-time average of the intensity, while PCS and TOA measure $I\langle\mathbf{K},t)\rangle$ integrated over all frequencies. In classical light scattering, the quantity of interest is the time-averaged scattered intensity $\langle I_s(\mathbf{K},t)\rangle = I_s(\mathbf{K})$. In the scattering of x-rays it is usually the time-averaged intensity for x-rays within a narrow energy range. Then, by means of near-ultraviolet radiation for light scattering and extension of x-ray scattering to very small angles, the gap in the momentum transfer for time-averaged scattered intensities can be filled (Chu, 1967), as demonstrated in Appendix 1.B. A striking feature of Fig. 1.1.1, as has been pointed out by Egelstaff (1967), is the extent of blank spaces that are not accessible by present-day techniques. Thus, it becomes important to explore the complementary nature of various radiation scattering processes.

A deeper understanding of the dynamical properties of systems often requires theoretical and experimental examinations of the scattering phenomena over wide ranges of momentum and energy changes and the combination of these results with measurements using other techniques such as ultrasonics, dielectric relaxation, and magnetic resonance.

1.2. TIME-AVERAGED AND TIME-DEPENDENT SCATTERED INTENSITY

In laser light scattering, we consider two aspects of the scattered intensity:

(1) *Angular distribution of time-averaged scattering intensity, $I_s(\mathbf{K})$.* From this quantity we can determine equilibrium properties such as the molecular weight, the size, and (at times) the shape of macromolecules in solution or colloidal particles in suspension. For $I_s(\mathbf{K})$, we can use different probing radiations, resulting in the complementary techniques of light scattering (intensity), SAXS, and SANS.

(2) *Spectral distribution of scattered intensity at different scattering angles, $I_s(\mathbf{K}, \Delta\omega)$.* Here we are more concerned with energy transfers that can be detected using an optical spectrometer (i.e., a grating spectrometer or an interferometer) or an optical mixing spectrometer. In the optical spectrometer, an optical dispersing element, such as a grating, is used. In the optical mixing spectrometer, we use the photomultiplier tube directly as a square-law detector (see also Fig. 3.2.1), The frequency ranges covered by PCS (including TOA), FP, and grating spectrometers are schematically represented by Fig. 1.1.2.

In Chapter II, the fundamental principles of light scattering, dealing with time-averaged scattered intensity and the light scattering spectrum, are discussed. The approach from the point of view of fluctuation phenomena is used in order to provide the reader with a quantitative derivation of how wavefront matching needs to be considered in optical mixing spectroscopy in the scattering volume, a very important practical problem for dynamic light scattering experiments. Chapter III deals with optical mixing spectroscopy, and Chapter IV with photon correlation spectroscopy. For completeness, discussions of Fabry–Perot interferometry are included in Chapter V. The practice of light scattering is summarized in Chapter VI, and methods of data analysis surveyed in Chapter VII. This basic book makes no attempt to include all the diverse applications of PCS, but it does single out the characterization of intractable polymers in dilute solution as a special demonstration of particle sizing in Chapter VIII. Additional summary on the topics covered in each chapter (I–VII) is presented in the introduction to Chapter VIII. The scattering behavior of complex fluids, as exemplified by Brillouin scattering,

Fabry–Perot interferometry, and photon–phonon interactions, has been excluded, even though the same light scattering technique is applicable to such studies.

APPENDIX 1.A.

Here is a listing of reviews, books, and proceedings covering some aspects of laser light scattering.

1. Spectral Distribution of Scattered Light in a Simple Fluid (Mountain, 1966);
2. Optical Mixing Spectroscopy, with Applications to Problems in Physics, Chemistry, Biology and Engineering (Benedek, 1969);
3. Light Beating Spectroscopy (Cummins and Swinney, 1970);
4. Quasi-elastic Light Scattering from Macromolecules (Pecora, 1972);
5. Laser Light Scattering (Chu, 1974)—first monograph on practice of laser light scattering;
6. Photon Correlation and Light Beating Spectroscopy (Cummins and Pike, eds., 1974)—NATO ASI Conference Proceedings;
7. Statistical Properties of Scattered Light (Crosignani *et al.*, 1975);
8. Dynamic Light Scattering (Berne and Pecora, 1976)—first monograph on theory of dynamic light scattering, and reviewed by Chu (1976);
9. Principles and Practice of Laser Doppler Anemometry (Durst *et al.*, 1976)—first monograph on laser Doppler velocimetry;
10. Photon Correlation Spectroscopy and Velocimetry (Cummins and Pike, eds., 1977)—NATO ASI Conference Proceedings;
11. Light Scattering in Liquids and Macromolecular Solutions (Degiorgio *et al.*, eds., 1980);
12. Scattering Techniques Applied to Supramolecular and Nonequilibrium Systems (Chen *et al.*, eds., 1981)—NATO ASI Conference Proceedings.
13. Biomedical Applications of Laser Light Scattering (Sattelle *et al.*, eds., 1982);
14. Application of Laser Light Scattering to the Study of Biological Motion (Earnshaw and Steer, eds., 1983)—NATO ASI Conference Proceedings;
15. Proceedings of the 5th International Conference on Photon Correlation Techniques in Fluid Mechanics (Schulz-DuBois, ed., 1983);
16. Measurement of Suspended Particles by Quasi-elastic Light Scattering (Dahneke, ed., 1983);
17. Dynamic Light Scattering: Applications of Photon Correlation Spectroscopy (Pecora, ed., 1985)—update of applications for Reference 8, reviewed by Chu (1986);
18. Light Scattering Studies of Polymer Solutions and Melts (Chu, 1985)—a review update;

19. Photoelectron Statistics (Saleh, 1978);
20. Light Scattering in Physics, Chemistry and Biology (Buckingham *et al.*, eds., 1980);
21. Modern Methods of Particle Size Analysis (Barth, ed., 1984);
22. Physics of Amphiphiles: Micelles, Vesicles and Microemulsions (Degiorgia and Corti, eds., 1985);
23. Particle Size Distribution (Provder, ed., 1987);
24. Comprehensive Polymer Science (Booth and Price, eds., 1989), Volume 1: Polymer Characterization, Chapters 7 and 8;
25. An Introduction to Dynamic Light Scattering (DLS) by Macromolecules (Schmitz, 1990)—an introduction to the applications of DLS in solutions;
26. Selected Papers on Quasielastic Light Scattering by Macromolecular, Supramolecular, and Fluid Systems (Chu, ed., 1990).

Remarks. Some early reviews are included for historical interest. The books tend to have continuity and provide an easier starting point. Uninitiated readers may find it more rewarding to first consult the books (e.g., this one for practice and Reference 8 for theory) before attempting to read the articles in the proceedings. Reviews (e.g., an extensive review on the determination of motile behavior of prokaryotic and eukaryotic cells by quasi-elastic light scattering by Chen and Hallett, 1982) have generally been omitted.

APPENDIX 1.B.

Table 1.B.1. gives typical magnitudes for momentum transfer and energy transfer in laser light scattering (LLS), small-angle x-ray scattering (SAXS), and small-angle neutron scattering (SANS).

Table 1.B.1

	LLS[a]			SAXS[b]		SANS[c]
Momentum transfer						
Representative wavelength (nm)	300			0.15		0.4
Scattering angle (rad)	Min. 5×10^{-2}	Typical $\pi/2$	Max. π	Min. 5×10^{-4}	Typical 10^{-2}	Typical 10^{-2}
K (nm^{-1})	1.05×10^{-3}	2.96×10^{-2}	4.19×10^{-2}	2.09×10^{-2}	0.419	0.157
K^{-1} (nm)	952	33.8	23.9	47.9	2.39	6.37
$d \; (= 2\pi/K, \text{nm})$	5.98×10^{3}	212	150	301	15.0	40.0
Energy transfer						
		Typical				Typical
$\Delta\bar{v}$ (wave number, cm^{-1})		8×10^{-9}				8×10^{-3}
$\hbar\Delta\omega$ (eV)		10^{-12}				10^{-6}
τ (μsec)		4×10^{3}				4×10^{-3}

(a) For light scattering, we have taken 5.00×10^{-2} rad or $2.86°$ as the lowest accessible scattering angle, while ordinary light scattering spectrometers can usually reach $\theta \geq \sim 15°$. The maximum scattering angle is assumed to be π rad or $180°$, while in practice $\theta < 180°$.

(b) For SAXS, a minimum scattering angle of 5×10^{-4} rad is assumed. A value of 0.5 mrad can be achieved using a Kratky block collimation system under favorable conditions (see references in Glatter and Kratky, 1982; also Chu, 1987) or a Bonse–Hart camera (see references in Glatter and Kratky, 1982; also Nave et al., 1986). By comparing K_{max} from LLS and K_{min} from SAXS, we see that the scattering curve can be made to overlap experimentally using the two techniques. The high scattering-angle limit for SAXS overlaps with x-ray diffraction for atomic and molecular structure studies.

(c) For SANS, the neutron wavelengths, which are longer than those of x-rays, permit easier study of larger structures. Thus, SANS covers comparable ranges to SAXS, and the scattering curves from LLS and SANS can also be made to overlap experimentally.

REFERENCES

Barth, H. G., ed. (1984). "Modern Methods of Particle Size Analysis," Wiley, New York.

Benedek, G. B. (1969). *In* "Polarization Matière et Rayonnement, Livre de Jubile en l'Honneur de Professeur A. Kastler," pp. 49–84, Presses Universitaires de France, Paris.

Berne, B. J. and Pecora, R. (1976). "Dynamic Light Scattering," Wiley-Interscience, New York.

Booth, C. and Price, C., eds. (1989). "Comprehensive Polymer Science, Volume 1: Polymer Characterization," Pergamon Press, New York.

Buckingham, A. D., Pike, E. R., and Poules, J. G., eds. (1980). "Light Scattering in Physics, Chemistry and Biology, A Royal Society Discussion held on 22 and 23 February, 1978," The Royal Society, London.

Chen, S. H., Chu, B., and Nossal, R., eds. (1981). "Scattering Techniques Applied to Supra-molecular and Nonequilibrium Systems," Plenum Press, New York.

Chen, S. H. and Hallett, F. R. (1982). *Quarterly Rev. Biophys.* **15**, 131.

Chu, B. (1967). "Molecular Forces Based on the Baker Lectures of Peter J. W. Debye," Wiley, New York.

Chu, B. (1968). *J. Chem. Educ.* **45**, 224.

Chu, B. (1974). "Laser Light Scattering," Academic Press, New York.

Chu, B. (1976). *Science* **194**, 1155.

Chu, B. (1985). *Polym. J. (Japan)* **17**, 225.

Chu, B. (1986). *J. Amer. Chem. Soc.* **108**, 7885.

Chu, B. (1990). "Selected Papers on Quasielastic Light Scattering by Macromolecular, Supramo-lecular, and Fluid Systems," SPIE Milestone Series Volume MS 12, SPIE Optical Engineering Press, Washington.

Chu, B., Wu, D.-Q., and Wu, C. (1987). *Rev. Sci. Instrum.* **58**, 1158.

Crosignani, B., DiPorto, P., and Bertolotti, M. (1975). "Statistical Properties of Scattered Light," Academic Press, New York.

Cummins, H. Z. and Pike, E. R., eds. (1974). "Photon Correlation and Light Beating Spectro-scopy," Plenum Press, New York.

Cummins, H. Z. and Pike, E. R., eds. (1977). "Photon Correlation Spectroscopy and Velocimetry," Plenum Press, New York.

Cummins, H. Z. and Swinney, H. L. (1970). *Progr. Opt.* **8**, 133.

Dahneke, B., ed. (1983). "Measurement of Suspended Particles by Quasi-elastic Light Scattering," Wiley, New York.

Degiorgio, V., and Corti, M. (1985). "Physics of Amphiphiles: Micelles, Vesicles and Microemul-sions," North-Holland, New York.

Degiorgio, V., Corti, M., and Giglio, M., eds. (1980). "Light Scattering in Liquids and Macro-molecular Solutions," Plenum Press, New York.

Dhadwal, H. S., Chu, B., and Xu, R. (1987). *Rev. Sci. Instrum.* **58**, 1494.

Durst, F., Melling, A., and Whitelaw, J. H. (1976). "Principles and Practice of Laser Doppler Anemometry," Academic Press, New York.

Earnshaw, J. C. and Steer, M. W., eds. (1983). "Application of Laser Light Scattering to the Study of Biological Motion," Plenum Press, New York.

Egelstaff, P. A. (1967). *Discuss. Faraday Soc.* **43**, 149.

Einstein, A. (1910). *Ann. Phys. (Leipzig)* **33**, 1275.

Glatter, O. and Kratky, O., eds. (1982). "Small Angle X-ray Scattering," Academic Press, New York.

Huglin, M., ed. (1972). "Light Scattering from Polymer Solutions," Academic Press, New York.

Huglin, M. (1977). *Topics in Current Chemistry* **73**, 141.

Kerker, M. (1969). "The Scattering of Light and Other Electromagnetic Radiation," Academic Press, New York and London.

Mountain, R. D. (1966). *Rev. Mod. Phys.* **38**, 205.

Nave, C., Diakun, G. P., and Bordas, J. (1986). *Nucl. Instrum. Methods Phys. Res.* **A246**, 609.

Ornstein, L. S. and Zernike, F. (1914). *Proc. Acad. Sci. Amsterdam* **17**, 793.

Ornstein, L. S. and Zernike, F. (1915). *Proc. Acad. Sci. Amsterdam* **18**, 1520.

Ornstein, L. S. and Zernike, F. (1916). *Proc. Acad. Sci. Amsterdam* **19**, 1312; **19**, 1321.

Ornstein, L. S. and Zernike, F. (1926). *Physik. Z.* **27**, 761.

Pecora, R. (1972). *Ann. Rev. Biophys. Bioeng.* **1**, 257.

Pecora, R., ed. (1985). "Dynamic Light Scattering: Applications of Photon Correlation Spectroscopy," Plenum Press, New York.

Provder, T. ed. (1987). "Particle Size Distribution; Assessment and Characterization," ACS Symposium Series 332, Washington D.C., 308.

Saleh, B. (1978). "Photoelectron Statistics," Springer-Verlag, Berlin.

Sattelle, D. B., Lee, W. I., and Ware, B. R., eds. (1982). "Biomedical Applications of Laser Light Scattering," Elsevier Medical Press, New York.

Schmitz, K. S. (1990). "An Introduction to Dynamic Light Scattering by Macromolecules," Academic Press, New York.

Schulz-DuBois, E. O., ed. (1983). "Proceedings of the 5th International Conference on Photon Correlation Techniques in Fluid Mechanics," Spinger-Verlag, New York.

Van de Hulst, H. C. (1957). "Light Scattering by Small Particles," Wiley, New York.

II

LIGHT SCATTERING THEORY

Light scattering theory can be approached from the single-particle analysis or Einstein's fluctuation viewpoint. The former is simpler to visualize and is a good starting point for studying the structure and dynamical motions of polymer molecules in solution or colloidal particles in suspension. The latter is applicable for liquids and polymer melts. We proceed with the scattered electric field from the simpler single-particle scattering approach (2.1) and then the fluctuation viewpoint (2.2). Details of the derivation are available in Chapter 3 of "Laser Light Scattering" (Chu, 1974; see also Lastovka, 1967 for original derivations) or Chapter 3 of "Dynamic Light Scattering" (Berne and Pecora, 1976) in which a more condensed version of the derivation has been presented.

The fluctuation viewpoint also permits us to consider how the signal-to-noise ratio of the measurements in dynamic light scattering can be optimized by taking into account considerations of the coherence solid angle. The reader should not be concerned with the exact definition of the coherence solid angle at this time, but only the fact that the finite size and geometry of the scattering volume can have a remarkable effect on the signal-to-noise ratio in dynamic light scattering measurements. For the reader who is not familiar with the classical aspect of light scattering, i.e., measurements of the angular distribution of time-averaged scattered intensity, the scattering volume v can be defined as the volume of the incident beam in the scattering medium of interest as viewed by the observer (an ideal point photodetector). For example, if an incident beam with a cross section of L_xL_z is propagating in the y-direction and is intersected by a plane-parallel sample of thickness L_y that is oriented perpendicular to the y-direction, the scattering volume $v\,(\equiv L_xL_yL_z)$ becomes independent of the scattering angle if the observer "sees" the same scattering volume over all scattering angles. We have also assumed, for simplicity, the incident beam to have a square cross section of L_xL_z, which is certainly not the case under most experimental conditions.

2.1. SINGLE-PARTICLE APPROACH TO TIME-AVERAGED SCATTERED INTENSITY

2.1.1. RAYLEIGH SCATTERING AND PHASE INTEGRAL

For a single small particle with polarizability α, the scattered electric field E_s is proportional to the polarizability, which is related to the volume of the particle (V_p). Thus, the intensity of scattered light for a single particle (i_s), which is proportional to E_s^2 (see Section 2.3.1), must be proportional to α^2, i.e., $i_s/I_{INC} \propto \alpha^2$ with I_{INC} being the intensity of the incident beam. In light scattering, the interaction of electromagnetic radiation with a small particle is a weak one, implying that most of incident light is transmitted and only a small amount of the light is scattered. The scattering from a single small particle (a molecule, in more precise terms) can be considered as the radiation from an induced dipole. The far-field radiation field observed at an angle perpendicular to the dipole, i.e., at $\sin\vartheta = 90°$ as shown in Fig. 2.1.1, is proportional to $1/R$ (see point (5) in Section 2.2). Thus, $I_s/I_{INC} \propto \alpha^2/R^2$. We know I_s/I_{INC} is dimensionless. Now $\alpha^2 \propto \text{length}^6$ and $R^2 \propto \text{length}^2$. Thus, $i_s/I_{INC} \propto (\alpha^2/R^2)(1/\text{length})^4$, which is the natural outcome for the Rayleigh law explaining why the sky is blue, i.e., $i_s/I_{INC} \propto 1/\lambda^4$.

By combining the above considerations, we get for a single-particle scattering

$$\boxed{\frac{I_s}{I_{INC}} = k_s^4 \frac{\sin^2\vartheta}{R^2}\alpha^2,}$$
(2.1.1)

where $k_s = 2\pi/\lambda$ and $\lambda = \lambda_0/n$ with $n = 1$ in free space.

For N particles in volume V, the scattered intensity per unit scattering volume becomes

$$\frac{I_s}{I_{INC}} = k_s^4 \frac{\sin^2\vartheta}{R^2}\alpha^2\frac{N}{V}.$$
(2.1.2)

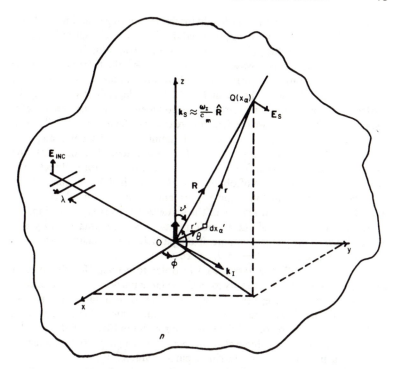

FIG. 2.1.1.   Relative position of the observer $Q(x_a)$ and the scattering volume $v(x'_a)$. (Index of refraction $= n$.) The scattered intensity is the square of the scattered electric field (Eq. (2.3.1)). The incident light field E_{INC} is polarized in the z-direction and is propagated in the k_I-direction, while the scattered electric field E_S under discussion is propagated in the R-direction. Note that $E_{INC} \perp k_I$. In Fig. 2.1.1, the vector R is defined from the origin of the scattering volume to the idealized point detector $Q(x_a)$, while in the particle-scattering approach, R is defined from the origin of the particle to the idealized point detector. θ is the scattering angle, and ϑ the direction of polarization of the incident beam and R. ϕ is the angle between x and k_I in the x–y plane. $\omega_I = 2\pi\nu_I$ and $\nu_I\lambda = c_m$, with c_m and ν_I being the speed of light and the frequency of light in the scattering medium, respectively. x_a is a running variable. For example, x_1, x_2, x_3 can be used to represent x, y, z. Thus, dx_a is like dr and represents a volume element in the scattering volume v. Here $r = R - r'$, and $\hat{R}$ is a unit vector in the direction of R.

In Eq. (2.1.2), the time-averaged scattered intensity I_s is proportional to the time-averaged incident intensity I_{INC} and is inversely proportional to λ_0^4 and R^2 with λ_0 being the wavelength of light in vacuo. Thus, for example, the blue light of 488-nm wavelength, i.e., $\lambda_0 = 488$ nm, from an argon ion laser is scattered more intensely than the red light of 632.8-nm wavelength from a He–Ne laser by a ratio of $(632.8/488)^4 \simeq 3$, implying that it is usually more efficient to use shorter-wavelength light for light scattering experiments. In estimating the factor of 3, we have neglected the dispersion change in Eq. (2.1.2) and the fact that both the polarizability and the electric inductive capacity are dependent on the incident wavelength. As $I_s \propto 1/R^2$, the total power of the scattered field through the surface of a sphere with radius R and centered at the scattering volume as shown in Fig. 2.1.1 is independent of R, i.e., the scattered intensity should remain unchanged when we increase the detector distance by ΔR and increase the detector area (A) by a factor of $(\Delta R)^2$. Thus, in light scattering, we shall be concerned mainly with the solid angle of acceptance. Equation (2.1.2) shows that I_s does not depend on the scattering angle θ but depends on the angle ϑ between the dipole vector (z-direction in Fig. 2.1.1) and the direction of observation ($\hat{\mathbf{R}}$). The $\sin \vartheta$ dependence reflects the fact that an oscillating dipole does not radiate in the direction of oscillation, i.e., $\sin \vartheta = 0$ when $\vartheta = 0$. The scattered intensity is proportional to α^2, or the sixth power of the particle radius, since the polarizability α is proportional to the particle volume. If the particle is in a vacuum instead of a dielectric medium, then $k = 2\pi/\lambda_0$, since $n = 1$.

A phase factor $e^{i\mathbf{K} \cdot \rho_r}$ is introduced in order to account for the angular dependence of scattered intensity for a single large particle in the presence of intraparticle interference; here $\rho_r = \mathbf{r}_1 - \mathbf{r}_2$. The usual notation $\mathbf{r} = \mathbf{r}_1 - \mathbf{r}_2$ with $|\mathbf{r}|$ being the distance between the two volume elements has been avoided in order to avoid the mixing of the definitions with $\mathbf{r}$ in Fig. 2.1.1. $d\mathbf{r}'_1$ and $d\mathbf{r}'_2$ represents *intraparticle* volume elements within the single-particle volume V_p, and the integrations are over the single-particle volume V_p.

The phase integral $P(K)$ in the scattered intensity due to a single large particle has the form

$$P(K) = \iint\limits_{V_P} \exp[i\mathbf{K} \cdot \rho'_r] \, d\mathbf{r}'_1 \, d\mathbf{r}'_2, \qquad (2.1.3)$$

where $\rho'_r = \mathbf{r}'_1 - \mathbf{r}'_2$ and we have dropped the polarizability by taking the local polarizability per volume element throughout the particle to be the same. $P(K)$ denotes the unnormalized (intra)particle scattering factor of a single particle of uniform density and finite size. The double integral is over volume elements in the same particle. (A variation of Eq. (2.1.3) is presented in Eq. (2.2.6) below.)

In Eq. (2.1.3), the physics depends on our interpretation of the terms. If there are N particles in volume V, the effect is N times the scattered intensity per volume V in the absence of interparticle interactions. Thus, we want to extrapolate the scattering measurements to infinite dilution and use Eq. (2.1.3) to determine the structure of a single particle. For concentrated solutions or condensed media where interparticle interactions must be taken into account, a more generalized expression, including intermolecular interactions, should be used. On the other hand, if we go to the other extreme, we can consider the scattered intensity from the fluctuation viewpoint.

2.1.2. EXCESS RAYLEIGH RATIO DUE TO SCATTERING BY SINGLE PARTICLES IN A DIELECTRIC MEDIUM

According to Eq. (2.1.2), the scattered intensity per unit scattering volume for N particles in volume V with $\vartheta = \pi/2$ has the form

$$\frac{I_s}{I_{INC}} = k_s^4 \frac{\alpha^2}{R^2} \frac{N}{V}. \tag{2.1.2}$$

Equation (2.1.2) represents the scattered intensity of dielectric spheres, each with a polarizability α in vacuo when $n = 1$. If we take the Clausius–Mosotti equation to be $(n^2 - 1)/(n^2 + 2) = \frac{4}{3}\pi(N/V)\alpha$ in cgs units with N/V being the total particle number per unit volume V and α the polarizability in units of volume per particle, then we have $n^2 - 1 = 4\pi(N/V)\alpha$ if $n \sim 1$. For N/V particles per unit volume in a dielectric solvent medium with solvent refractive index n_0, we then have $(n/n_0)^2 - 1 = 4\pi(N/V)\alpha_{ex}$, and $k_s = 2\pi/\lambda$. Here

$$\alpha_{ex} = \frac{2n_0^{-1}(n - n_0)}{4\pi C} \frac{CV}{N} = \frac{n_0^{-1}}{2\pi} \left(\frac{\partial n}{\partial C}\right)_T \frac{M}{N_a}, \tag{2.1.4}$$

is the excess polarizability for a single particle in suspension (or solution), where the weight concentration $C[\text{g-solute/cm}^3] = MN/(N_a V)$ with N_a being Avogadro's number. It should be noted that in Eq. (2.1.4) we have, assumed $n^2 + 2 \sim 3$. However, the error gets cancelled out in computing α_{ex}. The corresponding excess scattered intensity for N/V particles per unit volume per unit solid angle now becomes

$$\frac{I_{ex}}{I_{INC}} = \frac{4\pi^2}{\lambda_0^4} \frac{n_0^2}{R^2} \left(\frac{\partial n}{\partial C}\right)_T^2 \frac{CM}{N_a} \tag{2.1.5}$$

with $I_{ex} = I_s(\text{solution}) - I_s(\text{solvent})$. The excess Rayleigh ratio ($\mathcal{R}_{ex} = I_{ex} R^2/I_{INC}$; see also Eq. (2.3.4)) has the form

$$\boxed{\mathcal{R}_{ex} = \frac{4\pi^2 n_0^2}{\lambda_0^4 N_a} \left(\frac{\partial n}{\partial C}\right)_T^2 CM = HCM,} \tag{2.1.6}$$

where the optical constant H [mole cm^2 g^{-2}] for vertically polarized incident light is $4\pi^2 n_0^2 (\partial n/\partial C)_T^2/(\lambda_0^4 N_a)$. The depolarized scattered light intensity is assumed to be negligible (see Section 8.4.1.A). For unpolarized incident light,

$$\mathscr{R}_{ex,u} = \frac{I_{ex}R^2}{I_{INC}(1 + \cos^2\theta)} = \frac{2\pi^2\eta_0^2}{\lambda_0^4 N_a}\left(\frac{\partial n}{\partial C}\right)^2 CM, \qquad (2.1.7)$$

because by definition we have set $I_s/I_{INC} \propto (1 + \cos^2\theta)/2$. There is a factor of 2 difference between Eqs. (2.1.6) and (2.1.7) because of the nature of the polarization of the incident beam.

In the presence of intraparticle interference and interparticle interactions in the *dilute* solution regime, we have for polarized light

$$\frac{HC}{\mathscr{R}_{ex}} \simeq \frac{1}{MP(K)} + 2A_2 C, \qquad (2.1.8)$$

where $P(K)$ is the particle scattering factor $P(K) = \phi(K)$; see Eq. (2.2.6). The notation $P(K)$ is introduced in order to remind the reader that $P(K)$ is the usual symbol used in classical light scattering texts for the *single*-particle scattering factor. $\Phi(K)$ stands for the more general definition of the phase integral where the interference could come from a single particle, as now denoted by $P(K)$, or from a macroscopic scattering volume which is related to coherence considerations in optical mixing spectroscopy. Furthermore, when the particles are close to one another, the excess time-averaged scattered intensity I_{ex} is not only related to $P(K)$, but also to a structure factor $S(K)$ $(\equiv N^{-1}\sum_{i=1}^{N}\sum_{j=1}^{N}\langle \exp[i\mathbf{K}\cdot(\mathbf{r}_i - \mathbf{r}_j)]\rangle)$, which can be written in terms of the radical distribution function (see Section 4.2 of Chapter 4 by Pusey and Tough, 1985). $P(K) = 1 - K^2 R_g^2/3 + \cdots$, with the radius of gyration defined in Appendix 2.A. Equation (2.1.8) is the basic equation for monodisperse particles in light-scattering intensity experiments. We have

$$\lim_{C\to 0} \frac{HC}{\mathscr{R}_{ex}} \simeq \frac{1}{M}\left(1 + \frac{K^2 R_g^2}{3}\right), \qquad (2.1.9)$$

$$\lim_{K\to 0} \frac{HC}{\mathscr{R}_{ex}} \simeq \frac{1}{M} + 2A_2 C, \qquad (2.1.10)$$

$$\lim_{\substack{K\to 0 \\ C\to 0}} \frac{HC}{\mathscr{R}_{ex}} = \frac{1}{M}, \qquad (2.1.11)$$

where we can determine the radius of gyration R_g, the second virial coefficient A_2, and the molecular weight M.

In the very dilute solution regime, we are primarily interested in measuring the excess scattered intensity, I_{ex}, and then using Eq. (2.1.11) for molecular-weight determinations if the particles are monodisperse. However, at finite concentrations, there could be intermolecular interactions. Extrapolation to infinite dilution by means of Eq. (2.1.8) becomes the preferred procedure.

The Rayleigh ratio was first surveyed and reviewed critically by Kratohvil *et al.* (1962). Post-1962 average literature values for the Rayleigh ratios tend to show better agreement (Bender *et al.*, 1986). However, averages of mostly poor values do not make a better estimate of the correct answer. Thus, only the more recent $\mathcal{R}_{vv}$-values are listed in Table 2.1.1. If we take benzene as a reference standard and consider the two recent values at 488 nm to be reasonable estimates, then $\mathcal{R}_{vv}$ (at 25°C and $\theta = 90°$) are much lower than earlier determinations. The value by Finnigan and Jacobs (1970) is in good agreement with $\mathcal{R}_{vv}$(benzene) = 32 m^{-1} at 30°C as determined by Ehl *et al.* (1964). We tend to believe the 31.4-m^{-1} value for $\mathcal{R}_{vv}$(benzene) at 25°C and 488 nm. The temperature dependence of $\mathcal{R}_{vv}$(benzene) can be computed from the expression

$$\mathcal{R}_{vv}^{t} = \mathcal{R}_{vv}^{25}[1 + 0.368 \times 10^{-2}(t - 25°C)], \qquad (2.1.12)$$

Table 2.1.1

Wavelength (nm)	\multicolumn{4}{c}{Rayleigh ratio $\mathcal{R}_{vv}$ at 25°C and $\theta = 90°$ [a]}
	Benzene	Toluene	Carbon tetrachloride	Cyclohexane
\multicolumn{5}{c}{Moreels *et al.* (1987) (estimated 1–2% error)}
488	29.1	31.0	17.1	15.3
514.5	23.0	24.2	13.5	11.7
647.1	7.81	8.50	4.97	4.27
\multicolumn{5}{c}{Finnigan and Jacobs (1970) (converted from $\mathcal{R}_{v}$) [b]}
488	32.2	29.7	19.7	—
\multicolumn{5}{c}{Bender *et al.* (1986) (converted from $\mathcal{R}_{v}$) [b]}
488	27.5	29.5	16.8	14.9

	\multicolumn{4}{c}{Rayleigh ratio $\mathcal{R}_{u,90}$ and ρ_u at 633 nm, 22°C, and $\theta = 90°$}
	Benzene	Toluene	Carbon tetrachloride	Water
\multicolumn{5}{c}{Pike *et al.* (1975)}
$\mathcal{R}_{u,90}$ (10^{-4} m^{-1})	8.51	10.4	2.95	0.49
ρ_u	0.432	0.528	0.042	0.076

(a) See text for discussion and also Section 8.4.1.C.
(b) $\mathcal{R}_u = \frac{1}{2}\mathcal{R}_v(1 + \rho_u) = \frac{1}{2}\mathcal{R}_{vv}(1 + 3\rho_v)$, where ρ_u and ρ_v are the depolarization ratios for unpolarized and vertically polarized incident radiation, respectively, and $\mathcal{R}_{vv} = [2\mathcal{R}_u(6 - 3\rho_u)]/[6 + 6\rho_u]$.

where t is temperature expressed in °C. The measure $\mathscr{R}_{ex}$ of a solution can be computed according to

$$\mathscr{R}_{ex} = \frac{I_{ex}}{I_{benzene}} \mathscr{R}_{vv}(benzene) \tag{2.1.13}$$

provided that the refractive index of the solution and that of benzene are the same. Otherwise, there is a refraction correction depending on the optical geometry of the light-scattering instrument because one wants to "see" the same scattering volume from the solution and the reference standard, i.e., benzene in this case. In Fig. 2.1.1, if we take the scattering plane to be the $x-y$ plane with $\theta = 90°$, the correction factor is $n_{solution}/n_{benzene}$ if the observer sees all the incident beam in the z-direction, as would be the case with a slit geometry. The correction factor becomes $(n_{solution}/n_{benzene})^2$ if the observer sees only the *middle* portion of the scattered beam in the z-direction at $\theta = 90°$. In in-between cases, correction factors of the form $(n_{solution}/n_{benzene})^x$ with $1 < x < 2$ need to be used. The best approach would be to use a slit so that one sees the entire incident beam cross section. Then $x = 1$ (see Section 8.4.1.C for more details).

2.1.3. SMALL-ANGLE X-RAY SCATTERING

In view of our emphasis on the complementary nature of scattering techniques, we shall briefly touch upon small-angle x-ray scattering (SAXS) for comparison purposes. For a single electron scattering by x-rays,

$$\frac{i_e}{I_{INC}} \sim \frac{e^4}{(mc_0^2)^2 R^2} \left(\frac{1 + \cos^2 \theta}{2}\right), \tag{2.1.14}$$

where e and m are the electron charge and the electron mass, respectively, and the incident x-ray beam is unpolarized. Equation (2.1.14) is known as the Thomson formula. For small-angle x-ray scattering, i_e depends only slightly on the scattering angle, since the polarization factor $(1 + \cos^2 \theta)/2 \sim 1$. For an ensemble of N/V widely separated small particles per unit volume, each with v electrons and at small scattering angles where $\theta \simeq 0$, the observable scattered intensity over an average dimension $(V_P^{1/3})$ with $K V_P^{1/3} \ll 1$ becomes

$$\frac{I_e}{I_{INC}} \sim \frac{e^4}{(mc_0^2)^2 R^2} v^2 \frac{N}{V}. \tag{2.1.15}$$

The magnitude of K in SAXS is given in Appendix I.B of Chapter I. In a dilute solution, with v electrons per particle being replaced by $v - v_0$ excess electrons, we then have for the excess scattered intensity I_{ex}

$$\frac{I_{ex}}{I_{INC}} \sim \frac{e^4}{(mc_0^2)^2 R^2} (v - v_0)^2 \frac{N}{V} = \frac{e^4}{(mc_0^2)^2 R^2} \left(\frac{N(v - v_0)}{VC}\right)^2 \frac{CM}{N_a}. \tag{2.1.16}$$

Thus, the excess scattered x-ray intensity (Eq. (2.1.16)) resembles Eq. (2.1.5) with I_{ex} depending largely on the electron density difference (instead of the refractive-index difference) between the solute and the solvent.

2.2. SCATTERED ELECTRIC FIELD FROM THE FLUCTUATION VIEWPOINT

We now want to express the scattered electric field $E_s(R, t)$ in a fluctuating dielectric medium *over a finite scattering volume* (as shown schematically in Fig. 2.1.1) rather than in the simpler approach of single-particle scattering whereby $|E_s(R, t)| = |E_a(t) \exp(iK \cdot R)|$ with E_a being a time-dependent amplitude factor and R the vector from the origin of the particle to the observer position. For a fluid undergoing local dielectric-constant fluctuations, the phase integral $\left[\int_v \exp(iK \cdot r') \, dr' \right]$ over a finite scattering volume v (for example $v = L_x L_y L_z$) is computed (see Eq. (2.2.10)). This is an important development, as we shall note that the radiation field at the observer position but originating from the scattering volume must satisfy a wavefront matching condition. How the wavefront can be matched is the key to obtaining optimal signal-to-noise ratios in dynamic light scattering experiments. In a fluctuating dielectric medium, the source of scattered electric field is related to the local susceptibility (χ_e) fluctuations, which can be Fourier decomposed (see Appendix 2.B). From the Fourier decomposition of propagating sound waves, the conservations of energy (Eq. (2.B.11)) and momentum (Eq. (2.B.9)) give rise to the expression for the scattered electric field in a fluctuating dielectric medium in terms of its Fourier components (Eq. (2.B.8)).

In the remaining sections of this chapter, the intensity of scattered light is further explored from the fluctuation viewpoint (2.3), and the spectrum (2.4) of scattered light, based on the expression for the scattered electric field derived in Section 2.2, is discussed.

When a beam of linearly polarized light is passed through a nonabsorbing dielectric medium with an average dielectric constant $\langle \kappa_e \rangle$ (which is related to the refractive index $n = \sqrt{\langle \kappa_e \rangle}$), the incident electric field induces an oscillating dipole of the same frequency at each point along its path. Each oscillating dipole radiates energy in all directions. So the net field at any point in space is the vector sum of the induced dipole fields. In our calculations, we treat the dielectric medium as a continuum even for dimensions small compared with the wavelength of light and take the electric susceptibility (χ_e) to be a scalar.

The scattered electric field $E_s(x_\alpha, t)$, as shown in Fig. 2.1.1, can be computed either by summing the induced dipole fields reaching the point x_α in such a way so as to include the relative phase between waves originating from spatially separated points in the scattering volume v, or by demanding that the total field E ($= E_{INC} + E_s$ with E_{INC} being the incident electric field) satisfies the

Maxwell equations throughout all space. The second approach avoids the dipolar sum associated with the time–space average susceptibility $\langle \chi_e \rangle$, which is independent of the space coordinates in the scattering volume v if the system is at equilibrium. The effect corresponds to changing the velocity of light c_0 in free space to c_m in the scattering medium with $c_m = c_0/n$.

Let us consider the scattering in a dielectric medium that has a local fluctuating dielectric constant κ_e ($= \epsilon/\epsilon_0$ with ϵ and ϵ_0 being the electric inductive capacity and the electric inductive capacity in free space, respectively):

$$\kappa_e(\mathbf{r}', t) = \langle \kappa_e \rangle + \Delta\kappa_e(\mathbf{r}', t), \tag{2.2.1}$$

where $\kappa_e(\mathbf{r}', t)$ is the time-dependent dielectric constant at position $\mathbf{r}'$ and time t, and $\Delta\kappa_e(\mathbf{r}', t)$ represents local fluctuations of the dielectric constant at $\mathbf{r}'$ and t. Here $\kappa_e(\mathbf{r}', t) = 1 + \chi_e(\mathbf{r}', t)$, where χ_e is the electric susceptibility. For the reader who is not familiar with the electric susceptibility, it is acceptable to consider the source of scattering as due to local dielectric constant fluctuations, since $\Delta\kappa_e(\mathbf{r}', t) = \Delta\chi_e(\mathbf{r}', t)$. Thus, we could delete the electric susceptibility term and think of only the dielectric constant. The electric susceptibility is introduced here in order to make a connection with the first edition of this book. Berne and Pecora (1976) discussed the local fluctuations in terms of the dielectric constant.

For a plane-wave incident electric field,

$$\mathbf{E}_{\text{INC}}(\mathbf{R}, t) = \mathbf{E}_I^0 \exp[i(\mathbf{k}_I \cdot \mathbf{R} - \omega_I t)] \tag{2.2.2}$$

where $\mathbf{E}_{\text{INC}}$ is a simple sinusoidal traveling wave of frequency ω_I in the direction $\hat{\mathbf{k}}_I$ with phase velocity c_m ($= \omega_I/|\mathbf{k}_I|$). Equation (2.2.2) is equivalent to Eq. (5.1.1). If the reader is not familiar with Eq. (2.2.2), a reading of Chapter V up to Eq. (5.1.6) could be helpful. The periodic fluctuations are represented by complex exponentials for convenience. Only the real part has physical significance, and $\mathbf{E}_I^0$ is real.

It has been shown (Chu, 1974) that the component of the scattered electric field at a large distance R from the scattering volume v with electric susceptibility fluctuation $\Delta\chi_e$ can be expressed in vector form as

$$\mathbf{E}_s(\mathbf{R}, t) = \mathbf{k}_s \times (\mathbf{k}_s \times \mathbf{E}_I^0) \left(\frac{\epsilon_0}{\epsilon}\right) \frac{\exp[i(k_s R - \omega_I t)]}{4\pi R}$$

$$\times \int_v \Delta\chi_e(\mathbf{r}', t) \exp[i(\mathbf{k}_I - \mathbf{k}_s) \cdot \mathbf{r}'] \, d\mathbf{r}'. \tag{2.2.3}$$

The reader is likely to be discouraged by the complexity of this equation, as it contains two cross products, two functions of a complex variable, a volume integral, and 15 different physical quantities, not to mention its origin as a solution of Maxwell's equations. In addition, Eq. (2.2.3) is expressed in mks (rationalized) units, instead of the cgs units (see conversion chart in Panofsky and Phillips, 1955) used in Section 2.1. It is shown to the reader partly because it provides the linkage with the more detailed derivation in Chapter III of the first edition on this book, and partly because its components can be broken down and discussed, so that its essence can be understood physically without extensive mathematics.

(1) Readers who are not familiar with vector algebra should not worry too much about the mathematical cross product $\mathbf{k}_s \times (\mathbf{k}_s \times \mathbf{E}_I^0)$. The result is shown by $\mathbf{E}_\theta$ in Fig. 2.1.1 with a magnitude of $k_s^2 \sin \vartheta \, E_I^0$.

(2) In Eq. (2.2.3), we have assumed the equivalence of $\mathbf{E}$ and $\mathbf{E}_{INC}$, since the scattered intensity is usually only a very small fraction of the incident intensity. The coordinates of the scattering volume element dx_α' located at x_α' are shown in Fig. 2.1.1. Refractive effects at the interface of the scattering cell have been avoided by taking Q in the scattering medium.

(3) Equation (2.2.3) represents a summing of effective induced dipoles over the volume integral $\int dx_\alpha'$ $(\equiv \int d\mathbf{r}')$ with a phase factor that measures the interference between the wavelets emitted by different volume elements in the scattering volume v. It should be interesting to note that we can regard the scattering volume as the finite macroscopic scattering volume that one can perceive in an experiment, or the volume of a particle (as discussed in Section 2.1) having local electric-susceptibility (or dielectric-constant) fluctuations. In the particle scattering case, the phase integral produces the intraparticle interference effect whenever the particle size approaches the wavelength of light. In an idealized homogeneous medium where $\Delta\chi_e = 0$, there is no light scattering except in the direction of the incident beam. With the far-field approximation, r' is the microscopic distance from the origin 0 to the scattering center and r is the distance traveled by the scattered wave from the scattering center located at x_α' to the point of observation Q located at x_α. The relative phase of the field at $\mathbf{R}$ due to a source at $\mathbf{r}'$ is

$$\begin{aligned} \exp(i\mathbf{k}_s \cdot \mathbf{r}) &= \exp[i\mathbf{k}_s \cdot (\mathbf{R} - \mathbf{r}')] \\ &= \exp(ik_s R) \cdot \exp(-i\mathbf{k}_s \cdot \mathbf{r}'). \end{aligned} \tag{2.2.4}$$

$\mathbf{R}$ is in the same direction as $\mathbf{k}_s$ and can be taken outside the integral, since it does not depend on $\mathbf{r}'$. Thus, we have $\exp[i(\mathbf{k}_I - \mathbf{k}_s) \cdot \mathbf{r}']$ inside the integral.

(4) The factor $\exp[i(k_s R - \omega_I t)]/R$ represents the wave scattered from the origin (0), while $\exp[i(\mathbf{k}_I - \mathbf{k}_s) \cdot \mathbf{r}']$ is the phase factor that measures the interference between the wavelets emitted by different volume elements $d\mathbf{r}'$ $(= dx_\alpha')$.

(5) On closer examination of the electric dipole radiation, we note that only the factor involving $1/R$ contributes towards the radiating electric field

E_9 in the far-field approximation, with

$$E_9 = -\frac{k_s^2 \sin \vartheta}{4\pi R}\left(\frac{\epsilon_0}{\epsilon}\right)E_1^0 \exp[i(k_s R - \omega_1 t)]$$

$$\times \int_v \Delta\chi_e(\mathbf{r}', t)\exp[i(\mathbf{k}_1 - \mathbf{k}_s)\cdot\mathbf{r}']\,d\mathbf{r}'$$

(2.2.5)

and $\mathbf{k}_s = (\omega_1/c_m)\hat{\mathbf{R}}$.

The phase integral

$$\Phi = \int_v \exp(i\mathbf{K}\cdot\mathbf{r}')\,d\mathbf{r}'$$

(2.2.6)

forms the basis in coherence considerations, essential in experimental designs for optical mixing spectroscopy. We note that the integration can be performed over the macroscopic scattering volume v, which has an important effect in optical mixing of electromagnetic wavefronts. In order to impress on the reader the importance of coherence considerations in the mixing of wavefronts in dynamic light scattering, it is sufficient to mention here that the fluctuations at two different points in the scattering volume, as observed by the detector, will not be coincident unless the two points are very close together— or, more precisely, the angular divergence of the scattering angle $\Delta\theta$ is very small. But how small is small? What is the relationship of the scattering volume with respect to the scattering angle? These are some of the questions for which one hopes to get reasonable answers in the following discussions.

The phase factor $\langle e^{iw}\rangle = (\sin w)/w = g(w)$ (Appendix 2.A), where the function $g(w)$ peaks sharply at $w = 0$ and goes to unity as $w \to 0$, is shown in Fig. 2.2.1. The function $g(w)$ dies off in an oscillatory fashion for increasing w with an envelope of w^{-1}. The integral $\int[(\sin w)/w]\,dw$ can be evaluated analytically. For simplicity, Lastovka (1967) has approximated $g(w) = (\sin w)/w$ by a step function $g^*(w)$ with

$$g^*(w) = \begin{cases} 1, & -\Delta w \leq w \leq \Delta w, \\ 0 & \text{otherwise}. \end{cases}$$

The increment Δw is chosen by equating the areas under $g(w)$ and $g^*(w)$ so that the x, y, and z components of K satisfy the condition

$$\int_{-\infty}^{\infty}\frac{\sin w}{w}\,dw = \int_{-\pi/2}^{\pi/2}g^*(w)\,dw = \int_{-\pi/2}^{\pi/2}dw = \pi.$$

If we choose the shape of the scattering volume v to be a rectangular parallelepiped of dimensions L_x, L_y, and L_z, the phase integral as a function of

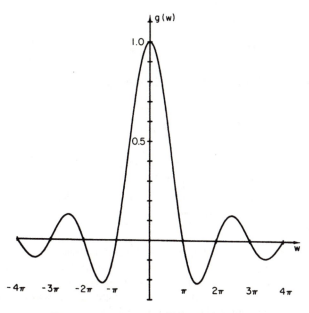

FIG. 2.2.1.   A plot of the function $g(w) = (\sin w)/w$.

$\mathbf{K}$ around $\mathbf{K} = 0$ has the form

$$\int_{-\infty}^{\infty} dK_x \, dK_y \, dK_z \left\{ \int_v \exp(i\mathbf{K} \cdot \mathbf{r}') \, d\mathbf{r}' \right\}$$

$$= \int_{-\infty}^{\infty} dK_x \, dK_y \, dK_z \, v \frac{\sin(\tfrac{1}{2}K_x L_x)}{\tfrac{1}{2}K_x L_x} \frac{\sin(\tfrac{1}{2}K_y L_y)}{\tfrac{1}{2}K_y L_y} \frac{\sin(\tfrac{1}{2}K_z L_z)}{\tfrac{1}{2}K_z L_z}$$

$$= \frac{8v}{L_x L_y L_z} \int_{-\infty}^{\infty} \frac{\sin(\tfrac{1}{2}K_x L_x)}{\tfrac{1}{2}K_x L_x} \, d(\tfrac{1}{2}K_x L_x) \int_{-\infty}^{\infty} \frac{\sin(\tfrac{1}{2}K_y L_y)}{\tfrac{1}{2}K_y L_y} \, d(\tfrac{1}{2}K_y L_y)$$

$$\times \int_{-\infty}^{\infty} \frac{\sin(\tfrac{1}{2}K_z L_z)}{\tfrac{1}{2}K_z L_z} \, d(\tfrac{1}{2}K_z L_z)$$

$$= 8\pi^3, \tag{2.2.7}$$

or

$$\int_v \exp(i\mathbf{K} \cdot \mathbf{r}') \, d\mathbf{r}' = (2\pi)^3 \, \delta(\mathbf{K}). \tag{2.2.8}$$

But for $\mathbf{K} \simeq 0$, we have

$$\int_{-\pi/L_x}^{\pi/L_x} dK_x \int_{-\pi/L_y}^{\pi/L_y} dK_y \int_{-\pi/L_z}^{\pi/L_z} dK_z = \frac{2\pi}{L_x} \times \frac{2\pi}{L_y} \times \frac{2\pi}{L_z} = \frac{8\pi^3}{v}. \quad (2.2.9)$$

Thus, by combining Eqs. (2.2.8) and (2.2.9), we get

$$\int_v \exp[i(\mathbf{k_l} - \mathbf{k_s} + \mathbf{K}) \cdot \mathbf{r}'] \, d\mathbf{r}'$$

$$= \begin{cases} v & \text{when } \begin{cases} -\pi/L_x \leq (k_l - k_s + K)_x \leq \pi/L_x, \\ -\pi/L_y \leq (k_l - k_s + K)_y \leq \pi/L_y, \\ -\pi/L_z \leq (k_l - k_s + K)_z \leq \pi/L_z, \end{cases} \\ 0 & \text{otherwise.} \end{cases} \quad (2.2.10)$$

From Eq. (2.2.10), we see that the radiation field observed at the point of observation $Q(\mathbf{R})$ comes from the plane-wave components of $\Delta\chi_e$ having a wave vector $\mathbf{K}$. In other words, in order to get light scattered in the $\mathbf{k_s}$-direction, $\Delta\chi_e(\mathbf{r}', t)$ must have a Fourier component whose wave vector is such that the phase integral satisfies Eq. (2.2.10).

In Eq. (2.2.10), we have changed the definition of $\mathbf{K} = \mathbf{k_l} - \mathbf{k_s}$ in the Introduction, i.e., $\mathbf{K}$ is now used for all values of the wave-vector index in the fluctuation amplitude; the old $\mathbf{K}$ is now called $\mathbf{K}'$:

$$\Delta\chi_e(\mathbf{r}, t) = \int_{\text{all } \mathbf{K}'} \Delta\chi_e(\mathbf{K}', t) \exp(i\mathbf{K}' \cdot \mathbf{r}) \, d\mathbf{K}', \quad (2.2.11)$$

where $\Delta\chi_e(\mathbf{K}, t) \equiv \Delta\chi_{e\mathbf{K}}$. Again, we shall drop the prime on $\mathbf{K}$ from here on. It should be noted that Eq. (2.2.10) represents an important result of this chapter.

2.3. INTENSITY OF SCATTERED LIGHT BASED ON SECTION 2.2

With Eq. (2.2.3) for the scattered electric field in a fluctuating dielectric medium, we proceed to compute the time-averaged intensity of scattered light $\langle I_s \rangle$ (2.3) and the spectrum of scattered light (2.4). We have used the particle-scattering approach to derive the expressions in $\langle I_s \rangle$. However, for the sake of completeness, we shall first write down the more general expression for the time-averaged scattered intensity (Eq. (2.3.3)) originating from a scattering volume v. In experiments related to the intensity of scattered light, the absolute scattered intensity, in terms of the Rayleigh ratio (2.3.3), is very important because it can be used in an absolute determination of the molecular weight of polymers in solution or particles in suspension. The reader is cautioned to distinguish particle scattering (2.1) from scattering in a condensed medium (2.3.1). In discussions of the Rayleigh ratio, the concept of correlation function is introduced, and the Rayleigh ratio is related to physical quantities which one can determine in a condensed medium (Sections 2.3.3.A and B).

2.3.1. SCATTERING IN CONDENSED MEDIA

The intensity of light scattered into the solid angle $d\Omega = \sin \vartheta \, d\vartheta \, d\phi$ around the field point $Q(\mathbf{R})$ as shown in Fig. 2.1.1 corresponds to the flow of power crossing a unit area along the direction of propagation $\mathbf{k}_s$:

$$I_s = \tfrac{1}{2}(\epsilon/\mu)^{1/2}\langle \mathbf{E}_s \cdot \mathbf{E}_s^* \rangle, \tag{2.3.1}$$

where ϵ and μ are the electric and magnetic inductive capacity, respectively. The scattered intensity I_s should be identified with the time-averaged scattered intensity $\langle I_s \rangle$, where the brackets are implicit since I_s is proportional to $\langle \mathbf{E}_s \cdot \mathbf{E}_s^* \rangle$. For simplicity, we have dropped the brackets around I_s. It should also be interesting to note that $(\mu/\epsilon)^{1/2}$ has the dimensions of resistance. In Eq. (2.3.1), the factor $\tfrac{1}{2}$ occurs because the scattered fields (and the incident waves) are taken as complex. The incident intensity has the form

$$I_I = \tfrac{1}{2}(\epsilon/\mu)^{1/2}|E_I^0|^2, \tag{2.3.2}$$

where I_I ($\equiv I_{INC}$) is the flow of power per unit area of the incoming light beam. By combining Eqs. (2.2.3), (2.3.1), and (2.3.2), we obtain for the time-averaged scattered intensity

$$I_s = \frac{1}{2}\left(\frac{\epsilon}{\mu}\right)^{1/2} \frac{|\mathbf{k}_s \times (\mathbf{k}_s \times \mathbf{E}_I^0)|^2}{(4\pi R)^2} \left(\frac{\epsilon_0}{\epsilon}\right)^2$$

$$\times \left\langle \iint_v \Delta\chi_e(\mathbf{r}_1', t)\, \Delta\chi_e(\mathbf{r}_2', t) \exp[i(\mathbf{k}_I - \mathbf{k}_s)\cdot(\mathbf{r}_1' - \mathbf{r}_2')] \, d\mathbf{r}_1' \, d\mathbf{r}_2' \right\rangle \tag{2.3.3}$$

$$= I_I \frac{k_s^4 \sin^2 \vartheta}{(4\pi R)^2} \left(\frac{\epsilon_0}{\epsilon}\right)^2 \iint_v \langle \Delta\chi_e(\mathbf{r}_1', t)\, \Delta\chi_e(\mathbf{r}_2', t) \rangle$$

$$\times \exp[i(\mathbf{k}_I - \mathbf{k}_s)\cdot(\mathbf{r}_1' - \mathbf{r}_2')] \, d\mathbf{r}_1' \, d\mathbf{r}_2'.$$

With the far-field approximation, there is no scattering along the z-axis, which is the direction of polarization of the incident wave. The $\sin^2 \vartheta$ angular dependence is characteristic of the dipole radiation.

We define the differential scattering cross section as the Rayleigh ratio:

$$\mathcal{R} = I_s R^2/I_I v, \tag{2.3.4}$$

where v is the scattering volume. The Rayleigh ratio has the dimensions of reciprocal length. The total scattering cross section, known as the turbidity τ^*, is related to the Rayleigh ratio by integration of $\mathscr{R}$ over the whole solid angle. Then,

$$\tau^* = \int_\Omega \mathscr{R}\, d\Omega(\vartheta, \phi). \tag{2.3.5}$$

By substituting Eqs. (2.3.3) and (2.3.4) into Eq. (2.3.5), we have

$$\tau^* = \frac{k_s^4}{6\pi v}\left(\frac{\epsilon_0}{\epsilon}\right)^2 \iint_v \langle \Delta\chi_e(\mathbf{r}'_1, t)\, \Delta\chi_e(\mathbf{r}'_2, t)\rangle$$
$$\times \exp[i(\mathbf{k}_I - \mathbf{k}_s)\cdot(\mathbf{r}'_1 - \mathbf{r}'_2)]\, d\mathbf{r}'_1\, d\mathbf{r}'_2, \tag{2.3.6}$$

where τ^* represents the total energy loss per second per unit scattering volume v and unit incident intensity. In the absence of absorption,

$$\tau^* = d^{-1}\ln(I_{\mathrm{INC}}/I_t), \tag{2.3.7}$$

where I_t and d are the transmitted intensity and the optical path length through the scattering medium, respectively.

2.3.2. SCATTERED ELECTRIC FIELD FROM PARTICLE-SCATTERING APPROACH

According to Eq. (2.2.3), the scattered electric field for a single small particle with polarizability α is

$$\mathbf{E}_s(\mathbf{R}, t) = \mathbf{k}_s \times (\mathbf{k}_s \times \mathbf{E}_I^0)\left(\frac{\epsilon_0}{\epsilon}\right)\frac{\exp[i(k_s R - \omega_I t)]}{4\pi R}\alpha e^{i\mathbf{K}\cdot\mathbf{r}'}. \tag{2.3.8}$$

Equation (2.3.8) implies that the point particle is located at $\mathbf{r}'$ and that the integral of the susceptibility over the particle volume (V_P) is equal to the total polarizability per particle, α. In Eq. (2.2.3), $\Delta\chi_e(\mathbf{r}', t)$ corresponds to a time-dependent local fluctuation of the electric susceptibility. In Eq. (2.3.8), $\mathbf{r}'(t)$ corresponds to a moving particle. Thus, the meanings of the symbols have changed.

In the presence of intraparticle interference, the phase shift due to the propagation of the scattered electric field from position $\mathbf{r}'$ to the position $\mathbf{R}$ is $\mathbf{k}_s\cdot(\mathbf{R} - \mathbf{r}')$, and that of the incident field at position $\mathbf{r}'$ is $\mathbf{k}_I\cdot\mathbf{r}'$. Thus, by combining the two terms, we get for the phase of the scattered electric field

$$\phi_s = \mathbf{k}_I\cdot\mathbf{r}' - \omega_I t + \mathbf{k}_s\cdot(\mathbf{R} - \mathbf{r}')$$
$$= -\omega_I t + \mathbf{k}_s\cdot\mathbf{R} + \mathbf{r}'\cdot(\mathbf{k}_I - \mathbf{k}_s) \tag{2.3.9}$$

with $e^{i\mathbf{K} \cdot \mathbf{r}'}$ being the phase factor and $\mathbf{K} = \mathbf{k}_{l} - \mathbf{k}_{s}$. For small particles, $e^{i\mathbf{K} \cdot \mathbf{r}'} \simeq 1$. Thus, the intensity of scattered light for a single small particle at $\mathbf{R}$ in a dielectric medium is

$$\frac{i_s}{I_1} = k_s^4 \left(\frac{\epsilon_0}{\epsilon}\right)^2 \frac{\sin^2 \vartheta}{(4\pi R)^2} \alpha^2, \qquad (2.3.10)$$

which is equivalent to Eq. (2.1.1), except for the factor $(\epsilon_0/4\pi\epsilon)^2$ due to a difference in unit convention.

2.3.3. RAYLEIGH RATIOS FROM THE FLUCTUATION VIEWPOINT

The scattered intensity as expressed in Eq. (2.3.3) can be calculated if we know the behavior of local fluctuations of the electric susceptibility $\Delta\chi_e(\mathbf{r}, t)$. We shall limit our present discussion to the time and position dependence of $\Delta\chi_e(\mathbf{r}, t)$ arising from the thermal motions of molecules. These thermal motions or thermal fluctuations are local disturbances in the "thermodynamic" coordinates of the system about their respective equilibrium values. Thus, fluctuations in the electric susceptibility can be represented by statistical fluctuations in local density, temperature, pressure, entropy, etc. In general, changes in the electric susceptibility could also depend upon other effects, such as molecular orientation and vibrations. Coupling between fluctuations in these quantities and $\Delta\chi_e$ (or $\Delta\kappa_e$) results in other light scattering phenomena. In the present section, we want to provide the physical basis for the source of scattered intensity in fluids. In light scattering *intensity* experiments of dilute polymer solutions (or colloidal suspensions), the scattered intensity due to the solvent is usually subtracted from that due to the solution (or suspension). Only the excess scattered intensity becomes pertinent. However, we must be aware of the scattering behavior of a solvent. For example, in concentrated polymer solutions, the presence of large amounts of polymer solute makes the solution behave like a complex fluid. Then, it often becomes appropriate to consider the scattering behavior based on local fluctuations of properties of interest rather than from the particle-scattering approach (Patterson, 1983). It should also be noted that in the above subtraction one has, in fact, only subtracted the incoherent scattering component for the solvent. The fact becomes especially pertinent in neutron scattering and has been discussed by Kops-Werkhoven *et al.* (1982) in applying the idea to self-diffusion measurements in light scattering experiments.

In a one-component system, there are two independent thermodynamic variables, such as the pairs (ρ, T), (P, T), and (V, T), which determine the thermodynamic state of the system. The symbols ρ, P, V, T denote the density, pressure, volume, and temperature of the system, respectively. In other words, if we know the values of two thermodynamic properties, then the values of all other thermodynamic properties of the closed homogeneous system are fixed. This does not mean that we know the values of all the other thermodynamic properties; rather, if we fix the values of any two of the properties and repeatedly measure the values of all other properties, we always get the same answers within the limits of experimental error.

In considering local electric-susceptibility fluctuations, we shall first take the entropy–pressure pair and derive the Brillouin–Mandel'shtam components of scattered light; and then take the density–temperature pair and show that local density fluctuations are responsible for the scattered intensity in a simple fluid, as derived by Einstein (1910).

The reader may wonder about the reason for the choice of an entropy–pressure pair to account for the Brillouin scattering in fluids. After all, the concept of entropy is already difficult to grasp. It would be even harder to visualize an entropy fluctuation (see Eqs. (2.4.22)–(2.4.27)). Nevertheless, the entropy–pressure pair must be used because we have already discussed how sound waves, which are mechanical pressure waves, propagate in fluids (see Appendix 2.B), and entropy–pressure forms a statistically independent pair, i.e., the cross terms $\Delta S \Delta P = 0$ (see Appendix 2.C).

2.3.3.A. RAYLEIGH RATIO DUE TO ENTROPY–PRESSURE FLUCTUATIONS

Thermal sound waves, responsible for the Brillouin–Mandel'shtam components, are adiabatic pressure disturbances. Thus, the fluctuations in the electric susceptibility can naturally be represented by local fluctuations in the entropy and the pressure around their equilibrium values s_0 and P_0, which are the average entropy per unit volume and pressure of the system:

$$\Delta\chi_e(\mathbf{r}, t) = (\partial\chi_e/\partial s)_{P_0}[\Delta s_P(\mathbf{r}, t)] + (\partial\chi_e/\partial P)_{s_0}[\Delta P_s(\mathbf{r}, t)], \qquad (2.3.11)$$

where $\Delta s_P(\mathbf{r}, t)$ are the local changes in entropy per unit volume from the equilibrium entropy per unit volume at constant pressure, and $\Delta P_s(\mathbf{r}, t)$, the local changes in pressure from the equilibrium pressure at constant entropy. The choice of s and P as the pair of state variables to represent the fluctuations in the electric susceptibility turns out to be an appropriate one because s and P are statistically independent, i.e., the cross terms involving $\Delta s_P \Delta P_s$ are zero (see Appendix 2.C and Benedek, 1968). The problem then is to find the mean squared fluctuations in Δs_P and ΔP_s.

We need to evaluate $\langle\Delta\chi_e(\mathbf{r}_1, t)\,\Delta\chi_e(\mathbf{r}_2, t)\rangle$ in Eq. (2.3.3). By substituting

Eq. (2.3.11) into $\langle\Delta\chi_e(\mathbf{r}_1,t)\,\Delta\chi_e(\mathbf{r}_2,t)\rangle$, we have

$$\begin{aligned}\langle\Delta\chi_e(\mathbf{r}'_1,t)\,\Delta\chi_e(\mathbf{r}'_2,t)\rangle &= (\partial\chi_e/\partial s)^2_{P_0}\langle\Delta s_P(\mathbf{r}'_1,t)\,\Delta s_P(\mathbf{r}'_2,t)\rangle\\
&\quad + (\partial\chi_e/\partial P)_{s_0}(\partial\chi_e/\partial s)_{P_0}[\langle\Delta P_s(\mathbf{r}'_1,t)\,\Delta s_P(\mathbf{r}'_2,t)\rangle\\
&\quad + \langle\Delta s_P(\mathbf{r}'_1,t)\,\Delta P_s(\mathbf{r}'_2,t)\rangle]\\
&\quad + (\partial\chi_e/\partial P)^2_{s_0}\langle\Delta P_s(\mathbf{r}'_1,t)\,\Delta P_s(\mathbf{r}'_2,t)\rangle.\end{aligned}\tag{2.3.12}$$

The time average in the right-hand side of Eq. (2.3.12) is equivalent to an ensemble average in which the time-averaged single sample is replaced by a large number of identical samples, and then the product $(\Delta P_s(\mathbf{r}_1,t)\,\Delta P_s(\mathbf{r}_2,t)$, etc.) is computed for each member of the ensemble. The prime on $\mathbf{r}$ is reintroduced here to remind us of the coordinates of Fig. 2.1.1. Again, we shall delete the prime hereafter and use $\mathbf{r}$ for some general position vector. As the time t is of no consequence and the statistical or thermodynamic averages of an isotropic medium are independent of the origins of space coordinates as well as time, the correlation function, which is the average value of the product of two simultaneous fluctuations of a thermodynamic variable, can only be a function of the separation distance ρ_r between the two fluctuations at $\mathbf{r}_1$ and $\mathbf{r}_2$. The spatial correlation function is a mathematical representation of the extension of local fluctuations in a system. By combining Eq. (2.3.3) with Eq. (2.D.5) we have for the scattered intensity

$$I_s(\mathbf{R}) = I_1\frac{k_s^4\sin^2\vartheta}{(4\pi R)^2}\left(\frac{\epsilon_0}{\epsilon}\right)^2 v\left[\left(\frac{\partial\chi_e}{\partial s}\right)^2_{P_0}\langle\Delta s_P^2\rangle v_F^* + \left(\frac{\partial\chi_e}{\partial P}\right)^2_{s_0}\langle\Delta P_s^2\rangle v_G^*\right],\quad (2.3.13)$$

where v_F^* and v_G^* are the correlation volumes. The final result for the Rayleigh ratio is shown in Eq. (2.3.20), together with the Rayleigh ratio derived from local density and temperature fluctuations for comparison purposes.

2.3.3.B. RAYLEIGH RATIO DUE TO DENSITY AND TEMPERATURE FLUCTUATIONS

If we take the density ρ and the temperature T as the state variables, then the fluctuations in the electric susceptibility are represented by local fluctuations in ρ and T around the average equilibrium values ρ_0 and T_0 of the system:

$$\Delta\chi_e(\mathbf{r},t) = \left(\frac{\partial\chi_e}{\partial p}\right)_{T_0}[\Delta\rho(\mathbf{r},t)] + \left(\frac{\partial\chi_e}{\partial T}\right)_{\rho_0}[\Delta T(\mathbf{r},t)],\tag{2.3.14}$$

where $\Delta\rho(\mathbf{r},t)$ are the local changes in density from the equilibrium density at constant temperature, and ΔT, the local changes in temperature from the equilibrium temperature at constant density. Similarly, as in Eqs. (2.D.2) and

(2.D.3), we may write

$$\langle \Delta\rho(\mathbf{r}_1, t)\,\Delta\rho(\mathbf{r}_2, t)\rangle = \langle(\Delta\rho)^2\rangle M(\rho_r), \tag{2.3.15}$$

$$\langle \Delta T(\mathbf{r}_1, t)\,\Delta T(\mathbf{r}_2, t)\rangle = \langle(\Delta T)^2\rangle N(\rho_r). \tag{2.3.16}$$

The computation of the mean squared fluctuations of the density and the temperature was first accomplished by Einstein (1910). Discussions on this topic have been presented by Landau and Lifshitz (1958) and by Benedek (1968). A derivation of the mean squared fluctuations of the fundamental thermodynamic quantities pertaining to a part of the system of interest is available in Appendix 2.C. With $(\delta v/v)^2 = (\delta\rho/\rho)^2$, the results are

$$\langle(\Delta\rho)^2\rangle = k_B T K_T \rho^2/v_m^*, \tag{2.3.17}$$

$$\langle(\Delta T)^2\rangle = (k_B T^2)v_n^*/C_V, \tag{2.3.18}$$

where $v_m^* = 4\pi\int_0^{r_\rho} \rho_r^2 M(\rho_r)\,d\rho_r$, $v_n^* = 4\pi\int_0^{r_T} \rho_r^2 N(\rho_r)\,d\rho_r$ with r_ρ and r_T being the corresponding correlation ranges; C_V is the specific heat at constant volume; K_T is the isothermal compressibility; and k_B is the Boltzmann constant, but we shall from here on drop the subscript B. The reader may recall that we have already used the symbol $\mathbf{k}(|\mathbf{k}| = 2\pi/\lambda)$ for the wave vector. By substituting Eqs. (2.3.17) and (2.3.18) into Eqs. (2.3.3) and (2.3.4), we have for the differential scattering cross section of an isotropic dielectric medium due to density and temperature fluctuations

$$\mathscr{R} = \frac{k_s^4 \sin^2\vartheta}{16\pi^2}\left(\frac{\epsilon_0}{\epsilon}\right)^2\left[\rho^2\left(\frac{\partial\chi_e}{\partial\rho}\right)_T^2 kTK_T + \left(\frac{\partial\chi_e}{\partial T}\right)_\rho^2 \frac{kT^2}{C_V'}\right]. \tag{2.3.19}$$

The first term on the right-hand side of Eq. (2.3.19) corresponds to the expression first derived by Einstein (1910), where we have neglected the temperature fluctuations, and C_V' is the specific heat at constant volume per unit volume.

Instead of V (or ρ) and T, we can choose P and s as the two independent thermodynamic variables (Section 2.3.3.A). The differential scattering cross section of an isotropic dielectric medium due to entropy and pressure fluctuations is given by

$$\mathscr{R} = \frac{k_s^4 \sin^2\vartheta}{16\pi^2}\left(\frac{\epsilon_0}{\epsilon}\right)^2\left[\left(\frac{\partial\chi_e}{\partial s}\right)_P^2 kC_P' + \left(\frac{\partial\chi_e}{\partial P}\right)_s^2 \frac{kT}{K_S}\right], \tag{2.3.20}$$

where C_P' is the specific heat at constant pressure per unit volume, and $K_s(= -(1/V)(\partial V/\partial P)_s)$ is the adiabatic compressibility. It should be noted that we may define the isothermal bulk modulus as $-V(\partial P/\partial V)_T$ and the adiabatic bulk modulus as $-V(\partial P/\partial V)_s$. Equation (2.3.20), like Eq. (2.3.19), is again independent of the direction of observation except for the dipole radiation factor $\sin^2 \vartheta$, because we have assumed rapidly decreasing $F(\rho_r)$ and $G(\rho_r)$ as ρ_r increases beyond the respective correlation ranges. Thus, the angle-independent Rayleigh ratio is due to spatial correlations in $F(\rho_r)$, $G(\rho_r)$, $M(\rho_r)$, and $N(\rho_r)$ whose range is short compared with the probing light wavelength. If the spatial extension of local density fluctuations becomes comparable to the wavelength of light as for fluids near the critical point, the Rayleigh ratio $\mathscr{R}$ will depend on $\mathbf{K}$.

The Rayleigh ratio may also be defined as (Lastovka, 1967)

$$\mathscr{R}^* = \mathscr{R}_s^* + \mathscr{R}_P^* = \frac{R^2}{vI_1}\left[(2\pi)^{-1}\int_0^{2\pi} I_s(\theta = 90°, \phi, \vartheta)\,d\vartheta\right], \quad (2.3.21)$$

which differs from our definition in Eq. (2.3.4). Then, with

$$\frac{1}{2\pi}\int_0^{2\pi}\sin^2\vartheta\,d\vartheta = \frac{1}{2\pi}\left[-\tfrac{1}{2}\cos\vartheta\sin\vartheta + \tfrac{1}{2}\vartheta\right]_0^{2x} = \frac{1}{2\pi}[\pi] = \frac{1}{2},$$

we have

$$\mathscr{R}_s^* = \left(\frac{\omega_1}{c_0}\right)^4\frac{1}{2(4\pi)^2}\left(\frac{\partial\chi_e}{\partial s}\right)_P^2 kC_P' \quad (2.3.22)$$

and

$$\mathscr{R}_P^* = \left(\frac{\omega_1}{c_0}\right)^4\frac{1}{2(4\pi)^2}\left(\frac{\partial\chi_e}{\partial P}\right)_s^2\frac{kT}{K_S}. \quad (2.3.23)$$

From Eq. (2.3.20), if we take $\vartheta = 90°$, i.e., polarization perpendicular to the plane of observation, we have

$$\mathscr{R}_s = \left(\frac{\omega_1}{c_0}\right)^4\frac{1}{(4\pi)^2}\left(\frac{\partial\chi_e}{\partial s}\right)_P^2 kC_P' \quad (2.3.24)$$

and

$$\mathscr{R}_P = \left(\frac{\omega_1}{c_0}\right)^4\frac{1}{(4\pi)^2}\left(\frac{\partial\chi_e}{\partial P}\right)_s^2\frac{kT}{K_S}, \quad (2.3.25)$$

where we have used $\epsilon = \epsilon_0(1 + \langle\chi_e\rangle)$ and $\kappa_e = n^2$. Equations (2.3.22)–(2.3.25) differ by a factor of $\tfrac{1}{2}$, so we should be careful in our understanding of the definition of Rayleigh ratios.

We may express the scattered intensity in terms of $\mathscr{R}$ or $\mathscr{R}^*$ (or values based on other definitions):

$$I_s(\mathbf{R}) = \frac{I_1 v}{R^2}\mathscr{R} = \frac{2I_1 v}{R^2}\mathscr{R}^*\sin^2\vartheta. \quad (2.3.26)$$

By writing $A \times L =$ (beam cross section) $\times$ (length of beam as seen by the observer) $=$ (volume of illuminated region as seen by the observer) $= v =$ scattering volume, we have for the power scattered into a solid angle Ω

$$P = 2P_1\mathscr{R}*L(\sin^2 \vartheta)\Omega = P_1\mathscr{R}\Omega L. \tag{2.3.27}$$

Thus, $\mathscr{R}$ is related to the fraction of the incident power P_1 scattered into a solid angle Ω per unit length of the scattering volume.

2.4. SPECTRUM OF SCATTERED LIGHT (LASTOVKA, 1967)

While the intensity of scattered light provides information on the mean squared amplitude of local fluctuations in the thermodynamic variables, the spectrum of scattered light tells us something about the time dependence of the fluctuations in the electric susceptibility $\Delta\chi_e(\mathbf{r}, t)$. By means of Eqs. (2.B.6) and (2.B.8), we have, from Eq. (2.2.3), the scattered electric field in the form

$$\begin{aligned}
\mathbf{E}_s(\mathbf{R}, t) &= \mathbf{k}_s \times (\mathbf{k}_s \times \mathbf{E}_I^0)\left(\frac{\epsilon_0}{\epsilon}\right)\frac{\exp[i(k_sR - \omega_I t)]}{4\pi R}\int_v \Delta\chi_e(\mathbf{r}', t) \\
&\quad \times \exp[i(\mathbf{k}_I - \mathbf{k}_s)\cdot\mathbf{r}']\,d\mathbf{r}' \\
&= \mathbf{f}(\mathbf{R})\exp[i(k_sR - \omega_I t)] \\
&\quad \times \frac{1}{2}\left\{\int_v\left(\sum_K \Delta\chi_{e\mathbf{K}}(t)\exp[i(\mathbf{k}_I - \mathbf{k}_s + \mathbf{K})\cdot\mathbf{r}']\right)d\mathbf{r}'\right. \tag{2.4.1} \\
&\quad \left. + \int_v\left(\sum_K \Delta\chi_{e\mathbf{K}}^*(t)\exp[i(\mathbf{k}_I - \mathbf{k}_s - \mathbf{K})\cdot\mathbf{r}']\right)d\mathbf{r}'\right\} \\
&= v\mathbf{f}(\mathbf{R})\exp[i(k_sR - \omega_I t)]\tfrac{1}{2}[\Delta\chi_{e_{k_s - k_I}} + \Delta\chi_{e_{k_I - k_s}}^*],
\end{aligned}$$

where $\mathbf{f}(\mathbf{R})$ $(= \mathbf{k}_s \times (\mathbf{k}_s \times \mathbf{E}_I^0)(\epsilon_0/\epsilon)/4\pi R)$ is independent of time, and $\mathbf{K}$ is fixed by the direction of $\mathbf{R}$ through $|\mathbf{K}| = 2k_I\sin(\frac{1}{2}\theta)$ with $k_I(= n(2\pi/\lambda_0))$ being the incident wave vector in the scattering medium.

Equation (2.4.1) represents a general expression for the scattered electric field in terms of Fourier components of the local electric susceptibility fluctuations and is the starting point for computing the spectrum of scattered light in a fluctuating dielectric medium, with v being the finite scattering volume as seen by the observer. The essential feature in the expression for the electric field scattered by a single particle is similar to the simplification expressed in Eq. (2.3.8) except for the fact that the amplitude factor α is now a time-dependent quantity. The polarizability amplitude is generally a tensor, i.e., the component of the dipole moment induced by the electric field is not generally parallel to the applied field. Time-dependent scattering of model systems of spherical molecules, optically anisotropic molecules, and very large molecules has been discussed elsewhere (Chapters 5, 7, and 8 of Berne and Pecora, 1976). We shall complete the development for the time-dependent scattering in a fluid so that the reader can have a better perception of the overall picture in the light-scattering spectrum.

By substituting Eq. (2.3.13) into Eq. (2.4.1), we have expressed the scattered electric field due to entropy and pressure fluctuations as

$$\mathbf{E}_s(\mathbf{R}, t) = \mathbf{v}\mathbf{f}(\mathbf{R}) \exp[i(k_s R - \omega_1 t)]$$

$$\times \left[\frac{1}{2}\left(\frac{\partial \chi_e}{\partial s}\right)_P [\Delta s_{k_s - k_1}(t) + \Delta s^*_{k_1 - k_s}(t)] \right.$$

$$\left. + \frac{1}{2}\left(\frac{\partial \chi_e}{\partial P}\right)_s [\Delta P_{k_s - k_1}(t) + \Delta P^*_{k_1 - k_s}(t)] \right]. \qquad (2.4.2)$$

The spectrum of scattered light, expressible in terms of the scattered optical intensity per unit frequency interval, is related to the ordinary Fourier time transform of $\mathbf{E}_s(\mathbf{R}, t)$. Unfortunately, $\mathbf{E}_s(\mathbf{R}, t)$ is not mean-square integrable over the infinite time domain (Margenau and Murphy, 1956). Hence, we cannot use the ordinary Fourier integral. Furthermore, even if we could expand $\mathbf{E}_s$ in a large but finite domain in a Fourier series, the exact time behavior of $\Delta s_K(t)$ and $\Delta P_K(t)$ could not be specified because both are random variables. How can we describe the time dependence of such random fluctuations? To proceed, we first discuss the time correlation function and the power spectral density in (2.4.1) before we return to the time correlation (2.4.2) and the spectrum (2.4.3) of the scattered electric field in terms of entropy and pressure fluctuations.

2.4.1. DEFINITIONS OF THE TIME CORRELATION FUNCTION AND THE POWER SPECTRAL DENSITY

We have discussed the spatial correlation function in Section 2.3, in which we have deliberately delayed our consideration of the time dependence of the correlation function. The time dependence of the correlation (or probability) function becomes important in problems involving random signals and noise.

Let x_1 and x_2 be random variables that refer to the possible values that can be assumed at time instants t_1 and t_2, respectively, by the sample function $x(\mathbf{r}, t)$ of a given random process. In our case, the random variables are (real-valued) functions defined on a sample at point $\mathbf{r}$ for which a probability is defined. We could make measurements at N instants of time, say $t_1 - t_N$, giving sample values of the random variables $x_1 - x_N$, and obtain a measure of the joint probability density function $w(x_1, x_2, \ldots, x_N)$. A continuous-parameter random process is defined by the specification of all such sets of random variables and of their joint probability distribution functions for all values of N. For example, the output current of a photodetector as a function of time is a sample function of our continuous-parameter random process as shown in Fig. 2.4.1. The measured bulk property of x at equilibrium is a time-averaged value

$$\overline{x(\mathbf{r}, t)} = \frac{1}{2T} \int_{t_0}^{t_0 + 2T} x(\mathbf{r}, t)\, dt, \qquad (2.4.3)$$

where t_0 is the time at which the measurement is initiated and $2T$ is the time period over which it is averaged. If x is stationary,

$$\langle x \rangle = \lim_{T \to \infty} \frac{1}{2T} \int_{-T}^{T} x(\mathbf{r}, t)\, dt, \qquad (2.4.4)$$

where $\langle x \rangle$ is the time average of x and is independent of t_0. The integration can be from $-T$ to T or 0 to $2T$. In Eq. (2.4.4), we have normalized the period based on $2T$. We define the time correlation function of $x(t)$ as

$$R_x(\tau) = \langle x(t)x(t + \tau) \rangle = \lim_{T \to \infty} \frac{1}{2T} \int_{-T}^{T} x(t)x(t + \tau)\, dt, \qquad (2.4.5)$$

where τ is the delay time. $x(t)x(t + \tau) \equiv x(t_i)x(t_j)$ with $\tau = t_j - t_i$.

The integral indicates a time average over all starting times t inside the periodogram of length $2T$. Furthermore, we have taken the random process to be stationary, so that the joint probability distribution of x_1 and x_2 depends only on the time difference and not on the particular values of t_i and t_j. The

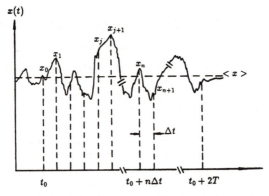

FIG. 2.4.1. A typical sample function of a continuous-parameter random process with periodograms of length $2T$. Here $t_n = n \Delta t$, and x_n is the value of x at $t_0 + n \Delta t$.

limit exists provided that the average "power" in $x(t)$, $R_x(0)$ ($\equiv \langle x^2(t) \rangle$), is finite. On the other hand, the requirement on $x(t)$ for the existence of a Fourier integral demands that $x(t)$ be mean-square integrable in the limit $T \to \infty$, or correspondingly that the total energy in $x(t)$ be bounded. We may consider the time correlation function as a mathematical representation of the persistence in time of a particular fluctuation before it dies out to zero. The time correlation function is obtained by computing $x(t)x(t + \tau)$ and then averaging the results over all starting times inside the periodogram, and it has similar properties to Eq. (2.D.1).

For time-invariant random processes, the correlation function

$$\langle x(t + \tau)x(t) \rangle = \langle x(\tau)x(0) \rangle$$

is independent of t. Then

$$\tau \to 0, \qquad \langle x(t + \tau)x(t) \rangle \to \langle |x(t)|^2 \rangle; \qquad (2.4.6)$$

$$\tau \to \infty, \qquad \langle x(t + \tau)x(t) \rangle \to \langle x(\tau) \rangle \langle x(0) \rangle = \langle x \rangle^2. \qquad (2.4.7)$$

The autocorrelation function of a nonperiodic property x decays from its initial value $\langle |x(t)|^2 \rangle$ as $\tau \to 0$ and becomes $\langle |x(t)| \rangle^2$ as $\tau \to \infty$. Figure 2.4.2 shows a schematic plot of the time correlation function $\langle x(0)x(\tau) \rangle$.

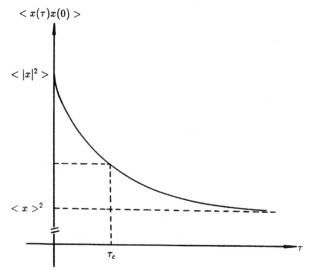

FIG. 2.4.2. Time correlation function $\langle x(\tau)x(0) \rangle$ for a nonconserved, nonperiodic property x. The function starts at $\langle |x|^2 \rangle$ at $\tau = 0$ and decays to $\langle x \rangle^2$ when $\tau \gg \tau_c$ with τ_c being the characteristic decay time for x.

If the time axis is divided into discrete time intervals Δt as shown in Fig. 2.4.1, such that $t_j = j\,\Delta t$, $t_n = n\,\Delta t$, $T = N\,\Delta t$, and $\tau_n = t_j + t_n - t_j = (j + n)\,\Delta t - j\,\Delta t$, Eqs. (2.4.4) and (2.4.5) can be approximated by

$$\langle x \rangle \cong \lim_{N \to \infty} \frac{1}{2N} \sum_{j=1}^{2N} x_j, \tag{2.4.8}$$

$$R_x(\tau_n) = \langle x(0)x(\tau_n) \rangle \cong \lim_{N \to \infty} \frac{1}{2N} \sum_{j=1}^{2N} x_j x_{j+n}, \tag{2.4.9}$$

where the sums become better approximations to the infinite-time averages as $\Delta t \to 0$. If $R_x(\tau)$ exists and is absolutely integrable, the correlation function and the power spectral density $S_x(\omega)$ are connected through Fourier transforms.

In Fig. 2.4.1, we break the infinitely random fluctuations into equal periods of length $2T$, i.e., periodograms, and regard $x(t) = 0$ outside the interval. Then, x_{2T} can be decomposed rigorously into its Fourier components

$$x_{2T}(t) = \int_{-\infty}^{\infty} \hat{x}_{2T}(\omega)\exp(-i\omega t)\,d\omega \tag{2.4.10}$$

and

$$\hat{x}_{2T}(\omega) = \frac{1}{2\pi} \int_{-\infty}^{\infty} x_{2T}(t)\exp(i\omega t)\,dt \tag{2.4.11}$$

with

$$2\pi\,\delta(\omega - \omega') = \int_{-\infty}^{\infty} \exp[i(\omega - \omega')t]\,dt. \tag{2.4.12}$$

For complex x, we have

$$
\begin{aligned}
\langle x_{2T}(t + \tau)x_{2T}^*(t) \rangle &= \lim_{T \to \infty} \frac{1}{2T} \int_{-T}^{T} x_{2T}(t + \tau)x_{2T}^*(t)\,dt \\
&= \lim_{T \to \infty} \frac{1}{2T} \left\{ \int_{-T}^{T} dt \int_{-\infty}^{\infty} d\omega \int_{-\infty}^{\infty} d\omega'\,\hat{x}_{2T}(\omega)\hat{x}_{2T}^*(\omega') \right. \\
&\qquad\qquad \left. \times \exp[-i\omega(t + \tau)]\exp(i\omega' t) \right\} \\
&= \int_{-\infty}^{\infty} d\omega \int_{-\infty}^{\infty} d\omega' \lim_{T \to \infty} \left(\frac{\hat{x}_{2T}(\omega)\hat{x}_{2T}^*(\omega')}{2T} \right) \\
&\qquad\qquad \times \exp(-i\omega\tau) \int_{-T}^{T} \exp[i(\omega' - \omega)t]\,dt.
\end{aligned}
$$

But from Eq. (2.4.12) we have

$$\int_{-\infty}^{\infty} \hat{x}_{2T}^*(\omega')\, d\omega' \lim_{T\to\infty} \int_{-T}^{T} \exp[i(\omega' - \omega)t]\, dt$$

$$= \int_{-\infty}^{\infty} \hat{x}_{2T}^*(\omega')\, d\omega' [2\pi\, \delta(\omega' - \omega)]$$

$$= 2\pi\hat{x}_{2T}^*(\omega)$$

and

$$\langle x(t + \tau)x^*(t)\rangle = \int_{-\infty}^{\infty} d\omega \left(2\pi \lim_{T\to\infty} \frac{|\hat{x}_{2T}(\omega)|^2}{2T} \right) \exp(-i\omega\tau),$$

which gives

$$\boxed{\langle x(t + \tau)x^*(t)\rangle = \int_{-\infty}^{\infty} S_x(\omega) \exp(-i\omega\tau)\, d\omega,} \tag{2.4.13}$$

i.e., the correlation function is the inverse Fourier transform of the power spectral density $S_x(\omega)$ with

$$\boxed{S_x(\omega) = \frac{1}{2\pi} \int_{-\infty}^{\infty} \langle x(t + \tau)x^*(t)\rangle \exp(i\omega\tau)\, d\tau,} \tag{2.4.14}$$

and the power spectral density is the time Fourier transform of the correlation function. Equations (2.4.13) and (2.4.14) are called the Wiener–Khintchine theorem (Wiener, 1930; Khintchine, 1934). At a topical meeting on "Photon Correlation Techniques and Applications" in 1988, J. Abbiss mentioned an interesting earlier work by Einstein (1914), who had indeed explored Eqs. (2.4.13) and (2.4.14). Thus, the Wiener–Khintchine theorem can be attributed to Einstein.

We may note that the total power contained under the spectral density curve is $\langle |x(t)|^2\rangle$, i.e.,

$$\int_{-\infty}^{\infty} S_x(\omega)\, d\omega = \frac{1}{2\pi} \int_{-\infty}^{\infty} \langle x(t + \tau)x^*(t)\rangle \int_{-\infty}^{\infty} \exp(i\omega\tau)\, d\omega\, d\tau$$

$$= \int_{-\infty}^{\infty} \langle x(t + \tau)x^*(t)\rangle\, \delta(\tau)\, d\tau = \langle |x(t)|^2\rangle. \tag{2.4.15}$$

Let $S_{x,N}(\omega)$ be the normalized power spectral density. Then we have

$$S_{x,N}(\omega) = \frac{1}{2\pi} \int \frac{\langle x(t+\tau)x^*(t)\rangle}{\langle |x(t)|^2 \rangle} \exp(i\omega\tau)\, d\tau, \qquad (2.4.16)$$

where

$$\int_{-\infty}^{\infty} S_{x,N}(\omega)\, d\omega = 1. \qquad (2.4.17)$$

Some typical power spectra and their corresponding time correlation functions are illustrated in Fig. 2.4.3. Lastovka (1967) has noted that the normalization is performed in terms of "power" $\langle |x(t)|^2 \rangle$ rather than "energy," which is "power" times time. This normalization permits us to treat the scattered field as a random variable, since the average "power" is finite even though the total energy is infinite over infinite time. In addition, when we compute $R_x(\tau)$, we do not need a precise knowledge of the time evolution of

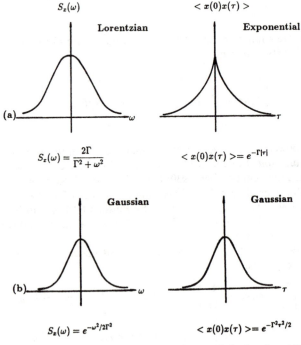

$S_z(\omega)$ $< x(0)x(\tau) >$

Lorentzian   Exponential

(a)

$$S_z(\omega) = \frac{2\Gamma}{\Gamma^2 + \omega^2} \qquad\qquad < x(0)x(\tau) >= e^{-\Gamma|\tau|}$$

Gaussian   Gaussian

(b)

$$S_z(\omega) = e^{-\omega^2/2\Gamma^2} \qquad\qquad < x(0)x(\tau) >= e^{-\Gamma^2\tau^2/2}$$

FIG. 2.4.3. Power spectra and their corresponding time correlation functions: (a) Lorenzian; (b) Gaussian; (c) rectangular; (d) white light; (e) double Lorenzian.

Rectangular

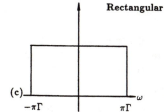

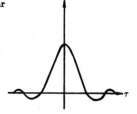

(c)

$S_x(\omega) = 1 \quad -\pi\Gamma < \omega < \pi\Gamma$

$\quad\quad\quad = 0 \text{ otherwise}$

$< x(0)x(\tau) > = \pi\Gamma sinc(\Gamma\tau)$
with $sinc(x)$ being $sin(\pi x)/\pi x$; same form as in Fig. 2.2.1

White Light

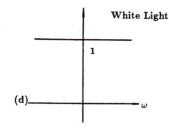

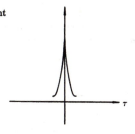

(d)

$S_x(\omega) = 1 \quad everywhere$

$< x(0)x(\tau) > = \delta(\tau)$

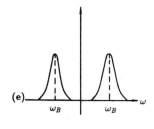

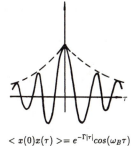

(e)

$S_x(\omega) = \dfrac{\Gamma}{\Gamma^2 + (\omega - \omega_B)^2}$

$\quad\quad + \dfrac{\Gamma}{\Gamma^2 + (\omega + \omega_B)^2}$

$< x(0)x(\tau) > = e^{-\Gamma|\tau|} cos(\omega_B\tau)$

FIG. 2.4.3. (*continued*)

$x(t)$. Rather, we need only to express an ensemble average of the time evolution of $x(t)$ from some fixed instant.

Following Eq. (2.D.1), we can define the time-dependent property $x(t)$ to consist of an average property $\langle x \rangle$ and a deviation from the instantaneous value $x(t)$, $\delta x(t)$ $[\equiv \Delta x(t)]$, such that $\delta x(t) \equiv x(t) - \langle x \rangle$. Then,

$$\langle x(0)x(\tau) \rangle = \langle [\langle x(0) \rangle + \delta x(0)][\langle x(\tau) \rangle + \delta x(\tau)] \rangle$$

$$= \langle x \rangle^2 + \langle \langle x \rangle [\delta x(0) + \delta x(\tau)] \rangle + \langle \delta x(0)\,\delta x(\tau) \rangle \quad (2.4.18)$$

$$= \langle x \rangle^2 + \langle \delta x(0)\,\delta x(\tau) \rangle,$$

where $\langle \delta x(0) \rangle = \langle \delta x(t) \rangle = 0$, since $\langle x \rangle \,(= \langle x(0) \rangle = \langle x(\tau) \rangle)$ is a constant and $\langle |x|^2 \rangle = \langle x \rangle^2 + \langle |\delta x|^2 \rangle$ as $\tau \to 0$.

If the time correlation function decays like a single exponential,

$$\langle x(0)x(\tau) \rangle = \langle x \rangle^2 + (\langle |x|^2 \rangle - \langle x \rangle^2) \exp(-\tau/\tau_c), \quad (2.4.19)$$

where τ_c is the characteristic decay time of property x. By combining Eqs. (2.4.19) and (2.4.18), we have retrieved

$$\langle \delta x(0)\,\delta x(\tau) \rangle = \langle |\delta x|^2 \rangle \exp(-\tau/\tau_c). \quad (2.4.20)$$

The autocorrelation function of *fluctuations of* x has a simpler structure. It decays to zero as $\tau \to \infty$.

2.4.2. TIME CORRELATION FOR THE SCATTERED ELECTRIC FIELD IN TERMS OF ENTROPY AND PRESSURE FLUCTUATIONS

We can obtain the time correlation function for the scattered electric field by substituting Eq. (2.4.1) into Eq. (2.4.5):

$$R_E(\tau) = \langle \mathbf{E}_s^*(\mathbf{R}, t) \cdot \mathbf{E}_s(\mathbf{R}, t + \tau) \rangle$$

$$= v^2 |\mathbf{f}(\mathbf{R})|^2 \left\{ \left(\frac{\partial \chi_e}{\partial s} \right)_P^2 \exp(-i\omega_I \tau) \langle \Delta s_K^*(t)\,\Delta s_K(t + \tau) \rangle \right.$$

$$\left. + \left(\frac{\partial \chi_e}{\partial P} \right)_s^2 \exp(-i\omega_I \tau) \langle \Delta P_K^*(t)\,\Delta P_K(t + \tau) \rangle + \text{cross terms} \right\}, \quad (2.4.21)$$

where we have used the relation $\Delta \chi_{e, k_I - k_s}^* = \Delta \chi_{e, K}^* = \Delta \chi_{e, K}$. The cross terms are zero because the entropy and the pressure terms are uncorrelated.

The forms of the correlation functions for the pressure and the entropy fluctuations were first proposed by Brillouin (1914, 1922) and by Landau and Placzek (1934), respectively.

The time correlation functions for the entropy and the pressure fluctuations are

$$R_{\Delta s}(K, \tau) = \langle \Delta s_K^*(t)\,\Delta s_K(t + \tau) \rangle$$

and

$$R_{\Delta P}(K, \tau) = \langle \Delta P_K^*(t)\, \Delta P_K(t + \tau) \rangle,$$

which can be evaluated in terms of their respective statistical averages. We shall first evaluate the local entropy fluctuations.

The time dependence of the local entropy fluctuations at constant pressure is related to that of local temperature fluctuations:

$$\Delta s_P(\mathbf{r}, t) = (C_P'/T)\, \Delta T_P(\mathbf{r}, t). \tag{2.4.22}$$

The space–time dependence of the local temperature fluctuations obeys the heat-flow equation of Fourier:

$$\int_v \frac{dq'}{dt}\, dv = \Lambda \int_S \nabla T(\mathbf{r}, t) \cdot d\mathbf{S}, \tag{2.4.23}$$

where $dq\,(= T\, ds = C_P'\, dT)$ is the heat absorbed by the system per unit volume, and Λ is the thermal conductivity. $d\mathbf{S}$ is the surface element (not ds), with $\mathbf{S}$ being the surface encompassing volume v. By application of Gauss's divergence theorem, we have

$$\int_v \frac{C_P'\, dT}{dt}\, dv = \Lambda \int_v \nabla \cdot \nabla T(\mathbf{r}, t)\, dv = \Lambda \int_v \nabla^2 T(\mathbf{r}, t)\, dv, \tag{2.4.24}$$

which gives us the equation

$$\frac{d(\overline{\Delta T_P(\mathbf{r}, t)})}{dt} = \left(\frac{\Lambda}{C_P'} \nabla^2 \overline{(\Delta T_P(\mathbf{r}, t)}) \right) \tag{2.4.25}$$

governing the relaxation to equilibrium of the ensemble average in $\Delta T_P(\mathbf{r}, t)$. In Eq. (2.4.25), Λ/C_P' is the thermal diffusivity. By taking the spatial Fourier transform of Eq. (2.4.25), we find

$$\frac{d(\overline{\Delta T_{P,K}(t)})}{dt} = \frac{\Lambda}{C_P'}(-K^2)\, \overline{\Delta T_{P,K}(t)}. \tag{2.4.26}$$

The initial condition requires that $\Delta T_K(0) = \Delta T_K^\circ$. Thus, the solution to Eq. (2.4.26) is

$$\overline{\Delta T_{P,K}(t)} = \overline{\Delta T_{P,K}(0)} \exp(-\Gamma_c t) = \Delta T_{P,K}^\circ \exp(-\Gamma_c t) \tag{2.4.27}$$

where

$$\Gamma_c = \Lambda K^2 / C_P'. \tag{2.4.28}$$

For local entropy fluctuations,

$$\Delta s_{P,K}(t) = \Delta s_{P,K}^\circ \exp(-\Gamma_c t), \tag{2.4.29}$$

i.e., the ensemble average entropy fluctuation relaxes back to equilibrium exponentially. Furthermore, it does not propagate. From Eq. (2.3.28), we have

$$\langle |\Delta s_K(t)|^2 \rangle = \overline{|\Delta s_K(0)|^2} = kC'_P/v, \qquad (2.4.30)$$

where v is the illuminated scattering volume. By using Eq. (2.4.20), we then have

$$\boxed{R_{\Delta s}(K, \tau) = (kC'_P/v)\exp(-\Gamma_c \tau),} \qquad (2.4.31)$$

where Γ_c is τ_c^{-1}

We shall now consider the pressure fluctuations, which also contribute to the scattered electric field:

$$R_{\Delta P}(K, \tau) = \langle \Delta P_K(t) \Delta P_K^*(t + \tau) \rangle.$$

Local pressure fluctuations at constant entropy can be represented, to a good approximation, by propagating sound waves. We take

$$\Delta P_{s, K}(t) = \Delta P_{s, K}(0)\exp(\pm i\bar{\omega}_K t)\exp(-\Gamma_P t), \qquad (2.4.32)$$

where $\bar{\omega}_K$ is the frequency of the sound wave having wave vector K, and Γ_P is the damping constant for the sound-wave amplitude. The $\pm$ sign corresponds to two sound waves with the same frequency propagating in opposite directions. The correlation function for the pressure fluctuations has the form

$$\langle \Delta P_K(t) \Delta P_K^*(t + \tau) \rangle$$
$$= \langle |\Delta P_K(t)|^2 \rangle \tfrac{1}{2}[\exp(i\bar{\omega}_K \tau) + \exp(-i\bar{\omega}_K \tau)]\exp(-\Gamma_P \tau) \qquad (2.4.33)$$
$$= \overline{|\Delta P_K(0)|^2}\tfrac{1}{2}[\exp(i\bar{\omega}_K \tau) + \exp(-i\bar{\omega}_K \tau)]\exp(-\Gamma_P \tau),$$

where the factor $\tfrac{1}{2}$ is introduced to take into account that half of the mean squared pressure fluctuations belong to each of the two degenerate sound waves.

From Eq. (2.3.28), we also have

$$\overline{|\Delta P_K|^2} = kT/vK_s. \qquad (2.4.34)$$

Again, v^{-1} is introduced to express the quantities per unit scattering volume.

In liquids, the hydrodynamic equations can be linearized for small fluctuations. Furthermore, the propagation of thermal sound waves produces negligible heat-conduction effect, and only longitudinal waves exist. For the pressure–time correlation function $R_{\Delta P}(K, \tau)$, we have

$$R_{\Delta P}(K, \tau) = (kT/vK_S)\tfrac{1}{2}\exp(-\Gamma_P \tau)[\exp(i\bar{\omega}_K \tau) + \exp(-i\bar{\omega}_K \tau)], \quad (2.4.35)$$

where we define a damping rate Γ_P:

$$\Gamma_P = (K^2/2\rho)(\tfrac{4}{3}\eta + \eta') = (\bar{\omega}_K^2/2\rho v_s^2)(\tfrac{4}{3}\eta + \eta') \qquad (2.4.36)$$

with a phase velocity

$$v_s = (\bar{\omega}_K/K) = (\rho K_S)^{-1/2} \tag{2.4.37}$$

and a sound-wave frequency

$$\bar{\omega}_K = (\rho K_S)^{-1/2}K. \tag{2.4.38}$$

η and η' are the shear and compressional viscosities, respectively. A more detailed derivation is available elsewhere (Chu, 1974).

2.4.3. SPECTRUM OF SCATTERED ELECTRIC FIELD

The normalized power spectral density of the scattered radiation has the form

$$S_{E_s,N}(\omega) = \frac{1}{2\pi} \int_{-\infty}^{\infty} \frac{\langle \mathbf{E}_s^*(\mathbf{R},t) \cdot \mathbf{E}_s(\mathbf{R},t+\tau) \rangle}{\langle |\mathbf{E}_s(\mathbf{R},t)|^2 \rangle} \exp(i\omega\tau)\, d\tau \tag{2.4.39}$$

by using Eq. (2.4.16). By substituting Eqs. (2.4.21), (2.4.31), and (2.4.35) into Eq. (2.4.39), we get

$$
\begin{aligned}
S_{E_s,N}(\omega) = \frac{1}{2\pi} \int_{-\infty}^{\infty} & \left\{ \left(\frac{\partial \chi_e}{\partial s}\right)_P^2 \frac{kC_P'}{v} \exp\left(\frac{-\Lambda K^2}{C_P'}\tau\right) + \left(\frac{\partial \chi_e}{\partial P}\right)_s^2 \frac{kT}{vK_S} \right. \\
& \times \frac{1}{2}\left\{ \exp\left[i\frac{K\tau}{(\rho K_S)^{1/2}} \right] + \exp\left[-i\frac{K\tau}{(\rho K_S)^{1/2}} \right] \right\} \\
& \times \left. \exp\left[-\frac{K^2\tau}{2\rho}(\tfrac{4}{3}\eta + \eta') \right] \right\} \left[\left(\frac{\partial \chi_e}{\partial S}\right)_P^2 \frac{kC_P'}{v} + \left(\frac{\partial \chi_e}{\partial P}\right)_s^2 \frac{kT}{vK_S} \right]^{-1} \\
& \times \exp[i(\omega - \omega_1)\tau]\, d\tau.
\end{aligned}
\tag{2.4.40}
$$

The normalized power spectral density produced by Eq. (2.4.40) has three Lorentzians. Since the correlation function must satisfy the condition

$$R(\tau) = R(-\tau),$$

we obtain the power spectral density by taking the Fourier cosine transform (the real part of the Fourier transform of the time correlation function):

$$
\begin{aligned}
\int_{-\infty}^{\infty} \exp(i\omega\tau)\exp(-\Gamma\tau)\, d\tau &= \int_{-\infty}^{\infty} \exp(-\Gamma\tau)\cos\omega\tau\, d\tau \\
&= 2\int_0^{\infty} \exp(-\Gamma\tau)\cos\omega\tau\, d\tau \\
&= 2\left(\frac{\Gamma}{\Gamma^2 + \omega^2}\right).
\end{aligned}
$$

So the normalized power spectral density consists of three Lorentzians symmetrically split around the center:

$$
\begin{aligned}
S_{E_s, N}(\omega) =& \frac{\left(\dfrac{\partial \chi_e}{\partial s}\right)_P^2 \dfrac{C_P' k}{v}}{\left(\dfrac{\partial \chi_e}{\partial s}\right)_P^2 \dfrac{C_P' k}{v} + \left(\dfrac{\partial \chi_e}{\partial P}\right)_s^2 \dfrac{T}{K_S} \dfrac{k}{v}} \frac{1}{\pi} \left(\frac{\dfrac{\Lambda K^2}{C_P'}}{(\omega - \omega_1)^2 + \left(\dfrac{\Lambda}{C_P'} K^2\right)^2} \right) \\
&+ \frac{\left(\dfrac{\partial \chi_e}{\partial P}\right)_s^2 \dfrac{T}{K_S}}{\left(\dfrac{\partial \chi_e}{\partial s}\right)_P^2 C_P' + \left(\dfrac{\partial \chi_e}{\partial P}\right)_s^2 \dfrac{T}{K_S}} \frac{1}{2\pi} \\
&\times \left\{ \frac{\dfrac{1}{2\rho}(\tfrac{4}{3}\eta + \eta') K^2}{\left(\omega - \omega_1 - \dfrac{K}{(\rho K_S)^{1/2}}\right)^2 + \left(\dfrac{1}{2\rho}(\tfrac{4}{3}\eta + \eta') K^2\right)^2} \right. \\
&\left. + \frac{\dfrac{1}{2\rho}(\tfrac{4}{3}\eta + \eta') K^2}{\left(\omega - \omega_1 + \dfrac{K}{(\rho K_S)^{1/2}}\right)^2 + \left(\dfrac{1}{2\rho}(\tfrac{4}{3}\eta + \eta') K^2\right)^2} \right\}.
\end{aligned}
\tag{2.4.41}
$$

Equation (2.4.41) is a heuristic formula for the spectrum of light scattered by a simple fluid where the central and the Brillouin components do not overlap, so that the cross terms, e.g., $\langle \Delta P_K(t) \Delta S_K(t + \tau) \rangle = 0$, vanish for all θ, and we have neglected depolarization effects. Figure 2.4.4 shows a theoretical spectrum of scattered light according to Eq. (2.4.41). There is a pair of Doppler-shifted lines, called Brillouin doublets, located at frequencies $\omega \pm \bar{\omega}$, as we discuss in Fig. 2.B.1.

The frequency of sound responsible for Brillouin scattering has the magnitude ($v_s/c_m \ll 1$)

$$
\bar{\omega} \approx \omega_1 (v_s/c_m) 2 \sin(\tfrac{1}{2}\theta).
\tag{2.4.42}
$$

With $v_s \approx 10^5$ cm/sec for liquids, $\bar{\omega}$ ranges from 0 for $\theta = 0$ to $\sim 10^{10}$ Hz for $\theta = \pi$. Such high frequencies are orders of magnitude larger than those of ultrasonic measurements. Furthermore, the light scattering measurements require no external acoustic excitation. Table 2.4.1 shows estimates of the linewidth of the central component and Doppler shifts for some liquids using v_s from ultrasonic measurements. Since Γ_c has a K^2 dependence, the linewidths for the central component of simple fluids can be measured only with

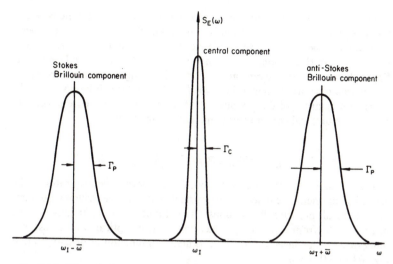

FIG. 2.4.4. Theoretical spectrum for the light scattered by thermal fluctuations in liquids according to Eq. (2.4.41). $\bar{\omega}_K = v_s K = K/\rho_d K_s)^{1/2}$; $\Gamma_c = (\Lambda/C_P')K^2$; $\Gamma_P = \alpha_p v_s = (\bar{\omega}K^2/2\rho_d v_s^2)(\frac{4}{3}\eta + \eta') = (K^2/2\rho_d)(\frac{4}{3}\eta + \eta')$.

Table 2.4.1

Computed Linewidths and Splittings for Some Liquids

Liquid	Λ (W/cm °C)	$\Gamma_c \, (\theta = \pi)^{(a)}$ (MHz)	v_s (cm/sec)	$\bar{\omega}_K \, (\theta = \pi)^{(b)}$ (MHz)
H_2O	6.18×10^{-3}	16.4	1.49×10^5	6.28×10^3
CS_2	1.61×10^{-3}	21.2	1.16×10^5	5.94×10^3
Benzene	1.32×10^{-3}	12.4	1.32×10^5	6.23×10^3
CCl_4	1.06×10^{-3}	10.5	0.94×10^5	4.33×10^3
Toluene	1.38×10^{-3}	13.3	1.32×10^5	6.21×10^3

(a) $\Gamma_c = (\Lambda/C_P')K^2$, Eq. (2.4.28).
(b) $\bar{\omega}_K = v_s K$, Eq. (2.4.37).

difficulty by conventional spectroscopy even at $\theta = \pi$. At smaller values of K, it becomes essential to use optical mixing spectroscopy (Lastovka and Benedek, 1966). The computation of Γ_P using Eq. (2.4.36) is less certain, since K_s, η, and η' may be functions of frequency for some liquids. It is in the range of 100–1000 MHz. Similarly, v_s can change with frequency. More generally, the shifted Brillouin lines are not Lorentzians as they would be for scattering from a freely decaying system.

Theory and measurements have been extended to relaxing liquids, where the interpretation of the spectrum is more complicated. Instrumental effects on Brillouin line shapes have also been discussed. The reader is referred to recent reviews (Patterson, 1983), since we are concerned mainly with the Rayleigh linewidth of the central component in the light-scattering spectrum. The discussions are presented here so that the reader will be aware of the order of magnitude of the characteristic frequencies that arise from local entropy, pressure, and concentration fluctuations for fluids.

2.4.4. THE LANDAU–PLACZEK INTENSITY RATIO

The Landau–Placzek intensity ratio, $I_C/2I_B$, is the intensity ratio of the central component due to local entropy fluctuations to the Brillouin doublets due to local pressure fluctuations. In computing the Landau–Placzek intensity ratio from Eq. (2.4.41), we assume that the optical electric susceptibility (in fact, the dielectric constant) is mainly a function of density, so that $(\partial\chi_e/\partial T)_\rho = 0$. Then,

$$(\partial\chi_e/\partial P)_s = (\partial\chi_e/\partial\rho)_T(\partial\rho/\partial P)_s \qquad (2.4.43a)$$

and

$$(\partial\chi_e/\partial s)_P = (\partial\chi_e/\partial\rho)_T(\partial\rho/\partial s)_P. \qquad (2.4.43b)$$

We recall

$$\frac{1}{\rho}\left(\frac{\partial\rho}{\partial}\right)_P = -\frac{1}{V}\left(\frac{\partial V}{\partial}\right)_P.$$

So,

$$\frac{1}{\rho}\left(\frac{\partial\rho}{\partial s}\right)_P = -\frac{1}{V}\left(\frac{\partial V}{\partial s}\right)_P = -\left(\frac{\partial V}{\partial S}\right)_P = \left(\frac{\partial T}{\partial P}\right)_S$$

$$= \frac{T}{C_P}\left(\frac{\partial V}{\partial T}\right)_P = \frac{T\alpha}{C_P},$$

where $s = S/V$ with S being the entropy. Finally,

$$\left(\frac{\partial\chi_e}{\partial P}\right)_s^2 = \left[\left(\frac{\partial\chi_e}{\partial\rho}\right)_T\left(\frac{\partial\rho}{\partial P}\right)_s\right]^2$$

$$= \left[\rho\left(\frac{\partial\chi_e}{\partial\rho}\right)_T\right]^2\left(\frac{1}{\rho}\left(\frac{\partial\rho}{\partial P}\right)_s\right)^2 \qquad (2.4.44)$$

$$= \left[\rho\left(\frac{\partial\chi_e}{\partial\rho}\right)_T\right]^2 K_S^2$$

and

$$\left(\frac{\partial \chi_e}{\partial s}\right)_P^2 = \left[\rho\left(\frac{\partial \chi_e}{\partial \rho}\right)_T\right]^2 \frac{T^2 \alpha^2}{C_P'^2} = \left[\rho\left(\frac{\partial \chi_e}{\partial \rho}\right)_T\right]^2 \left(\frac{C_P' - C_V'}{C_P'^2}\right) TK_T. \quad (2.4.45)$$

In Eq. (2.4.45), we have used the general thermodynamic relation $C_P' - C_V' = T\alpha^2/K_T$, where C_P' and C_V' are the heat capacities per unit volume (Bearman and Chu, 1967). With $C_P/C_V = K_T/K_S$ (Bearman and Chu, 1967), Eq. (2.4.41) gives

$$
\begin{aligned}
S_{E_s, N}(\omega) &= \frac{[(C_P - C_V)/C_P]K_T}{K_S + [(C_P - C_V)/C_P]K_T} \frac{1}{\pi} \frac{\Gamma_c}{(\omega - \omega_l)^2 + \Gamma_c^2} \\
&\quad + \frac{K_S}{K_S + [(C_P - C_V)/C_P]K_T} \frac{1}{2\pi} \\
&\quad \times \left\{ \frac{\Gamma_P}{(\omega - \omega_I - \bar{\omega}_K)^2 + \Gamma_P^2} + \frac{\Gamma_P}{(\omega - \omega_I + \bar{\omega}_K)^2 + \Gamma_P^2} \right\} \quad (2.4.46) \\
&= \frac{(C_P - C_V)/C_V}{1 + [(C_P - C_V)/C_V]} \mathscr{L}_C \\
&\quad + \frac{1}{2\{1 + [(C_P - C_V)/C_V]\}} \{\mathscr{L}_P(\bar{\omega}) + \mathscr{L}_P(-\bar{\omega})\} \\
&= \frac{C_P - C_V}{C_P} \mathscr{L}_C + \frac{1}{2(C_P/C_V)} \{\mathscr{L}_P(\bar{\omega}) + \mathscr{L}_P(-\bar{\omega})\},
\end{aligned}
$$

where

$$\mathscr{L}_C = \frac{1}{\pi} \frac{\Gamma_c}{(\omega - \omega_l)^2 + \Gamma_c^2} \quad (2.4.47)$$

and

$$\mathscr{L}_P(\bar{\omega}) = \frac{1}{\pi} \frac{\Gamma_P}{(\omega - \omega_I + \bar{\omega}_K)^2 + \Gamma_P^2}. \quad (2.4.48)$$

The Lorentzian functions $\mathscr{L}_C$ and $\mathscr{L}_P$ have been normalized so that

$$\int_{-\infty}^{\infty} \mathscr{L}_C(\Gamma_c, \omega - \omega_l)\, d\omega = \int_{-\infty}^{\infty} \mathscr{L}_P(\bar{\omega}, \Gamma_P, \omega - \omega_l)\, d\omega = 1, \quad (2.4.49)$$

which agrees with the earlier normalization for $S_{E_s}(\omega)$,

$$\int_{-\infty}^{\infty} S_{E_s, N}(\omega)\, d\omega = \frac{C_P - C_V}{C_P} + \frac{C_V}{C_P} = 1.$$

From Eq. (2.4.46), we see that the intensity of the central component is related to $(C_P - C_V)/C_P$, while that of the Brillouin component varies with $C_V/(2C_P)$. Thus,

$$I_c/2I_B = (C_P - C_V)/C_V, \qquad (2.4.50)$$

where I_c is the relative intensity of the central component and I_B is that of one of the two Brillouin components. Equation (2.4.50) is the Landau–Placzek ratio, and may be written in the form

$$(I_c + 2I_B)/2I_B = C_P/C_V. \qquad (2.4.51)$$

In comparing Eq. (2.4.51) with experiments, it is essential to take into account the dispersion effects. Then, the Landau–Placzek ratio is in substantial agreement with experiments (Cummins and Gammon, 1966).

APPENDIX 2.A. RADIUS OF GYRATION

The radius of gyration, R_g, for a particle may be defined as

$$R_g = \frac{\displaystyle\int_v \rho(r) r^2 \, dv}{\displaystyle\int \rho(r) \, dv}, \qquad (2.A.1)$$

where $\rho(r)$ is the mass density at r. According to Eq. (2.1.3), the particle scattering factor $P(K)$ can be related to measurements of the radius of gyration of monodisperse particles as follows:

$$P(K) = \iint_{V_P} \exp[i\mathbf{K} \cdot \mathbf{r}] \, d\mathbf{r}_1 \, d\mathbf{r}_2, \qquad (2.A.2)$$

where we have changed the symbol ρ_r to $\mathbf{r}$ and dropped the primes in the volume elements for simplicity. In the phase integral, the evaluation of $\langle e^{i\mathbf{K} \cdot \mathbf{r}} \rangle$ over the particle volume V_P represents an average taken over all directions of $\mathbf{r}$.

Evaluation of $\langle e^{i\mathbf{K} \cdot \mathbf{r}} \rangle = (\sin Kr)/(Kr)$ (Debye, 1915): Let us define the angle between $\mathbf{K}$ and $\mathbf{r}$ as φ. To calculate the average, we have $\langle e^{i\mathbf{K} \cdot \mathbf{r}} \rangle = \langle \cos(\mathbf{K} \cdot \mathbf{r}) \rangle$ and

$$\int_0^\pi \cos(Kr \cos\varphi) \frac{\sin\varphi}{2} \, d\varphi = -\frac{1}{Kr} \int_{Kr}^0 \cos u \, du \qquad (2.A.3)$$
$$= \frac{\sin Kr}{Kr},$$

leading to the classical result

$$\langle e^{i\mathbf{K}\cdot\mathbf{r}} \rangle = \frac{\sin Kr}{Kr}, \tag{2.A.4}$$

$$\int_{V_P} \langle e^{i\mathbf{K}\cdot\mathbf{r}} \rangle \, dv = \int_0^R 4\pi r^2 \frac{\sin Kr}{Kr} \, dr, \tag{2.A.5}$$

where R is the particle radius. At small values of KR and with the mass density dependent on r, we then have for the normalized particle scattering factor

$$P(K) = \frac{\phi(K)}{\phi(0)} \simeq 1 - \frac{K^2}{3} \frac{\displaystyle\int_{V_P} \rho(r) r^2 \, dv}{\displaystyle\int_{V_P} \rho(r) \, dv} = 1 - \frac{K^2 R_g^2}{3}, \tag{2.A.6}$$

where $\langle R_g^2 \rangle = \int_{V_P} \rho(r) r^2 \, dv / \int_{V_P} \rho(r) \, dv$ as defined in Eq. (2.A.1), and we have taken $\sin X \simeq X - X^3/3! + \cdots$.

For a uniform solid sphere,

$$\langle R_g^2 \rangle = \frac{\displaystyle\int_0^R 4\pi r^4 \, dr}{\displaystyle\int_0^R 4\pi r^2 \, dr} = \tfrac{3}{5} R^2. \tag{2.A.7}$$

For a rigid rod of length L, the number of mass elements at a distance between r and $r + dr$ is proportional to dr. Then,

$$\langle R_g^2 \rangle = \frac{\displaystyle\int_0^{L/2} r^2 \, dr}{\displaystyle\int_0^{L/2} dr} = \frac{L^2}{12}. \tag{2.A.8}$$

APPENDIX 2.B. FOURIER DECOMPOSITION
OF SUSCEPTIBILITY FLUCTUATIONS
(AFTER LASTOVKA, 1967)

Consideration of the phase integral (2.2.10) leads us to the realization that we must make a Fourier decomposition of the susceptibility fluctuation. In an isotropic medium, the electric susceptibility becomes a scalar quantity. The polarization field per unit volume may also be expressed in terms of the polarizability α:

$$\mathbf{P} = N\epsilon_0 \alpha \mathbf{E}_{\text{eff}} / V, \tag{2.B.1}$$

where N is the total number of molecules in volume V, and $\mathbf{E}_{\text{eff}}$ is the effective field, which is the sum of the macroscopic external field and an internal field owing to the polarization of neighboring molecules. For example, in an isotropic medium, α is again a scalar quantity and $\mathbf{E}_{\text{eff}} = \mathbf{E} + \mathbf{P}/3\epsilon_0$, which leads to the Clausius–Mossotti relation applicable to a wide class of solids and liquids.

The plane-wave fluctuation of the local electric susceptibility $\Delta\chi_e(\mathbf{r}, t)$ $(=\Delta\chi_e^0 \exp[i(\mathbf{K} \cdot \mathbf{r} - \bar{\omega}t)])$ in terms of a Fourier series expansion has the form

$$\Delta\chi_e(\mathbf{r}, t) = \int_{\text{all } \mathbf{K}} \Delta\chi_e(\mathbf{K}, t) \exp(i\mathbf{K} \cdot \mathbf{r}) \, d\mathbf{r}, \qquad (2.\text{B}.2)$$

where $\Delta\chi_e^0$ is the amplitude, $\bar{\omega}$ is the frequency of the sinusoidal traveling wave for the fluctuating electric susceptibility with phase velocity $v_s \, (=\bar{\omega}/|\mathbf{K}|)$, and the allowed $\mathbf{K}$-values are found by applying the cyclic boundary conditions on the surface bounding v. Equation (2.B.2) is Eq. (2.2.11), and the relaxed definition of $\mathbf{K}$ is again being used. For waves of different $\mathbf{K}$, we have from Eq. (2.2.10)

$$\int_v \exp[i(\mathbf{K}_1 - \mathbf{K}_2) \cdot \mathbf{r}'] \, d\mathbf{r}' = v\,\delta(\mathbf{K}_1 - \mathbf{K}_2). \qquad (2.\text{B}.3)$$

The Fourier amplitudes $\Delta\chi_{e\mathbf{K}}(t)$ can be obtained in terms of $\Delta\chi_e(\mathbf{r}, t)$:

$$\begin{aligned}
\int_v \Delta\chi_e(\mathbf{r}', t)&\exp(-i\mathbf{K}_2 \cdot \mathbf{r}') \, d\mathbf{r}' \\
&= \int_v \sum_{\mathbf{K}} \Delta\chi_{e\mathbf{K}}(t) \exp(i\mathbf{K} \cdot \mathbf{r}') \exp(-i\mathbf{K}_2 \cdot \mathbf{r}') \, d\mathbf{r}' \\
&= \int_v \sum_{\mathbf{K}} \Delta\chi_{e\mathbf{K}}(t) \exp[i(\mathbf{K} - \mathbf{K}_2) \cdot \mathbf{r}'] \, d\mathbf{r}' \\
&= \sum_{\mathbf{K}} \Delta\chi_{e\mathbf{K}}(t) v\,\delta(\mathbf{K} - \mathbf{K}_2) \\
&= \Delta\chi_{e\mathbf{K}_2}(t) v \qquad \text{for } \mathbf{K} = \mathbf{K}_2,
\end{aligned} \qquad (2.\text{B}.4)$$

or

$$\Delta\chi_{e\mathbf{K}}(t) = \frac{1}{v} \int_v \Delta\chi_e(\mathbf{r}', t) \exp(-i\mathbf{K} \cdot \mathbf{r}') \, d\mathbf{r}'. \qquad (2.\text{B}.5)$$

Since $\Delta\chi_e(\mathbf{r}, t)$ is real, we can rewrite Eq. (2.B.2) as a sum of explicitly real terms:

$$\begin{aligned}
\Delta\chi_e(\mathbf{r}, t) &= \tfrac{1}{2}[\Delta\chi_e(\mathbf{r}, t) + \Delta\chi_e^*(\mathbf{r}, t)] \\
&= \tfrac{1}{2}\left[\sum_{\mathbf{K}} \Delta\chi_{e\mathbf{K}}(t) \exp(i\mathbf{K} \cdot r) + \sum_{\mathbf{K}} \Delta\chi_{e\mathbf{K}}^*(t) \exp(-i\mathbf{K} \cdot \mathbf{r}) \right] \\
&= \tfrac{1}{2}\sum_{\mathbf{K}} [\Delta\chi_{e\mathbf{K}}(t) \exp(i\mathbf{K} \cdot \mathbf{r}) + \Delta\chi_{e\mathbf{K}}^*(t) \exp(-i\mathbf{K} \cdot \mathbf{r})]
\end{aligned} \qquad (2.\text{B}.6)$$

By substituting Eq. (2.B.6) into Eq. (2.2.3) and neglecting the amplitude factor $\mathbf{k}_s \times (\mathbf{k}_s \times \mathbf{E}_I^0)(\epsilon_0/\epsilon)\exp[i(k_s R - \omega_I t)]/4\pi R$, the phase integral in Eq. (2.2.3) becomes

$$\mathbf{E}_s \propto \frac{1}{2}\sum_K \Delta\chi_{eK}(t)\int_v \exp[i(\mathbf{k}_I - \mathbf{k}_s + \mathbf{K})\cdot\mathbf{r}']\,d\mathbf{r}'$$

$$+ \frac{1}{2}\sum_K \Delta\chi_{eK}^*(t)\int_v \exp[i(\mathbf{k}_I - \mathbf{k}_s - \mathbf{K})\cdot\mathbf{r}']\,d\mathbf{r}'. \quad (2.B.7)$$

By combining Eqs. (2.B.7) and (2.2.10) with Eq. (2.2.3), we finally have

$$\mathbf{E}_s(\mathbf{R}, t) = \mathbf{k}_s \times (\mathbf{k}_s \times \mathbf{E}_I^0)\left(\frac{\epsilon_0}{\epsilon}\right)\frac{\exp[i(k_s R - \omega_I t)]}{4\pi R}$$

$$\times \frac{1}{2}v[\Delta\chi_{e_{\mathbf{k}_I - \mathbf{k}_s}}^*(t) + \Delta\chi_{e_{\mathbf{k}_s - \mathbf{k}_I}}(t)]. \quad (2.B.8)$$

We shall now examine the physics associated with the subscripts in $\Delta\chi_e$, i.e., $k_I - k_s$ and $k_s - k_I$ in Eq. (2.B.8). From Eq. (2.B.3), we know that the first integral in Eq. (2.B.7) vanishes unless $\mathbf{K} = -(\mathbf{k}_I - \mathbf{k}_s)$. Similarly, for the second integral, we have $\mathbf{K} = +(\mathbf{k}_I - \mathbf{k}_s)$, which is Eq. (1.1). Thus, the scattered electric field at $Q(\mathbf{R})$ is due to the two plane-wave components of $\Delta\chi_e(\mathbf{r}, t)$ with wave vectors

$$\mathbf{K} = \pm(\mathbf{k}_I - \mathbf{k}_s). \quad (2.B.9)$$

With the local electric susceptibility $\Delta\chi_e$ being

$$\Delta\chi_e(\mathbf{r}, t) = \Delta\chi_e^0 \exp[i(\mathbf{K}\cdot\mathbf{r} - \bar{\omega}t)],$$

Eq. (2.B.7) shows that the scattered electric field is being modulated by the local electric susceptibility fluctuations. The modulation produces a frequency shift whose magnitude is equal to the frequency of the scattering plane wave ($\pm\bar{\omega}$). In Fig. 2.B.1, we envisage light waves as being diffracted by propagating sound waves owing to local thermal fluctuations in the electric susceptibility, and the Bragg law holds. The two sound-wave vectors at $\mathbf{q}^+$ and $\mathbf{q}^-$, as shown in Fig. 2.B.1, must satisfy Eq. (2.B.9) and have magnitudes

$$|\mathbf{K}| = |\mathbf{q}^+| = |\mathbf{q}^-| = k_s s, \quad (2.B.10)$$

where we have $k = k_s \simeq k_I$ for the case $\bar{\omega} \ll \omega_I$, since the velocity of sound is much smaller than that of light; $k = 2\pi/\lambda$; $s = 2\sin(\frac{1}{2}\theta)$; and λ and θ are the wavelength of light in the scattering medium and the scattering angle, respectively. $|\mathbf{q}|$ is equivalent to $|\mathbf{K}|$. We introduce the symbol $\mathbf{q}$ and the $+$ and $-$ superscripts in order to emphasize the two directions of propagation of the traveling thermal sound waves that are responsible for the Bragg reflection of the electric field.

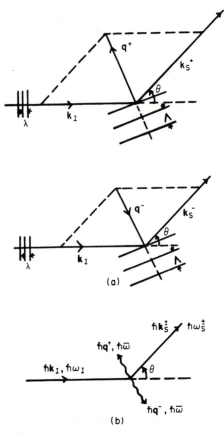

FIG. 2.B.1.   Brillouin scattering. (a) Classical description $(v_s/c_m \ll 1)$; $\mathbf{q}^+ = \mathbf{k}_s^+ - \mathbf{k}_I$, $\mathbf{k}_s^+ = \mathbf{k}_s^- = \hat{\mathbf{R}}, \mathbf{q}^- = \mathbf{k}_I - \mathbf{k}_s^-$. (b) Quantum description $(v_s/c_m \ll 1)$; $\hbar\mathbf{k}_s = \hbar\mathbf{k}_I \pm \hbar\mathbf{q}$, $\hbar\omega_s = \hbar\omega_I \pm \hbar\bar{\omega}$.

In our example, if the grating were fixed in space, the scattered light would have the incident frequency. However, our grating moves with a phase velocity v_s $(= \bar{\omega}/|\mathbf{q}|)$, so that the frequency of the back-reflected light is reduced by a Doppler shift of $(2v_s/c_m)\omega_I$. An analogous example of the Doppler shift is to measure the speed of a train by measuring the pitch of its whistle as it moves with respect to a fixed listener. Quantum-mechanically, we may consider the scattering process as inelastic photon–phonon collisions. The incident photon has an energy $\hbar\omega_I$ and a momentum $\hbar k_I$, and the sound wave may be described as a quasiparticle of energy $\hbar\bar{\omega}$ and quasimomentum $\hbar q$. Conservations of energy and momentum in the scattering process require

$$\hbar \mathbf{k}_s = \hbar \mathbf{k}_I \pm \hbar \mathbf{q} \qquad \text{(momentum conservation)}, \qquad (2.B.11)$$

$$\hbar \omega_s = \hbar \omega_I \pm \hbar \bar{\omega} \qquad \text{(energy conservation)}. \qquad (2.B.12)$$

The situation $\mathbf{q} = \mathbf{k}_s - \mathbf{k}_I$ and $\bar{\omega} = \omega_s - \omega_I$ corresponds to $\mathbf{q}^+ = \mathbf{k}_s^+ - \mathbf{k}_I$ and $\bar{\omega} = \omega_s^+ - \omega_I$, respectively, and represents *anti-Stokes* radiation or phonon annihilation, while the situation $\mathbf{q} = \mathbf{k}_I - \mathbf{k}_s^-$ and $\bar{\omega} = \omega_I - \omega_s$ corresponds to $\mathbf{q}^- = \mathbf{k}_I - \mathbf{k}_s^-$ and $\bar{\omega} = \omega_I - \omega_s^-$, respectively, and represents *Stokes* radiation or phonon creation. In phonon creation, the scattered photon has a lower frequency than that of the incident photon, so the shift is toward the red. The opposite is true for phonon annihilation, where the shift is toward the blue.

The idea of diffraction of light by ultrasonic waves was first proposed by Debye (1932), on the basis of earlier work of Brillouin (1914, 1922). Tunable lasers have been constructed successfully using this principle. In addition to propagating sound waves, overdamped plane waves also give rise to light scattering.

In the more general discussion of the scattered intensity in Section 2.3, the origin of light scattering from a simple fluid is discussed.

APPENDIX 2.C. MEAN-SQUARED FLUCTUATIONS OF FUNDAMENTAL THERMODYNAMIC QUANTITIES (BENEDEK, 1968)

Consider a subsystem I that is small compared with the whole, yet it is large enough to contain a sufficient number of molecules so that it is a macroscopic body. Subsystem I is subject to continuous interaction with the remaining parts of the system II that is the medium. From the second law of thermodynamics, subsystem I is in a state of equilibrium if (Kirkwood and Oppenheim, 1961)

$$dS_I \le \delta q_I / T_0, \qquad (2.C.1)$$

where T_0 is the uniform temperature of medium II and δq_I is the heat absorbed by the subsystem. For systems with only PV work, we have

$$\delta E_I + P_0 \, \delta V_I - T_0 \, \delta S_I \ge 0. \qquad (2.C.2)$$

Equation (2.C.2) is the general criterion for equilibrium in a closed system. The criterion for equilibrium in a closed system at constant energy E and volume V is

$$(\delta S_1)_{E,V} \leq 0. \tag{2.C.3}$$

The subsystem may be considered as a quasiclosed system, and it interacts in various ways with the medium of the system in a complex and intricate manner. From Eq. (2.C.2), we take the probability w_1 for finding the system in a state that may be reached reversibly from the equilibrium state to the nonequilibrium state by performing a minimum amount of work δW_{min} such that

$$w_1 = \text{const} \times \exp[-(\delta W_{min}/kT)], \tag{2.C.4}$$

where $\delta W_{min} = \delta E_1 + P\,\delta V_1 - T\,\delta S_1$. Equation (2.C.4) could also be derived from the following key concepts. The first point is to recognize that the fluctuations in thermodynamic quantities are governed by Gaussian statistics. The second point is that the relative probability of a particular state of the system is given by $W([X_i])d[X_i] = \exp[S([X_i])/k]\,d[X_i]$, which is equivalent to Eq. (2.C.4). $[X_i]$ is the set of variables that determine the entropy S in that state.

The difference in energy between states in a continuous change can be expanded in a Taylor's series in the form

$$\begin{aligned}
\delta E = &\left(\frac{\partial E}{\partial V}\right)_S \delta V + \left(\frac{\partial E}{\partial S}\right)_V \delta S \\
&+ \frac{1}{2}\left[\left(\frac{\partial^2 E}{\partial S^2}\right)_V (\delta S)^2 + \left(\frac{\partial^2 E}{\partial V^2}\right)_S (\delta V)^2\right. \\
&+ \left. 2\left(\frac{\partial^2 E}{\partial S\,\partial V}\right)\delta S\,\delta V\right] + \text{higher-order terms.}
\end{aligned} \tag{2.C.5}$$

From thermodynamics, we have

$$\left(\frac{\partial E}{\partial V}\right)_S = -P_0 \quad \text{and} \quad \left(\frac{\partial E}{\partial S}\right)_V = T_0, \tag{2.C.6}$$

where the subscript zero denotes evaluation of the coefficient in the uniform medium II. However, for simplicity, we shall drop the subscript zero from now on.

By combining Eqs. (2.C.5) and (2.C.6), we get

$$\delta E \simeq \frac{1}{2}\left[\left(\frac{\partial^2 E}{\partial S^2}\right)_V (\delta S)^2 + \left(\frac{\partial^2 E}{\partial V^2}\right)_S (\delta V)^2 + 2\left(\frac{\partial^2 E}{\partial V\,\partial S}\right)\delta S\,\delta V\right]_1, \tag{2.C.7}$$

where we have neglected higher-order terms. The right-hand side of Eq. (2.C.7) can be simplified. In particular, we want to eliminate the cross term involving

$\delta S \, \delta V$ by choosing two statistically independent thermodynamic variables such as T and V. With the use of Eq. (2.C.6), we get

$$\left(\frac{\partial^2 E}{\partial V^2}\right)_S = -\left(\frac{\partial P}{\partial V}\right)_S, \qquad \left(\frac{\partial^2 E}{\partial S^2}\right)_V = \left(\frac{\partial T}{\partial S}\right)_V \qquad (2.C.8)$$

and

$$\left(\frac{\partial^2 E}{\partial S \, \partial V}\right) = \left(\frac{\partial T}{\partial V}\right)_S = -\left(\frac{\partial P}{\partial S}\right)_V. \qquad (2.C.9)$$

With

$$\delta S = \left(\frac{\partial S}{\partial V}\right)_T \delta V + \left(\frac{\partial S}{\partial T}\right)_V \delta T, \qquad (2.C.10)$$

we then have

$$\left(\frac{\partial^2 E}{\partial S^2}\right)_V (\delta S)^2 + \left(\frac{\partial^2 E}{\partial V^2}\right)_S (\delta V)^2 + 2\left(\frac{\partial^2 E}{\partial V \, \partial S}\right) \delta S \, \delta V$$

$$= \left(\frac{\partial S}{\partial T}\right)_V (\delta T)^2 + 2\left(\frac{\partial S}{\partial V}\right)_T \delta V \, \delta T + \left(\frac{\partial S}{\partial V}\right)_T^2 \left(\frac{\partial T}{\partial S}\right)_V (\delta V)^2 \qquad (2.C.11)$$

$$- \left(\frac{\partial P}{\partial V}\right)_S (\delta V)^2 + 2\left(\frac{\partial T}{\partial V}\right)_S \left(\frac{\partial S}{\partial T}\right)_V \delta T \, \delta V$$

$$- 2\left(\frac{\partial P}{\partial S}\right)_V \left(\frac{\partial S}{\partial V}\right)_T (\delta V)^2$$

Since $(\partial T/\partial V)_S (\partial S/\partial T)_V (\partial V/\partial S)_T = -1$, the cross term is zero. Therefore, V and T are indeed statistically independent Gaussian random variables. The term involving the mean squared fluctuations in V can be simplified by utilizing the relation

$$\left(\frac{\partial P}{\partial V}\right)_T = \left(\frac{\partial P}{\partial V}\right)_S + \left(\frac{\partial P}{\partial S}\right)_V \left(\frac{\partial S}{\partial V}\right)_T. \qquad (2.C.12)$$

More specifically, we have

$$\left(\frac{\partial S}{\partial V}\right)_T^2 \left(\frac{\partial T}{\partial S}\right)_V - 2\left(\frac{\partial P}{\partial S}\right)_V \left(\frac{\partial S}{\partial V}\right)_T - \left(\frac{\partial P}{\partial V}\right)_S$$

$$= -\left(\frac{\partial S}{\partial V}\right)_T \left(\frac{\partial T}{\partial V}\right)_S - 2\left(\frac{\partial P}{\partial S}\right)_V \left(\frac{\partial S}{\partial V}\right)_T - \left(\frac{\partial P}{\partial V}\right)_S \qquad (2.C.13)$$

$$= -\left(\frac{\partial P}{\partial V}\right)_S - \left(\frac{\partial P}{\partial S}\right)_V \left(\frac{\partial S}{\partial V}\right)_T = -\left(\frac{\partial P}{\partial V}\right)_T.$$

By definition,

$$C_V = T\left(\frac{\partial S}{\partial T}\right)_V \quad \text{and} \quad K_T = -\frac{1}{V}\left(\frac{\partial V}{\partial P}\right)_T.$$

We then get

$$w_1(\delta V, \delta T)\, d(\delta V)\, d(\delta T)$$

$$= \text{const} \times \exp\left[\left(-\frac{1}{VK_T}(\delta V)^2 - \frac{C_V}{T}(\delta T)^2\right)_1 \bigg/ 2kT\right] d(\delta V)\, d(\delta T).$$

(2.C.14)

From Eq. (2.C.4) we can obtain the mean squared values of $(\delta V)^2$ and $(\delta T)^2$ for the *subsystem*:

$$(\delta V)^2 = \frac{\displaystyle\int_{-\infty}^{\infty} (\delta V)^2 w_1(\delta V, \delta T)\, d(\delta V)\, d(\delta T)}{\displaystyle\int_{-\infty}^{\infty} w_1(\delta V, \delta T)\, d(\delta V)\, d(\delta T)} = kTK_T v_m^* \qquad (2.C.15)$$

and

$$\overline{(\delta T)^2} = kT^2/C_{V,1}, \qquad (2.C.16)$$

where $C_{V,1}$ is the specific heat at constant volume in subsystem I.

The reader may wish to refer to the derivations by Benedek (1968) and the famous text by Landau and Lifshitz (1958) for further details.

APPENDIX 2.D. SPATIAL CORRELATION FUNCTION AND SPATIAL VOLUME

The isotropic spatial correlation function has the form (Chu, 1967)

$$C(\rho_r) = \langle \Delta_1 \Delta_2 \rangle / \langle \Delta^2 \rangle, \qquad (2.D.1)$$

where Δ_i represents the local fluctuations in a random process at point $\mathbf{r}_i$. The bracket indicates an equilibrium ensemble average. The local deviations from the equilibrium value Δ could mean fluctuations in pressure P, entropy per unit volume s, concentration C, dielectric constant κ_e, electric field $\mathbf{E}$, etc., depending upon the problem of interest. Physically we may interpret the correlation function as follows.

Let us imagine that we can measure the instantaneous dielectric constant on a line passing through the interior of a liquid, as shown schematically in Fig. 2.D.1. The quantity $\Delta_1 \Delta_2$ represents a product of local fluctuations measured at $\mathbf{r}_1$ and $\mathbf{r}_2$ at a distance $\rho_r (= |\mathbf{r}_1 - \mathbf{r}_2|)$ apart. We may consider Δ_1 and Δ_2 as the observed local fluctuations measured at the two ends 1 and 2, located at $\mathbf{r}_1$ and $\mathbf{r}_2$, respectively, of a stick with length ρ_r. The ensemble aver-

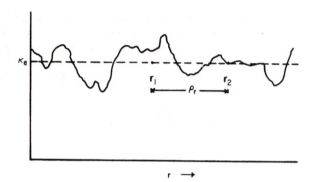

FIG. 2.D.1.   Instantaneous dielectric constant as a function of separation distance ρ_r.

age is obtained by throwing the stick at random into the liquid so as to allow the stick to take up all possible orientations and positions. The average square of the local fluctuations, $\langle \Delta^2 \rangle$, corresponds to the ensemble average of the square of local fluctuations measured with a stick of zero length. $C(\rho_r)$ starts at 1 when $\rho_r = 0$, which corresponds to perfect correlation, and eventually approaches zero for large values of ρ_r, as the fluctuations at $\mathbf{r}_1$ and at $\mathbf{r}_2$ are statistically independent because molecular interactions decrease with increasing distance. The correlation curve from a plot of $C(\rho_r)$ vs ρ_r can have different shapes.

The normalized spatial correlation function for entropy and pressure fluctuations may be defined as

$$F(\rho_r) = \langle \Delta s_P(\rho_r) \Delta s_P(0) \rangle / \langle \Delta s_P^2(0) \rangle, \tag{2.D.2}$$

$$G(\rho_r) = \langle \Delta P_s(\rho_r) \Delta P_s(0) \rangle / \langle \Delta P_s^2(0) \rangle, \tag{2.D.3}$$

where we have dropped the time reference.

Both $F(\rho_r)$ and $G(\rho_r)$ decrease rapidly to zero as ρ_r increases beyond correlation ranges r_s and r_P that are determined by the range of molecular pair correlation functions. In a liquid, the angular distribution of scattered x-ray intensity shows that the intermolecular distance is small compared to the wavelength of visible light, so that we may set

$$\exp[i(\mathbf{k}_I - \mathbf{k}_s) \cdot (\mathbf{r}_1' - \mathbf{r}_2')] = \exp[i(\mathbf{k}_I - \mathbf{k}_s) \cdot \rho_r'] \approx 1 \tag{2.D.4}$$

in Eq. (2.3.3) with $\rho_r' = \mathbf{r}_1' - \mathbf{r}_2'$. Thus, the phase integral in Eq. (2.3.3) becomes

$$v\left[\left(\frac{\partial \chi_e}{\partial s}\right)_{P_0}^2 \int 4\pi \rho_r^2 F(\rho_r) \langle \Delta s_P^2 \rangle \, d\rho_r + \left(\frac{\partial \chi_e}{\partial P}\right)_{s_0}^2 \int 4\pi \rho_r^2 G(\rho_r) \langle \Delta P_s^2 \rangle \, d\rho_r \right]$$

$$= v\left[\left(\frac{\partial \chi_e}{\partial s}\right)_{P_0}^2 \langle \Delta s_P^2 \rangle v_F^* + \left(\frac{\partial \chi_e}{\partial P}\right)_{s_0}^2 \langle \Delta P_s^2 \rangle v_G^* \right]. \tag{2.D.5}$$

In Equation (2.D.5), there was a change of variable in the integral of Eq. (2.3.3) resulting in the factor of the scattering volume v, and the limits of integral were changed from v to arbitrarily small distances. The dimensions defining the scattering volume could, however, still be used. The main point is that $\langle \Delta s_p^2 \rangle$ and $\langle \Delta P_s^2 \rangle$ are averaged over large enough dimensions to yield truly thermodynamic quantities:

$$v_F^* = 4\pi \int_0^{r_s} \rho_r^2 F(\rho_r) \, d\rho_r \tag{2.D.6}$$

and

$$v_G^* = 4\pi \int_0^{r_P} \rho_r^2 G(\rho_r) \, d\rho_r. \tag{2.D.7}$$

The correlation volumes represent the persistence of the regions over which essentially uniform values in Δs_P and ΔP_s take place. We have dropped the cross terms involving $\Delta P_s(\mathbf{r}_1, t) \Delta s_P(\mathbf{r}_2, t)$ and $\Delta s_P(\mathbf{r}_1, t) \Delta P_s(\mathbf{r}_2, t)$, remembering that the variables P and s are statistically independent.

REFERENCES

Bearman, R. J. and Chu, B. (1967). "Problems in Chemical Thermodynamics," pp. 77–78, 80–81, Addison-Wesley, Reading, Massachusetts.

Bender, T. M., Lewis, R. J., and Pecora, R. (1986). *Macromoecules* **19**, 244.

Benedek, G. B. (1968). *In* "Statistical Physics, Phase Transitions and Superfluidity" (M. Cretien, E. P. Gross, and S. Deser, eds.), Gordon and Breach, New York.

Berne, B. J. and Pecora, R. (1976). "Dynamic Light Scattering," Wiley-Interscience, New York.

Brillouin, L. (1914). *C. R. H. Acad. Sci.* **158**, 1331.

Brillouin, L. (1922). *Ann. Phys. (Paris)* **17**, 88.

Chu, B. (1967). "Molecular Forces Based on the Baker Lectures of Peter J. W. Debye," Wiley, New York.

Chu, B. (1970). *Ann. Rev. Phys. Chem.* **21**, 145.

Chu, B. (1974). "Laser Light Scattering," Academic Press, New York.

Cummins, H. Z. and Gammon, R. W. (1966). *J. Chem. Phys.* **44**, 2785.

Debye, P. (1915). *Ann. Physik* **46**, 809.

Debye, P. (1932). "The Collected Papers of Peter J. W. Debye," Interscience, New York, 1954.

Ehl, J., Loucheux, C., Reiss, C., and Benoit, H. (1964). *Makromol. Chem.* **75**, 35.

Einstein, A. (1910). *Ann. Phys. (Leipzig)* **38**, 1275.

Einstein, A. (1914). *Archives des Sci. Phys. Nat.* **37**, 254.

Finnigan, J. A. and Jacobs, D. J. (1970). *Chem. Phys. Lett.* **6**, 141.

Khintchine, A. (1934). *Math. Ann.* **109**, 604.

Kirkwood, J. G. and Oppenheim, I. (1961). "Chemical Thermodynamics," Chap. 6, McGraw-Hill, New York.

Kops–Werkhoven, M. M., Pathmamanoharan, C., Vrij, A., and Fijnaut, H. M. (1982). *J. Chem. Phys.* **77**, 5913.

Kratohvil, J. P., Dezelic, Gj., Kerker, M., and Matijevic, E. (1962). *J. Polym. Sci.* **57**, 59.

Landau, L. D. and Lifshitz, E. M. (1958). "Statistical Physics," Chapter XII, Addison-Wesley, Reading, Massachusetts.

Landau, L. and Placzek, G. (1934). *Phys. Z. Sowjetunion* **5**, 172.

Lastovka, J. B. (1967). "Light Mixing Spectroscopy and the Spectrum of Light Scattered by Thermal Fluctuations in Liquids," Ph.D. Thesis, MIT, Cambridge, Massachusetts.

Lastovka, J. B. and Benedek, G. (1966). *Phys. Rev. Lett.* **17**, 1039.

Margenau, H. and Murphy, G. M. (1956). "The Mathematics of Physics and Chemistry," pp. 247ff, Van Nostrand Reinhold, Princeton, New Jersey.

Moreels, E., DeCeuninck, W., and Finsy, R. (1987). *J. Chem. Phys.* **86**, 618.

Panofsky, W. K. H. and Phillips, N. (1955). "Classical Electricity and Magnetism," Addison-Wesley, Reading, Massachusetts.

Patterson, G. (1983). *Adv. Polym. Sci.* **48**, 125; *Ann. Rev. Mater. Sci.* **13**, 219.

Pike, E. R., Pomeroy, W. R. M., and Vaughan, J. M. (1975). *J. Chem. Phys.* **62**, 3188.

Pusey, P. N. and Tough, R. A. (1985). *In* "Dynamic Light Scattering" (R. Pecora, ed.), Plenum Press, New York, Chapter 4, pp. 85–179.

Wiener, N. (1930). *Acta Math.* **55**, 117.

III

OPTICAL MIXING
SPECTROSCOPY

In Chapter I, we emphasized the complementary nature of laser light scattering, small-angle x-ray scattering (SAXS), and small-angle neutron scattering (SANS), as shown in Fig. 1.1.1 and as listed in Table 1.B.1. Some basic aspects of light scattering theory, in terms of time-averaged intensity (Section 2.3) and spectrum (Section 2.4) of scattered light, were discussed in Chapter II. The approach to computing the time-averaged scattered intensity includes particle scattering as well as scattering by a fluctuating dielectric medium. By considering the phase integral (Eq. (2.2.10)) of a fluctuating dielectric medium over the finite macroscopic scattering volume v, the basic expression for optical wavefront matching has been introduced. In this chapter, we shall first consider the coherence solid angle (Section 3.1) as viewed by the observer (detector), because measurements of the spectrum of scattered light using optical mixing spectroscopy require wavefront matching, which controls the signal-to-noise ratio of our measured intensity time correlation function or power spectrum (see Eqs. (2.4.13) and (2.4.14) for the relationship between them and Eqs. (3.3.26) and (3.3.27)). The quality of signal in terms of intensity time correlation function is also controlled by the way the signal is accumulated. This point will be discussed in Section 4.8. Thus, from a practical experimental viewpoint, even the size and shape of the scattering volume have very important effects on measurement efficiency in optical mixing spectroscopy. The photoelectric detection of the scattered electric field is discussed in Section 3.2, and the different optical mixing schemes, as shown in Fig. 3.2.1, are described in Section 3.3.

Section 3.1 (Coherence Solid Angle) and Section 3.2 (Photoelectric Detection of the Scattered Electric Field) are not closely related. The reader may skip Section 3.1 until later, when its consideration for optimization of the signal-to-noise ratio of the intensity time correlation function at a given magnitude becomes important for designing experiments.

The theory of light-beating spectroscopy is developed in terms of the classical coherence functions (Mandel and Wolf, 1965) by Lastovka (1967), Cummins and Swinney (1970), and Schulz-DuBois (1983). The quantum-mechanical theory of optical coherence by Glauber (1963, 1964, 1965, 1969) provides an equivalent fundamental approach. However, the classical theory is adequate for our purposes.

3.1. COHERENCE SOLID ANGLE

3.1.1. INTRODUCTION

Optical coherence theory is concerned with the whole field of statistical optics. The correlations associated with optical fields can be measured by means of photoelectric detectors. Thus, in practice, the studies deal with correlations between *photoelectrons*, which can be related to the correlations between *photons*. Furthermore, such correlations can be treated semiclassically, i.e., we may use the classical electromagnetic wave theory even when the interactions have to be treated quantum-mechanically (Mandel *et al.*, 1964; Mandel and Wolf, 1965).

Glauber (1963, 1964) has developed quantum-mechanical analogs of the correlation functions. His quantum correlation functions indeed provide a more basic approach to optical coherence theory, since the radiation field should be treated as a quantum-mechanical system. In general, classical and quantum-mechanical descriptions can lead to different results. However, we shall make the valid assumption that the semiclassical theory provides us with correct predictions in laser light scattering. Sudarshan (1963a,b) has shown the existence of a phase-space representation of the density operator of the electromagnetic field that leads to a correspondence between the classical and quantum-mechanical descriptions, at least for a stationary, ergodic field and in the case of thermal light, in which the ensemble distribution of the complex field amplitude is Gaussian.

Coherence solid angle is a crucial property into which the reader should have some insight. In order to be successful in performing optical mixing spectroscopic experiments, we need to consider the relationship between the coherence solid angle and the scattering volume v as observed at $Q(x_\alpha)$ of Fig. 2.1.1. The intensity time correlation function is a two-photon process. The two photons need to come essentially from the same wavefront.

3.1.2. CALCULATION OF THE COHERENCE SOLID ANGLE (AFTER LASTOVKA, 1967)

In Eq. (2.2.10), we recall that the scattered electric field observed at the point of observation $Q(\mathbf{R})$ comes from the plane-wave components of $\Delta\chi_e$ having a wave vector near $\mathbf{K} = \mathbf{k}_s - \mathbf{k}_I$:

$$J(\mathbf{k}_I - \mathbf{k}_s, \mathbf{K}') = \int_v \exp[i(\mathbf{k}_I - \mathbf{k}_s + \mathbf{K}') \cdot \mathbf{r}'] \, d\mathbf{r}'$$

$$= \begin{cases} v, & \begin{cases} -\pi/L_x \leq (\mathbf{k}_I - \mathbf{k}_s + \mathbf{K}')_x \leq \dfrac{\pi}{L_x}, \\ -\pi/L_y \leq (\mathbf{k}_I - \mathbf{k}_s + \mathbf{K}')_y \leq L_y, \\ -\pi/L_z \leq (\mathbf{k}_I - \mathbf{k}_s + \mathbf{K}')_z \leq L_z, \end{cases} \\ 0 \text{ otherwise.} \end{cases} \quad (2.2.10)$$

This implies that the scattering occurs in the direction $\mathbf{R}$ as shown in Fig. 3.1.1, and $\mathbf{k}_\pm + \mathbf{K}$ falls inside a rectangular parallelepiped cell ξ^* centered at $\mathbf{k}_s$ and with dimensions $\Delta k_x, \Delta k_y, \Delta k_z$ ($= 2\pi/L_x, 2\pi/L_y, 2\pi/L_z$), i.e., $\mathbf{K}$ comes within $(\Delta k_x, \Delta k_y, \Delta k_z)$ of satisfying the condition $\mathbf{k}_I - \mathbf{k}_s + \mathbf{K} = 0$. Furthermore, $J(\mathbf{k}_I - \mathbf{k}_s, \mathbf{K}) \neq 0$ only for those orientations on $Q(\mathbf{R})$ around the direction $\mathbf{k}_I + \mathbf{K}$ for which $\mathbf{k}_s$ falls inside the cell ξ^*. Based on the above boundary conditions (Zinman, 1964), the lattice points in reciprocal space are given by

$$K_x = \frac{2\pi\bar{l}}{L_x}, \qquad K_y = \frac{2\pi\bar{m}}{L_y}, \qquad K_z = \frac{2\pi\bar{n}}{L_z} \quad (3.1.1)$$

with $\bar{l}, \bar{m}, \bar{n} = 0, \pm 1, \pm 2, \dots.$

The Ewald sphere, as shown in Fig. 3.1.1, has a finite thickness equivalent to the dimensions of the unit cell ξ^* in reciprocal space because of the finite size of the scattering volume, which produces the uncertainty in momentum $((\Delta k_x, \Delta k_y, \Delta k_z) = \pm(\pi/L_x, \pi/L_y, \pi/L_z))$. Alternatively, by examining Eq. (2.2.10), we note that the conservation-of-momentum condition is not sharp. The light scattered in the direction $Q(\mathbf{R})$ is produced by a single Fourier component $\mathbf{K}_{\overline{lmn}}$ of the fluctuations falling inside the unit cell ξ^*, whose dimensions in reciprocal space are controlled by the finite dimensions of the scattering volume. Furthermore, the unit cells associated with all possible vectors $\mathbf{k}_I + \mathbf{K}_{\overline{lmn}}$ form a nonoverlapping net filling all $\mathbf{k}$-space on the Ewald sphere. It should be noted that $Q(\mathbf{R})$ in Fig. 2.1.1 is defined as the point of observation at $\mathbf{R}$. The same $Q(\mathbf{R})$ is now transformed to Q_k on the Ewald sphere in $\mathbf{k}$-space, and Q_{k_s} falls inside ξ^*. In Fig. 3.1.1, $\mathbf{k}_I$ has only the appearance of being longer than $\mathbf{k}_s$. In fact, $k_s = k_I$ and both $\mathbf{k}_s$ and $\mathbf{k}_I$ are on Q_k.

If $\mathbf{k}_s$ is allowed to assume all possible orientations on the Ewald sphere Q_k, then a shell-like volume is swept out by the unit cell ξ^* as shown in Fig. 3.1.2.

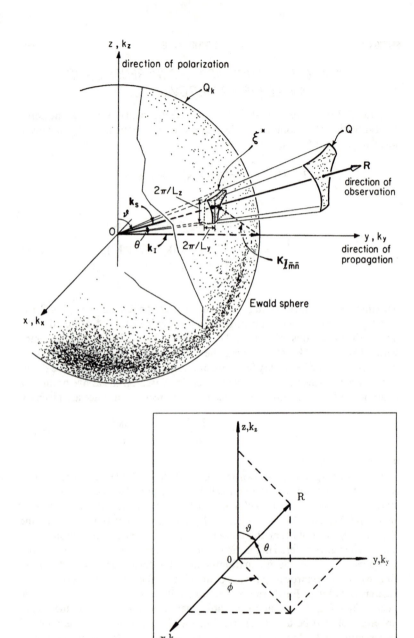

FIG. 3.1.1. Reciprocal space associated with Fourier series expansion of the electric susceptibility fluctuations (after Lastovka, 1967).

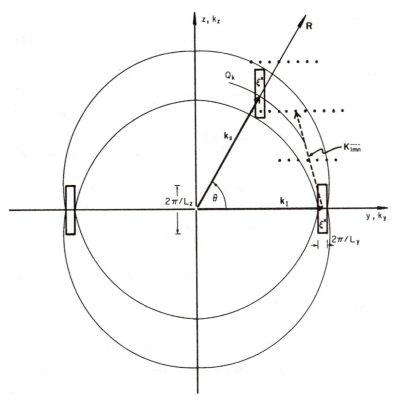

Fig. 3.1.2. Shell swept out by unit cell ξ^* if $\mathbf{k}_s$ takes up all orientations on the Ewald sphere Q_k (after Lastovka, 1967).

Since the momentum uncertainty gives the dimensions of the unit cell, we have

$$\Delta\mathbf{k} = \frac{2\pi}{L_x}\hat{\mathbf{x}} + \frac{2\pi}{L_y}\hat{\mathbf{y}} + \frac{2\pi}{L_z}\hat{\mathbf{z}}, \tag{3.1.2}$$

where the circumflex denotes the unit vector, i.e., $\hat{\mathbf{x}} = \mathbf{x}/|\mathbf{x}|$, etc. Thus, the shell thickness in spherical polar coordinates, as shown in the inset of Fig. 3.1.2, has the form

$$\Delta\mathbf{k} \cdot \frac{\mathbf{k}_s}{|k_s|} = 2\pi\left[\frac{\sin\vartheta\cos\phi}{L_x} + \frac{\sin\vartheta\sin\phi}{L_y} + \frac{\cos\vartheta}{L_z}\right], \tag{3.1.3}$$

and the volume of the shell in **k**-space is

$$v_\mathbf{k} = |\mathbf{k}_\mathrm{I}|^2 (\Delta \mathbf{k} \cdot \hat{\mathbf{k}}_\mathrm{s}) \Omega, \tag{3.1.4}$$

where Ω is a solid angle.

From Eq. (2.2.9), we get the density of cells in reciprocal **k**-space:

$$\rho_k = \left(\int_{-\pi/L_x}^{+\pi/L_x} dK_x \int_{-\pi/L_y}^{+\pi/L_y} dK_y \int_{-\pi/L_z}^{+\pi/L_z} dK_z \right)^{-1} = \frac{L_x L_y L_z}{(2\pi)^2} = \frac{v}{8\pi^3}. \tag{3.1.5}$$

Equation (3.1.5) suggests that the shape of the scattering volume has an effect on the coherence solid angle. Thus, the number of cells per unit solid angle is

$$\begin{aligned}
\frac{dN}{d\Omega} &= \frac{v}{8\pi^3} |\mathbf{k}_\mathrm{I}|^2 2\pi \left(\frac{\sin\vartheta\cos\phi}{L_x} + \frac{\sin\vartheta\sin\phi}{L_y} + \frac{\cos\vartheta}{L_z} \right) \\
&= \frac{v}{\lambda^2} \left(\frac{\sin\vartheta\cos\phi}{L_x} + \frac{\sin\vartheta\sin\phi}{L_y} + \frac{\cos\vartheta}{L_z} \right) \\
&= \frac{1}{\lambda^2} (L_y L_z \sin\vartheta\cos\phi + L_x L_z \sin\vartheta\sin\phi + L_x L_y \cos\vartheta), \quad (3.1.6)
\end{aligned}$$

and the coherence solid angle Ω_{coh} is

$$\Omega_{\mathrm{coh}} = \frac{\partial\Omega}{\partial N} = \frac{\lambda^2}{L_y L_z \sin\vartheta\cos\phi + L_x L_z \sin\vartheta\sin\phi + L_x L_y \cos\vartheta}. \tag{3.1.7}$$

We recall that the power scattered into a solid angle Ω has the form

$$P = 2P_\mathrm{I} \mathscr{R}^* L (\sin^2\vartheta) \Omega. \tag{2.3.27}$$

By substituting Eq. (3.1.7) into Eq. (2.3.27) and with $\mathbf{k}_\mathrm{I}$ in the y-direction, we obtain for the scattered power per coherence solid angle, P_{coh}:

$$P_{\mathrm{coh}} = 2P_\mathrm{I} \mathscr{R}^* L_y (\sin^2\vartheta) \frac{\lambda^2}{L_y L_z \sin\vartheta\cos\phi + L_x L_z \sin\vartheta\sin\phi + L_x L_y \cos\vartheta}. \tag{3.1.8}$$

For $\vartheta = 90°$, we then have

$$P_{\mathrm{coh}} = 2P_\mathrm{I} \mathscr{R}^* L_y \frac{\lambda^2}{L_z (L_x \sin\phi + L_y \cos\phi)}, \tag{3.1.9}$$

where ϑ is the angle between the incident polarization vector and the direction of observation. Equation (3.1.9) shows that P_{coh} depends upon the geometry of the scattering volume:

$$P_{\mathrm{coh}} \propto \{ L_z [\cos\phi + (L_x/L_y)\sin\phi] \}^{-1}.$$

By varying the dimensions of L_x, L_y, and L_z, we may maximize P_{coh}. In terms of the scattering angle θ, we have, according to Fig. 3.1.2 (inset),

$$P_{coh} = 2P_1\mathscr{R}^*L_y \frac{\lambda^2}{L_z(L_y \sin\theta + L_x \cos\theta)} \quad \text{for} \quad \vartheta = 90°, \quad (3.1.10)$$

since the angles ϕ and θ are defined from the x and y axes, respectively.

It is sufficient to mention here that as $P_{coh} \propto \lambda^2 \mathscr{R}^*$ and $\mathscr{R}^* \propto \lambda^{-4}$, the increase in P_{coh} on decreasing the incident wavelength is no longer as dramatic as $\mathscr{R}^* \propto \lambda^{-4}$. On closer examination, such a conclusion is reasonable, since wavefront matching in optical mixing spectroscopy becomes less critical with increasing wavelength. P_{coh} increases with increasing L_y, which is the dimension of the scattering volume in the direction of the incident beam ($\mathbf{k}_1$); but the behavior is more complex with respect to $L_x \cos\theta$ and $L_y \sin\theta$, including the presence of L_z in the denominator.

In Eq. (3.1.10), the power scattered into a coherence solid angle, P_{coh}, is shown to be dependent on $\lambda^2 \mathscr{R}^*$, on the dimensions of the scattering volume (L_x, L_y, L_z), and the scattering angle θ. As P_{coh} is not symmetric with respect to $\theta = 90°$, detector design at small and large scattering angles can be different with respect to angular apertures. We shall discuss this point later in connection with the practice of light scattering experiments.

It should be noted that Eq. (3.1.10) depends on the directions of polarization and the incident beam. If we take $\mathbf{k}_1$ to be in the z-direction and the polarization to be in the x-direction, then Eq. (3.1.10) is changed to

$$P_{coh} = 2P_1\mathscr{R}^*L_z \frac{\lambda^2}{L_x(L_z \sin\theta + L_y \cos\theta)} \quad \text{for} \quad \vartheta = 90°, \quad (3.1.11)$$

where we have only changed the axis designations ($x \to y$, $y \to z$, and $z \to x$), e.g., ϑ is the angle between x and $\mathbf{k}_s$, instead of z and $\mathbf{k}_s$. We shall follow the geometry for Eq. (3.1.11) in subsequent discussions. Let us first examine the half angle of coherence in the x–$\mathbf{k}_s$ plane, as shown in Fig. 3.1.3, for $\vartheta = 90°$:

$$(\Delta\vartheta)_{coh} = \frac{1}{2}\left(\frac{2\pi}{L_x}\right)\frac{1}{|\mathbf{k}_s|} = \frac{\lambda}{2L_x}. \quad (3.1.12)$$

Similarly, in the y–z plane, the half angle of coherence is

$$(\Delta\theta)_{coh} = \frac{1}{2}\frac{\lambda}{L_z \sin\theta + L_y \cos\theta} \quad (3.1.13)$$

as shown in Fig. 3.1.4. Thus, L_x and the ratio of L_y to L_z should be made as small as possible so as to maximize P_{coh}. To summarize, the scattering volume

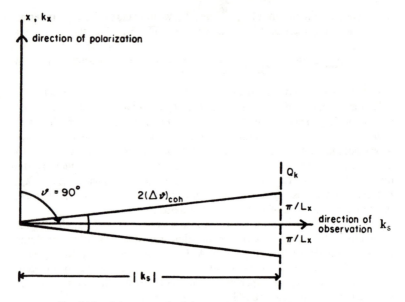

FIG. 3.1.3. Coherence angle of the scattering volume in the x–k_s plane.

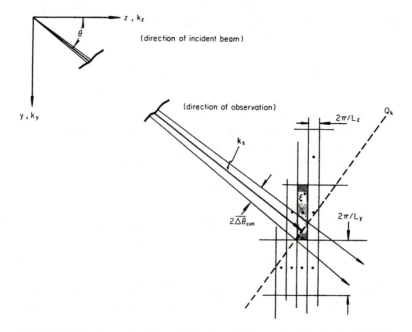

FIG. 3.1.4. Coherence angle of the scattering volume in the x–z plane (after Lastovka, 1967).

should have (1) small L_x with $(\Delta\vartheta)_{\text{coh}} = \lambda/2L_x$ in the direction of the incident polarization vector; (2) small L_y in the direction of the scattering plane; (3) large L_z in the direction of the incident beam.

A simple rule is to observe the scattering process by using a *thin, narrow* incident beam and a *long* scattering volume.

On closer examination of Eq. (3.1.13), we see that there are two limiting behaviors:

$$(\Delta\theta)_{\text{coh}} = \frac{\lambda}{2L_z\sin\theta}, \qquad \text{when} \quad L_z\sin\theta \gg L_y\cos\theta \text{ or } \tan\theta \gg L_y/L_z,$$

$$(3.1.14)$$

and

$$(\Delta\theta)_{\text{coh}} = \frac{\lambda}{2L_y\cos\theta}, \qquad \text{when} \quad L_y\cos\theta \gg L_z\sin\theta \text{ or } \tan\theta \ll L_y/L_z.$$

$$(3.1.15)$$

The dependence of $(\Delta\theta)_{\text{coh}}$ on $\sin\theta$ and $\cos\theta$ tells us that

$$P_{\text{coh}} \propto \begin{cases} \dfrac{1}{L_x\sin\theta} & \text{when} \quad \tan\theta \gg L_y/L_z, & (3.1.16) \\[2ex] \dfrac{L_z}{L_xL_y\cos\theta} & \text{when} \quad \tan\theta \ll L_y/L_z, & (3.1.17) \\[2ex] \dfrac{1}{2L_x\sin\theta} & \text{when} \quad \tan\theta_c = L_y/L_z. & (3.1.18) \end{cases}$$

From Eqs. (3.1.12) and (3.1.13), we retrieve the coherence solid angle for $\vartheta = 90°$:

$$\boxed{\Omega_{\text{coh}}(\vartheta = 90°) = 4(\Delta\theta)_{\text{coh}}(\Delta\vartheta)_{\text{coh}} = \frac{\lambda^2}{L_x(L_z\sin\theta + L_y\cos\theta)}.} \qquad (3.1.19)$$

Thus, *for fixed dimensions of L_x, L_y, and L_z*, we see from Eqs. (3.1.11) and (3.1.19) that $P_{\text{coh}} \propto L_z\Omega_{\text{coh}} \propto \Omega_{\text{coh}}$. Further clarification of Ω_{coh} for a scattering volume which changes with scattering angle is given in Fig. 6.9.2. The reader is advised to read Section 6.9.

Figure 3.1.5 shows a plot of $1/\{L_x[\sin\theta + (L_y/L_z)\cos\theta]\}$ and Ω_{coh} vs θ for two flat scattering cells whose scattering volumes have the following dimensions:

Cell	L_x (cm)	L_y (cm)	L_z (cm)	$\tan\theta_c$	θ_c
I	0.05	0.05	0.10	0.50	26°34′
II	0.05	0.05	1.0	0.05	2°52′

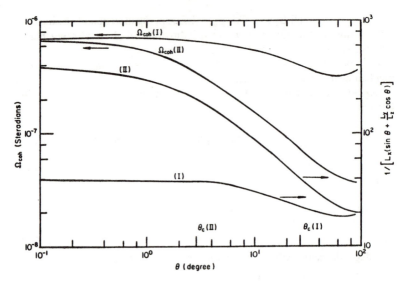

FIG. 3.1.5. A plot of $1/\{L_x[\sin\theta + (L_y/L_z)\cos\theta]\}$ and Ω_{coh} vs θ. Cell I has $L_x = L_y = 0.05$ cm, $L_z = 0.10$ cm; cell II has $L_x = L_y = 0.05$ cm, $L_z = 1.0$ cm.

With $\lambda_{air} = 632.8$ nm and $n = 1.5$, $(\Delta\theta)_{coh}$ varies from 0.02° to 0.01° for cell I and from 0.014° to 0.001° for cell II, between $\theta = 2°$ and 70°.

Thus we can draw the following conclusions:

(1) A dimension of 0.05 cm forms a reasonable upper limit for L_x and L_y. In fact, we should focus the laser beam down to smaller dimensions in order to reduce its cross section, which can be identified with L_x and L_y in the scattering volume. For example, a reduction of beam diameter from 0.05 cm down to 0.01 cm could increase $(\Delta\vartheta)_{coh}$ by a factor of 5. An angular aperture of about 0.25° for the detector is fairly easy to construct for a light-scattering spectrometer. Ordinary continuous-wave (cw) He–Ne or argon ion lasers have beam diameters of 0.1–0.3 cm, a factor of about 10 to 30 too large for our purposes. The reduction in beam diameter should *not* be performed with a diaphragm, but with a focusing lens, which increases the power density of the incident beam. In conventional light scattering, mercury arcs were used as light sources. Replacing the mercury arc lamp by a laser source *without focusing*, in a light-beating experiment, will result in a loss factor of at least about 30 in P_{coh} when compared with a focused laser incident beam. The point to be made here is that, because of wavefront matching, it pays to focus the laser beam in order to produce a small scattering volume that is convenient to

observe, and the light-beating experiment is different from the time-averaged experiment, where large beam cross sections resulting in large scattering volumes are acceptable.

(2) An increase in the dimension L_z (in the direction of the incident beam) becomes less important soon after $\theta < \theta_c$. A good criterion is $\tan \theta_c = L_y/L_z$ from Eq. (3.1.18). Then we need only to compare the sizes of effective scattering volumes between different L_x, L_y, and L_z.

(3) For fixed L_y, θ_c becomes smaller with increasing L_z.

If we have a long thin scattering volume (as in cell II), the most favorable measurements should be done at very small scattering angles ($\theta < \theta_c \approx 2°52'$) even though dust can then become a serious problem. At larger angles, the gain in P_{coh} decreases, as shown typically in Fig. 3.1.5. In the practice of light-beating experiments (see discussions in Section 3.2.1), we aim to maximize the signal-to-noise ratio, i.e., we want to measure the spectrum of scattered light to achieve the desired signal-to-noise ratio over the shortest measurement time period. For a given incident power, we want P_{coh} as high as possible over the range of scattering angles of interest. The equations already tell us that the optical geometry for the detector at small and large scattering angles should be different. In addition, we need to consider the counting statistics (Section 3.2). Thus, the two sources of maximization in P_{coh} and counting statistics are coupled and should be considered together, i.e., it does not make sense to maximize P_{coh} in such a way that the detector receives very little scattered intensity.

This is a subtle point, which deserves special attention from an uninitiated reader. It can be amplified as follows. We can focus the laser beam in order to provide easier wavefront matching conditions. Thus, with higher power density from the incident beam and smaller angular apertures as viewed by the observer (detector), most of the photons received by the observer will be capable of contributing toward the intensity time correlation function. However, we must keep sight of the total amount of scattered intensity received by the observer. It does not make sense, even if one gets perfect light beating, to reduce total intensity so greatly that the net total signal becomes less than that with somewhat imperfect light beating.

3.2. PHOTOELECTRIC DETECTION OF THE SCATTERED ELECTRIC FIELD

3.2.1. INTRODUCTION

We use photoelectric detectors to measure correlations in photoelectrons, which in turn can be related to correlations of the scattered electric field. In laser light scattering, we are concerned mainly with the UV–visible spectrum

Table 3.2.1

Wavelength, Frequency, Wave Number, Color

λ (nm)	$\tilde{v}\,(=1/\lambda)$ $(10^4\,\mathrm{cm}^{-1})$	$v\,(=c/\lambda)^{(a)}$ $(10^{14}\,\mathrm{Hz})$	$\omega\,(=2\pi v)$ $(10^{14}\,\mathrm{rad/sec})$	Color
380–480	2.63–2.08	7.90–6.25	49.6–39.3	violet
480–520	2.08–1.92	6.25–5.77	39.3–36.2	blue
520–560	1.92–1.78	5.77–5.36	36.2–33.7	green
560–610	1.78–1.64	5.36–4.92	33.7–30.9	yellow
610–630	1.64–1.59	4.92–4.76	30.9–29.9	orange
630–720	1.59–1.39	4.76–4.17	29.6–26.2	red

(a) $c = 3 \times 10^{10}$ cm/sec.

in the range from about 350 to 700 nm. Table 3.2.1 conveniently relates the various units of wavelength, frequency, wave numbers, and color.

Often, we are interested in knowing the bandwidth of a spectrometer, which is related to the percentage changes in frequency or wavelength that the instrument can measure. Since we have $v = c/\lambda$,

$$|\Delta v|/v \approx |\Delta \lambda|/\lambda, \tag{3.2.1}$$

which is the bandwidth, with c and v being the velocity and frequency of light. The best available grating spectrometer has a bandwidth of about 10^{-6}, while Fabry–Perot interferometers have bandwidths of about 10^{-7}. On the other hand, for example, the linewidth of the central Rayleigh component for a one-component system in the neighborhood of its critical point becomes very small (Chu, 1970), so we need a detection scheme with an extremely high resolving power ($v/|\Delta v| \approx \lambda/|\Delta \lambda|$), which is about a million times better than the narrowest bandpass filter in the optical frequencies, as has been illustrated schematically in Fig. 1.1.2.

The *direct*-detection receiver, in which the signal is detected with no prior translation of the incident signal energy to another frequency, is not appropriate. Instead, we should shift the frequency down to a sufficiently low value prior to filtering so that the desired bandpass filters, which are no longer optical filters, may be used at the new center frequency. This technique, called *light-beating detection*, can be subdivided into three categories: heterodyne detection, whereby a portion of the incident beam acting as a local oscillator but with a shifted frequency is optically mixed with the scattered light; homodyne detection (a special case of heterodyne detection), whereby the local-oscillator frequency and the input signal frequency have the same value; and self-beating, whereby the scattered intensity is optically mixed and the beat frequencies measured among themselves. A schematic representation of light-beating spectroscopy and optical spectroscopy is given in Fig. 3.2.1. It should be noted that in optical mixing spectroscopy we always have self-beating. While optical spectrometers filter at the optical frequency, the light-beating spectrometer filters only after the light carrier wave has been shifted down, by means of a nonlinear element (a mixer—such as a photomultiplier tube, which measures the *intensity* of light), to a much lower frequency, where very narrow bandpass filters are available. The photosensitive element (mixer) detects light intensity in terms of the number of photoelectrons—or photocurrent—which is proportional to the *square* of the electric field falling on the photosensitive surface. Thus, the photomultiplier (PMT) is a perfect nonlinear square-law detector.

Before describing the optical mixing spectrometers (as shown in Fig. 3.2.1), we need to know that the photoelectric detection process has its own

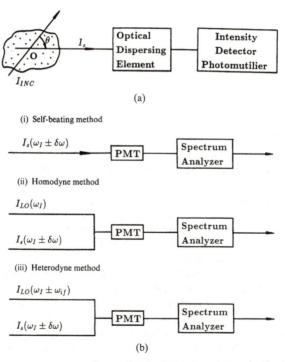

FIG. 3.2.1. Schematic representation of light-beating spectroscopic and optical scattering techniques. $I_s^c(K, \tau) = \langle I_s(K, 0) I_s(K, \tau) \rangle$; $S(K, \omega) = (1/2\pi) \int I_s^c(K, \tau) e^{iK\tau} d\tau$. (a) Optical spectroscopy; (b) optical mixing spectroscopy; (c) $S(K, \omega)$ from different methods of optical mixing spectroscopy.

(i) Self-beating method

(ii) Homodyne method

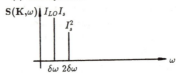

(iii) Heterodyne method

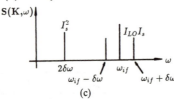

(c)

FIG. 3.2.1. (*continued*)

probabilistic nature. The relationship between the photoelectric current (or count) spectrum and the optical spectrum depends crucially on the statistical nature of optical signals as well as the detection process itself. Furthermore, photodetectors are not able to follow the optical frequency, but only the modulation due to motions of molecules in our scattering process, as shown pictorially in Fig. 3.2.2. The photodetector output signal contains information on the modulation frequency, which is the slowly varying envelope of a rapidly oscillating optical field.

We may consider the resulting time variation of the photocurrent (or photocount) as beats (differences) between closely spaced optical frequencies from a nonlinear (square-law) mixer. The net result is a transposition of the optical *field* spectrum centered at $v_l \approx 5 \times 10^{14}$ Hz to a *current* (or photoelectron count) spectrum at frequencies centered at 0 Hz, where very narrow-band (e.g. 1-Hz) filter is easily available. The advantages in the translation of the center frequency may be visualized as follows:

(1) In the self-beating technique, if the spectrum has a Lorentzian linewidth of Γ_c, and the band filter has $\Delta v_0 = 1$ Hz, then the effective resolving power

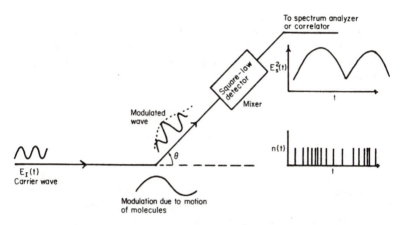

FIG. 3.2.2. Pictorial representation of the self-beating technique.

using an incident frequency ν of 5×10^{14} Hz is

$$\left(\frac{\omega_l}{\Delta\omega}\right)_{\text{eff}} = \left(\frac{\omega_l}{\Gamma_c}\right)\left(\frac{\Gamma_c}{\Delta\omega_0}\right) = \frac{\omega_l}{\Delta\omega_0} = 5 \times 10^{14},$$

where $\omega = 2\pi\nu$.

(2) In heterodyne detection, if the local oscillator has a frequency ω_{LO}, the beat signals between ω_{LO} and the frequency components of the input signal give a spectrum that is identical to the spectrum of the input signal except for a shift in the center frequency from ω_l to $\omega_{\text{if}} = \omega_l - \omega_{\text{LO}}$. Thus, the power spectrum can be centered around the zero frequency, as shown by using the configuration in Fig. 3.2.1(b)(ii) if $\omega_{\text{LO}} = \omega_l$, i.e., the local oscillator has the same frequency as the incident laser beam; or it can be centered around $\omega_{\text{if}} = |\omega_l - \omega_{\text{LO}}|$ as shown by using the configuration in Fig. 3.2.1(b)(iii). In Fig. 3.2.1(c), we have neglected the dc and shot-noise components and emphasized the frequency shift ω_{if} of the signal power spectrum. In addition, by shifting the local-oscillator frequency, we have separated the always present self-beating component of the power spectrum. $\Delta\omega_{\text{if}}$ is the bandwidth of the tuned filter in the intermediate frequency (if) range. The effective resolving power for $\Delta\nu_{\text{if}} = 1$ Hz is

$$\left(\frac{\omega_l}{\Delta\omega}\right)_{\text{eff}} = \left(\frac{\omega_l}{\omega_{\text{if}}}\right)\left(\frac{\omega_{\text{if}}}{\Delta\omega_{\text{if}}}\right) = \frac{\omega_l}{\Delta\omega_{\text{if}}} = 5 \times 10^{14}.$$

Figure 3.2.1 also shows that the time-dependent photocurrent $i_s(t)$ $[\propto E_s^2(t)]$ output can be expressed in terms of time-dependent photopulses $n(t)$. The deviation of the counting distribution from a purely Poisson distribution provides us with the same information on the spectrum of light. However, the information is in digital form. As the photomultiplier tube measures photoelectric pulses, it is easier to design a more precise instrument using the photoelectron count method.

3.2.2. PROBABILITY DISTRIBUTION AND FACTORIAL MOMENTS FOR EMITTED PHOTOPULSES

In the light-beating technique, the photomultiplier, acting as a nonlinear square-law detector for the intensity of light, is the key to optical mixing spectroscopic measurements, as illustrated in Fig. 3.2.1(b). Photons impinging on the photocathode of the photomultiplier produce amplified photoelectron current pulses at its anode output. However, only a fraction (α) of the photons falling on the photocathode are responsible for the output photoelectron pulses. Thus, the detection of photons is a statistical process. The quantum efficiency a of the photomultiplier plays an important role in making an appropriate choice for the photodetector in optical mixing spectroscopy. The probability distribution of registering n photoelectron pulses by an ideal detector in a sampling time interval T from t to $t + T$ is

$$p(n, t, T) = (1/n!)(aI_0T)^n \exp(-aI_0T), \tag{3.2.2}$$

where I_0 is a constant short-time intensity, and T, the sampling time period. The average number of photoelectron pulses emitted during the sampling period T is

$$\langle n(T) \rangle = \sum_{n=0}^{\infty} \frac{n(aI_0T)^n}{n!} \exp(-aI_0T) = \sum_{n=1}^{\infty} \frac{1}{(n-1)!} \exp(-aI_0T)(aI_0T)^n$$

$$= \exp(-aI_0T)(aI_0T) \sum_{n=1}^{\infty} \frac{(aI_0T)^{n-1}}{(n-1)!}$$

$$= \exp(-aI_0T)(aI_0T) \exp(aI_0T) = aI_0T,$$

or

$$\langle n(T) \rangle = aI_0T. \tag{3.2.3}$$

More generally, we redefine a so that it is proportional to the averaged gain of the photodetector as well as the quantum efficiency with which the photons are able to eject electrons, i.e., $a = $ (quantum efficiency) $\times$ (detector gain). By combining Eq. (3.2.3) with Eq. (3.2.2), we may write

$$p(n, t, T) = \frac{1}{n!} \exp[-\langle n(T) \rangle] \langle n(T) \rangle^n, \tag{3.2.4}$$

which is a form of the classic Poisson distribution and is normalized, since

$$\sum_{n=0}^{\infty} \frac{1}{n!} \exp[-\langle n(T)\rangle]\langle n(T)\rangle^n = \exp[-\langle n(T)\rangle] \sum_{n=0}^{\infty} \frac{\langle n(T)\rangle^n}{n!}$$

$$= \exp[-\langle n(T)\rangle + \langle n(T)\rangle] = 1.$$

Using Sterling's approximation for large n:

$$n! = (2\pi n)^{1/2} n^n e^{-n}(1 + 1/12n + \cdots),$$

we obtain for the probability of finding $\langle n(T)\rangle$ over the sample period T

$$p(\langle n(T)\rangle, t, T) = \frac{1}{\langle n(T)\rangle!} \exp[-\langle n(T)\rangle]\langle n(T)\rangle^{\langle n(T)\rangle}$$

$$\approx \frac{1}{\langle n(T)\rangle!} \frac{\langle n(T)\rangle!}{[2\pi\langle n(T)\rangle]^{1/2}} = \frac{1}{[2\pi\langle n(T)\rangle]^{1/2}}. \tag{3.2.5}$$

The mean squared fluctuation in the number of photopulses is equal to $\langle n(t)\rangle$:

$$\langle[n - \langle n(T)\rangle]^2\rangle_{T,\text{Poisson}} = \sum_{n=0}^{\infty} [n - \langle n(T)\rangle]^2 p(n, t, T)$$

$$= \sum_{n=0}^{\infty} [n^2 - 2n\langle n(T)\rangle + \langle n(T)\rangle^2] p(n, t, T)$$

$$= \sum_{n=0}^{\infty} [n^2 - \langle n(T)\rangle^2] p(n, t, T)$$

$$= \sum_{n=0}^{\infty} n^2 p(n, t, T) - \langle n(T)\rangle^2 \sum_{n=0}^{\infty} p(n, t, T)$$

$$= \left(\sum_{n=0}^{\infty} \frac{n}{(n-1)!} \langle n(T)\rangle^n \exp[-\langle n(T)\rangle]\right) - \langle n(T)\rangle^2$$

$$= \left[\sum_{n=1}^{\infty} \left(\frac{(n-1)}{(n-1)!} + \frac{1}{(n-1)!}\right) \right.$$

$$\left. \times \langle n(T)\rangle^n \exp[-\langle n(T)\rangle]\right] - \langle n(T)\rangle^2$$

$$= \sum_{n=2}^{\infty} \frac{\exp[-\langle n(T)\rangle]\langle n(T)\rangle^{n-2}\langle n(T)\rangle^2}{(n-2)!}$$

$$+ \sum_{n=1}^{\infty} \frac{\exp[-\langle n(T)\rangle](\langle n(T)\rangle^{n-1})\langle n(T)\rangle}{(n-1)!} - \langle n(T)\rangle^2$$

$$= \langle n(T)\rangle^2 + \langle n(T)\rangle - \langle n(T)\rangle^2 = \langle n(T)\rangle,$$

or

$$\langle [n - \langle n(T) \rangle]^2 \rangle_{T, \text{Poisson}} = \langle n(T) \rangle. \tag{3.2.6}$$

If the intensity I is a random variable, the integral

$$W = \int_t^{t+T} I(T') \, dT' \tag{3.2.7}$$

is the rate of events integrated over the counting interval T and is a random variable with a nonzero mean value. Then, the *probability distribution* $p(n, t, T)$ for registering n photoelectrons by an ideal detector in a time interval from t to $t + T$ is also a random variable, and its average over the probability density $P(W)$ of the light intensity is

$$p(n, t, T) = \int_0^{\infty} \frac{(aW)^n}{n!} e^{-aW} P(W) \, dW. \tag{3.2.8}$$

Equation (3.2.8) was first derived by Mandel (1958, 1959) using classical arguments. Jakeman and Pike (1968) have computed $P(W)$ under the conditions that the light field is a Gaussian random variable and that T is small compared to the coherence time τ_c, so that $W = IT$. For radiation fields produced from most of the available sources, we may take $P(W)$ as a probability function, with

$$P(W) = \frac{1}{\langle W \rangle} \exp\left(\frac{-W}{\langle W \rangle}\right) \tag{3.2.9}$$

where $\langle W \rangle = \langle I \rangle T \, (= I_0 T)$.

Equation (3.2.8) can be used to derive the expectation values $\langle n \rangle$ and $\langle n(n-1) \rangle$, whereby Eqs. (3.2.3) and (3.2.6), known as the first two factorial moments, are retrieved.

For the first factorial moment $\langle n \rangle$, the average number of photoelectron pulses emitted during the sampling period T, we have

$$\begin{aligned}
\langle n \rangle &= \sum n p(n, t, T) \\
&= \int \sum n \frac{(aW)^n}{n!} \exp(-aW) P(W) \, dW \\
&= \int aW \sum \frac{(aW)^{n-1}}{(n-1)!} \exp(-aW) P(W) \, dW \\
&= \int aW \sum \frac{(aW)^m}{m!} \exp(-aW) P(W) \, dW \\
&= \int aW \exp(aW) \exp(-aW) P(W) \, dW
\end{aligned}$$

$$= a \int W P(W) \, dW$$

$$= a\langle W \rangle = aT\langle I \rangle. \tag{3.2.10}$$

For the second factorial moment we have

$$\langle n(n-1) \rangle = \sum_n n(n-1) p(n, t, T)$$

$$= \int_0^\infty \sum n(n-1) \frac{(aW)^n}{n!} \exp(-aW) P(W) \, dW$$

$$= \int (aW)^2 \sum \frac{(aW)^{n-2}}{(n-2)!} \exp(-aW) P(W) \, dW$$

$$= \int (aW)^2 \sum \frac{(aW)^m}{m!} \exp(-aW) P(W) \, dW$$

$$= \int (aW)^2 P(W) \, dW = a^2 \langle W^2(t, T) \rangle,$$

or

$$\langle n(n-1) \rangle = a^2 \langle W^2(t, T) \rangle = a^2 T^2 \langle I^2 \rangle. \tag{3.2.11}$$

In deriving Eqs. (3.2.10) and (3.2.11), we started with the definition of expectation values, eliminated a common factor (n or $n(n-1)$) in the numerator and the factorial, and renamed the summation index. More generally, we can write the kth factorial moment as

$$\langle n(n-1) \cdots (n-k+1) \rangle = a^k T^k \langle I^k \rangle. \tag{3.2.12}$$

Several conclusions of importance to photon correlation may be drawn from the above equations (Schulz-DuBois, 1983):

(1) Equation (3.2.10) shows that the expectation value for the intensity of light, $\langle I \rangle$, and that of the photoelectron pulses, $\langle n \rangle$ ($= aT\langle I \rangle$), are equivalent, with the proportionality constant aT. Thus, $\langle I \rangle$ can be measured without error from $\langle n \rangle$.

(2) By combining Eqs. (3.2.10) and (3.2.11) with Eq. (3.2.6), we can write down the relative photon-count variance

$$\frac{\langle n^2 \rangle - \langle n \rangle^2}{\langle n \rangle^2} = \frac{\langle I^2 \rangle - \langle I \rangle^2}{\langle I \rangle^2}$$

$$= \frac{\langle n \rangle^2 + \langle n \rangle - \langle n \rangle^2}{\langle n \rangle^2} = \frac{1}{\langle n \rangle} = \frac{1}{aT\langle I \rangle}. \tag{3.2.13}$$

Equation (3.2.13) shows that the relative photon-count variance is finite and decreases with increasing sampling period T.

3.3. OPTICAL MIXING SPECTROMETERS

3.3.1. INTRODUCTION

B. Saleh (1978) discussed the technique of optical mixing spectroscopy in detail. The use of the photoelectric effect for nonlinear detection was first demonstrated in an experiment by Forrester *et al.* (1955) and then confirmed by the fluctuation-correlation experiments of Hanbury Brown and Twiss (1956). However, correlation experiments were very difficult because of the broad linewidth and low power density of conventional light sources. In 1961, Forrester sugggested the use of lasers as a light source and pointed out that photoelectric mixing is analogous to the mixing of ac electrical signals in nonlinear circuit elements. Thus, opticl mixing spectrometers are the analogs of the superheterodyne and the homodyne receivers. The nonlinear detector or mixer in an optical mixing spectrometer is a photoelectric device such as a photomultiplier tube or a photodiode, which produces a photocurrent proportional to the *square* of the total electric field falling on the photosensitive surface of the device.

With the advent of lasers, Javan *et al.* (1961) first obtained intermode beats of a He–Ne laser using a radio-frequency spectrum analyzer. When combined with lasers, photoelectric mixing is particularly suitable for quasielastic light scattering resulting from various time-dependent nonpropagating local thermodynamic fluctuations. Several closely related experimental approaches have been developed. These include spectrum analysis as well as signal correlation. In signal correlation, the emphasis has been on photon counting statistics.

According to the Wiener–Khintchine theorem (Eqs. (2.4.13) and (2.4.14)) the power spectrum $S_j(\omega)$ of the photocurrent density and the current-density time correlation function $R_j(\tau)$ *at one point of the photocathode surface* with perfect coherence are related through the equations

$$R_j(\tau) = \langle j(t + \tau)j(t) \rangle = \int_{-\infty}^{\infty} S_j(\omega) \cos \omega\tau \, d\omega, \qquad (3.3.1)$$

$$S_j(\omega) = \frac{1}{2\pi} \int_{-\infty}^{\infty} R_j(\tau) \cos \omega\tau \, d\tau, \qquad (3.3.2)$$

where $j(t)$ is the photocurrent density at time t. The total anode current from a photodetector, $i(t)$, is

$$i(t) = \int_{S} j(\mathbf{R}, t) \, dS$$

with S being the illuminated photocathode area. More rigorously speaking, correlation of the current output of the photodetector has the form

$$
\begin{aligned}
R_i(\tau) &= \langle\langle i(t + \tau)i(t)\rangle_{\text{st}}\rangle \\
&= \int_S \int_S \langle j(\mathbf{R}_2, t + \tau)j(\mathbf{R}_1, t)\rangle_{\text{st}} \, d^2R_2 \, d^2R_1,
\end{aligned}
\tag{3.3.3}
$$

where $d^2R_p = dS_p$ and the subscript st denotes short-time averages. Equation (3.3.3) requires us to consider the spatial characteristics of the mixing process in the double surface integral. Furthermore, the time behavior of $R_i(\tau)$ is contained in $R_j(\tau) = \langle\langle j(\mathbf{R}_2, t + \tau)j(\mathbf{R}_1, t)\rangle_{\text{st}}\rangle$, the current-density correlation function. The instantaneous intensity $I(t)$ ($\propto E^*(t)E(t)$) is responsible for the photoelectric current at a *single point* of the photocathode surface, or the photocurrent density $j(\mathbf{R}, t)$:

$$
j(\mathbf{R}, t) \propto I(\mathbf{R}, t) \propto n(\mathbf{R}, t, \delta T),
\tag{3.3.4}
$$

where δT represents a very short sampling time. Equation (3.3.4) implies that we get the same correlation information whether we consider photocurrent density (j), intensity of light (I), or photoelectron count (n). For two correlation intensities at $\mathbf{R}$, we have

$$
\begin{aligned}
\langle n(t)n(t + \tau)\rangle &= a^2T^2\langle I(t)I(t + \tau)\rangle \\
&= \alpha^2T^2\langle E^*(t)E(t)E^*(t + \tau)E(t + \tau)\rangle \\
&= a^2T^2\langle I(t)\rangle^2 g^{(2)}(\tau) \\
&= \alpha^2T^2\langle |E(t)|^2\rangle^2 g^{(2)}(\tau),
\end{aligned}
\tag{3.3.5}
$$

where

$$
\boxed{g^{(2)}(\tau) = \frac{\langle E^*(t)E(t)E^*(t + \tau)E(t + \tau)\rangle}{\langle E^*(t)E(t)\rangle^2} = \frac{\langle I(t)I(t + \tau)\rangle}{\langle I(t)\rangle^2}}
\tag{3.3.6}
$$

is the normalized correlation function; α and a are suitably defined quantum efficiencies with an appropriate amplification factor and unit conversion. Equation (3.3.5) shows that the intensity–intensity correlation function can be determined by means of the photon-count correlation function. It should be noted that for the photon-count correlation function at the "zeroth channel," i.e., $\langle n(t)n(t + \tau$ with $\tau \to 0)\rangle$ or $\langle n(t)n(t)\rangle$, we have according to Eq. (3.2.6)

$$
\begin{aligned}
\langle n^2\rangle &= \langle n\rangle^2 + \langle n\rangle \\
&= a^2T^2\langle I\rangle^2 + aT\langle I\rangle.
\end{aligned}
\tag{3.3.7}
$$

The term $aT\langle I \rangle$ is known as the shot-noise term; it leads to a discontinuous step change from the zeroth channel (Eq. (3.3.7)) to the first channel given by $\langle n(t)n(t + \Delta\tau) \rangle$, where $\Delta\tau$ is the delay-time increment based on summations of the product $n(t)n(t + \Delta t)$ with $\Delta t \, (= \Delta\tau)$ being the sampling time period (δT) for the $n(t)$ photocounts at time t and $n(t + \Delta t)$ photocounts at time $t + \Delta t$. Thus, the zeroth-channel value is usually not available in commercial correlators. The intensity–intensity time correlation function in Eq. (3.3.5) is incomplete. More generally, $\langle n(t)n(t + \tau) \rangle$ could also be considered to have two distinct contributions:

$$\langle n(t)n(t + \tau) \rangle = a^2 T^2 \langle I(t) \rangle^2 g^{(2)}(\tau) + aT\langle I(t) \rangle \delta(\tau), \qquad (3.3.8)$$

where the second term on the right-hand side appears if the same photocount occurs at t and $t + \tau$ (i.e., when $t = t + \tau$ or $\tau = 0$); $\delta(\tau) = 1$ when $\tau = 0$, and $\delta(\tau) = 0$ when $\tau \neq 0$. Similarly, the spectrum of a light wave is related to the time dependence of its electric field through a statistical average quantity, the autocorrelation function $R_E(\tau)$ defined by

$$\boxed{R_E(\tau) = \langle E^*(t)E(t + \tau) \rangle = \langle E^*(t)E(t) \rangle g^{(1)}(\tau),} \qquad (3.3.9)$$

where $g^{(1)}(\tau)$ is the normalized first-order correlation function for the electric field.

3.3.2. THE SELF-BEATING SPECTROMETER

The self-beating spectrometer measures the spectrum of the photocurrent with the detector illuminated only by the field under study, as shown schematically in Fig. 3.2.1. For an optical field that obeys *Gaussian* statistics, the normalized first- and second-order correlation functions $g^{(2)}(\tau)$ and $g^{(1)}(\tau)$ are related (Mandel, 1963):

$$\boxed{g^{(2)}(\tau) = 1 + |g^{(1)}(\tau)|^2.} \qquad (3.3.10)$$

This equation is known as the Siegert relation (Siegert, 1943) and can be derived as follows (Schulz-DuBois, 1983).

Let us consider the total scattered electric field $E(t)$ from N particles as an amplitude function with

$$E(t) = \sum_{k}^{N} E_k \exp[i\phi_k(t)], \qquad (3.3.11)$$

where the optical phases ϕ_k for the kth particle are distributed with equal probability between $-\pi$ and π, and the scattering amplitudes E_k for the kth particle may or may not depend on time. In our previous discussions, we have used the sumbol E_s for the scattered electric field. The subscript s is now dropped to avoid confusion with that on E_k. By definition, the intensity has the form

$$I(t) \propto E^*(t)E(t) = \sum_{k,l} E_k E_l \exp[i\phi_l(t) - i\phi_k(t)]. \qquad (3.3.12)$$

We can set $I(t) = (\alpha/a)E^*(t)E(t) = E^*(t)E(t)$ with $\alpha/a = 1$, i.e., ignore the factor $\frac{1}{2}(\epsilon/\mu)^{1/2}$ in Eq. (2.3.1). Then the amplitude autocorrelation function is

$$\begin{aligned} G^{(1)}(\tau) &= \langle E^*(t)E(t+\tau) \rangle \\ &= \sum_{k,l}^{N} \langle E_k E_l \exp[i\phi_l(t+\tau) - i\phi_k(t)] \rangle. \end{aligned} \qquad (3.3.13)$$

The corresponding intensity–intensity correlation function has the form

$$\begin{aligned} G^{(2)}(\tau) &= \langle I(t)I(t+\tau) \rangle \\ &= \left\langle \sum_{klmn}^{N} E_k E_l E_m E_n \exp[i(\phi_l(t) - \phi_k(t) + \phi_n(t+\tau) - \phi_m(t+\tau)] \right\rangle, \end{aligned} \qquad (3.3.14)$$

where $G^{(2)}(\tau)$ vanishes unless the indices k, l, m, n are pairwise equal as follows:

1. If $k = n$ and $l = m$ but $k \neq l$, the summation is equal to

$$N(N-1)\langle E_k^2 \rangle \langle \exp[i\phi_k(t+\tau) - i\phi_k(t)] \rangle \langle E_l^2 \rangle \langle \exp[i\phi_l(t) - i\phi_l(t+\tau)] \rangle$$

$$= \left(1 - \frac{1}{N}\right) \left| \sum_{k}^{N} E_k^2 \langle \exp[i\phi_k(t+\tau) - i\phi_k(t)] \rangle \right|^2$$

$$= \left(1 - \frac{1}{N}\right) |G^{(1)}(\tau)|^2.$$

2. If $k = l$ and $m = n$, but $k \neq m$, the summation is equal to

$$N(N-1)\langle E_k^2 \rangle \langle E_m^2 \rangle = \left(1 - \frac{1}{N}\right) \left| \sum_{k}^{N} E_k^2 \right|^2 = \left(1 - \frac{1}{N}\right) [G^{(1)}(0)]^2.$$

3. If $k = l = m = n$, the summation is equal to

$$N\langle E_k^4 \rangle = N\langle E_k^2 \rangle^2 = \frac{[G^{(1)}(0)]^2}{N}.$$

By summing (1), (2), and (3) in the limit $N \to \infty$, we get

$$\langle I(t)I(t+\tau) \rangle = \langle E^*(t)E(t) \rangle^2 + \langle E^*(t)E(t+\tau) \rangle^2, \qquad (3.3.15)$$

or

$$g^{(2)}(\tau) = 1 + |g^{(1)}(\tau)|^2, \tag{3.3.16}$$

where $g^2(\tau) \, (= G^{(2)}(\tau)/G^{(2)}(0))$ and $g^{(1)}(\tau) \, (= G^{(1)}(\tau)/G^{(1)}(0))$ are the normalized correlation functions.

Equation (3.3.16) is valid only for Gaussian signals with short sampling time such that $\delta T = \Delta\tau \ll \tau_c$, with τ_c being a characteristic time, as shown in Fig. 2.4.2, and with perfect optical mixing satisfying the condition discussed in Section 3.1. It is the essential formula linking the intensity–intensity time correlation function $G^{(2)}(\tau)$ with the first-order field (or amplitude) time correlation function $G^{(1)}(\tau)$ for the self-beating technique in optical mixing spectroscopy. It should also be noted that there is no phase information in the self-beating intensity–intensity correlation function. By combining Eq. (3.3.16) with Eq. (3.3.8), the photon-count time correlation function in the self-beating mode becomes

$$\boxed{\langle n(t)n(t+\tau)\rangle = \langle n\rangle\,\delta(\tau) + \langle n\rangle^2(1 + |g^{(1)}(\tau)|^2),} \tag{3.3.17}$$

where $\langle n\rangle = aT\langle I\rangle$ and $\langle n\rangle^2 = a^2T^2\langle I\rangle^2$, and where $\delta(\tau) = 1$ for $\tau = 0$ and 0 otherwise.

Instead of the photon-count time correlation function, we may express Eq. (3.3.17) in terms of photocurrent density j with

$$
\begin{aligned}
R_j(\tau) &= \langle j(t)j(t+\tau)\rangle \\
&= \langle j\rangle\,\delta(\tau) + \langle j\rangle^2(1 + |g^{(1)}(\tau)|^2).
\end{aligned}
\tag{3.3.18}
$$

If $g^{(1)}(\tau)$ has the form

$$g^{(1)}(\tau) = \exp(-i\omega_0\tau)\exp(-\Gamma|\tau|),$$

the optical spectrum of a *field* described by Eq. (3.3.9) is

$$
\begin{aligned}
S_E(\omega) &= \frac{1}{2\pi}\int_{-\infty}^{\infty} R_E(\tau)e^{i\omega\tau}\,d\tau \\
&= \frac{a}{\alpha}\langle I(t)\rangle\frac{1}{2\pi}\int_{-\infty}^{\infty}\exp[i(\omega-\omega_0)\tau]\exp(-\Gamma|\tau|)\,d\tau \\
&= \frac{a}{\alpha}\langle I(t)\rangle\frac{\Gamma/\pi}{(\omega-\omega_0)^2 + \Gamma^2},
\end{aligned}
\tag{3.3.19}
$$

where $\omega_0 = \omega_1$ is the angular frequency of the incident light. Equation (3.3.19) is a Lorentzian with a half width at half maximum $\Delta\omega_{1/2} = \Gamma$, centered at $\omega = \omega_0$. However, it does not represent the spectrum we measure using a square-law detector, such as a photomultiplier tube, which detects the intensity $I(t) \, (\propto E^*(t)E(t))$, not the electric field. Therefore, we should consider the power spectrum $S_j(\omega)$ of the photocurrent at a single point of the photo-cathode surface using Eqs. (2.4.14) and (3.3.18):

$$S_j(\omega) = \frac{1}{2\pi} \int_{-\infty}^{\infty} e^{i\omega\tau} [e\langle j \rangle \delta(\tau) + \langle j \rangle^2 (1 + e^{-2\Gamma|\tau|})]\, d\tau$$

$$= \frac{1}{2\pi} e\langle j \rangle + \langle j \rangle^2 \delta(\omega) + \langle j \rangle^2 \frac{2\Gamma/\pi}{\omega^2 + (2\Gamma)^2}. \tag{3.3.20}$$

The power spectrum $S_j(\omega)$ is symmetric about $\omega = 0$. However, we can obtain a power spectrum for positive frequencies only. By combining the positive- and negative-frequency parts, we get

$$S_j^+(\omega) \underset{\omega \geq 0}{=} \frac{e\langle j \rangle}{\pi} + \langle j \rangle^2 \delta'(\omega) + 2\langle j \rangle^2 \frac{2\Gamma/\pi}{\omega^2 + (2\Gamma)^2},$$

$$S_j^-(\omega) \underset{\omega < 0}{=} 0, \tag{3.3.21}$$

where the δ-function is normalized for positive frequencies:

$$\int_0^{\infty} \delta'(\omega)\, d\omega = 1.$$

Equation (3.3.21) has three components: a shot-noise term $e\langle j \rangle/\pi$, a dc component $\langle j \rangle^2 \delta'(\omega)$, and a light-beating spectrum, which, with our present form for $g^{(1)}(\tau)$, is a Lorentzian of half width $\Delta\omega_{1/2} = 2\Gamma$, centered at $\omega = 0$, with total *power* density $\langle j \rangle^2$. Figure 3.3.1 shows the general features of the

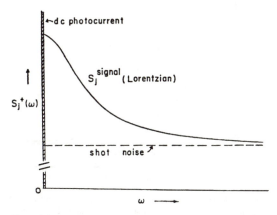

Fig. 3.3.1. General features of the photocurrent power spectrum, showing the dc photo-current $\langle j^2 \rangle \delta'(\omega)$ and the power spectrum of the signal, which includes the background shot noise $\propto \langle j \rangle/\pi$. (Aftr Chu, 1970.)

photocurrent power spectrum. It should be emphasized that Eq. (3.3.15) is valid only for optical fields with Gaussian statistics, and Eq. (3.3.21) is a special case in which the correlation function has the form $e^{-\Gamma|\tau|}$.

We may relax the form for $g^{(1)}(\tau)$ and express $S_j(\omega)$ as a convolution of the optical spectrum with itself:

$$
S_j(\omega) = \frac{\langle j \rangle}{2\pi} + \langle j \rangle^2 \delta(\omega)
$$

$$
\qquad\qquad + \alpha^2 \int_{-\infty}^{\infty} S_E(\omega')[S_E(\omega' + \omega) + S_E(\omega' - \omega)]\, d\omega',
$$

(3.3.22)

where we have taken $S_E(\omega)$ to be a symmetric function. However, for non-Gaussian fields, there is no simple connection between the optical spectrum $S_E(\omega)$ and the photocurrent power spectrum $S_j(\omega)$. In contrast with a spectrum analyzer, which measures $S_j(\omega)$, the photocurrent time-dependent autocorrelation function $R_j(\tau)$, as expressed in Eq. (3.3.18), can be related directly to measurements from a signal correlator.

Figure 3.3.2 shows the general features of the photocurrent (density) correlation function, which has a shot-noise term, a dc component, and the

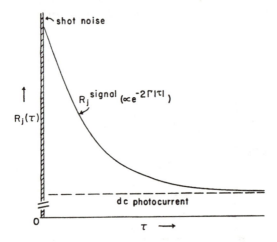

FIG. 3.3.2. General features of the photocurrent autocorrelation function, showing the dc photocurrent, the shot noise, and the current correlation function. Note: the shot noise is now represented by a δ-function, and the dc current forms the background. The R_j-signal is represented by an exponential form, in agreement with our special consideration of a Lorentzian power spectrum of the photocurrent signal S_j^+ as shown in Fig. 3.3.1. (After Chu, 1970.)

desired correlation signal. In the time correlation function, the shot-noise term, instead of the dc component, becomes the δ-function. Signal correlation is a more efficient way of obtaining the desired information, since the rate of data collection with a correlation-function computer is usually faster than that of data collection with a spectrum analyzer. On the other hand, fast-Fourier-transform (FFT) spectrum analyzers are available commercially. Thus, the two methods are complementary.

Here, the signal correlator is an *analog* correlator, the type that one finds in noise, sound, or vibration analysis. A signal correlator (or a FFT spectrum analyzer) accepts an analog signal (not digital photoelectron pulses). By means of an analog-to-digital (A/D) converter, the analog signal is converted to digital numbers and then processed in terms of $R(\tau)$ or $S(\omega)$. Unfortunately, most A/D converters have a fixed rate of conversion, and the analog signal is not time-averaged optimally over the sampling time. Therefore, there is often an unnecessary loss of signal in using a signal correlator or a FFT spectrum analyzer, since the output signal from a photomultiplier tube is digital in nature and can best be processed using a digital correlator, which counts the photoelectron pulses per sampling time directly. However, if the signals are so strong that photoelectron pulses overlap and the output signals are in photocurrents, then FFT spectrum analyzers can be an efficient way to process such time-dependent signals.

3.3.3. THE HOMODYNE-HETERODYNE SPECTROMETER

In homodyne detection, the photomultiplier is illuminated simultaneously by the field under study, E_s, and by a constant *coherent* local-oscillator signal, $\mathbf{E}_{LO}(\mathbf{R}, t) = \mathbf{E}_{LO}^0 \exp(-i\omega_{LO}t)$. A more precise definition is in terms of whether the local-oscillator frequency has been shifted from the incident-beam frequency. If so, we speak of heterodyne detection; if not, of homodyne, as shown schematically in Fig. 3.2.1. Alternatively, one may define a heterodyne technique as any one in which a local oscillator is used, as is often done in the literature. Henceforth, we shall use the word in the latter sense.

Now $E(t) = E_s(t) + E_{LO}^0 \exp(-i\omega_{LO}t)$, so Eq. (3.3.8) takes on a very complex form and for $I_{LO} (\equiv \langle I_{LO} \rangle) \gg \langle I_s \rangle$,

$$
\begin{aligned}
\langle n(t)n(t+\tau)\rangle = {} & aTI_{LO}\,\delta(\tau) + a^2T^2I_{LO}^2 \\
& + a^2T^2I_{LO}\langle I_s\rangle[\exp(i\omega_{LO}\tau)g^{(1)}(\tau) \\
& + \exp(-i\omega_{LO}\tau)g^{(1)*}(\tau)],
\end{aligned} \tag{3.3.23}
$$

where $g^{(1)}(\tau) = \langle E_s^*(t)E_s(t+\tau)\rangle/\langle E_s^*(t)E_s(t)\rangle$ and

$$
g^{(1)*}(\tau) = \langle E_s(t)E_s^*(t+\tau)\rangle/\langle E_s(t)E_s^*(t)\rangle.
$$

The corresponding photocurrent power spectrum at a single point of the photocathode surface resulting from heterodyne detection for $j_{LO} \gg \langle j_s \rangle$ (with the subscripts LO and s denoting local oscillator and scattering, respectively) has the form

$$S_j(\omega) \cong \frac{j_{LO}}{2\pi} + j_{LO}^2 \delta(\omega)$$

$$+ \frac{j_{LO}\langle j_s \rangle}{2\pi} \int_{-\infty}^{\infty} \exp(i\omega\tau)[\exp(i\omega_{LO}\tau)g^{(1)}(\tau) \quad (3.3.24)$$

$$+ \exp(-i\omega_{LO}\tau)g^{(1)*}(\tau)]\,d\tau,$$

which consists of a shot-noise term ($j_{LO}/2\pi$), a dc term ($j_{LO}^2 \delta(\omega)$), and the heterodyne light-beating spectrum.

A very important fact comes into play in Eqs. (3.3.23) and (3.3.24), i.e., unlike the $g^{(2)}$ in Eq. (3.3.8), we need not assume Gaussian statistics for the signal field. Substituting $g^{(1)}(\tau) = \exp(-i\omega_s\tau)\exp(-\Gamma|\tau|)$ into Eq. (3.3.24), we obtain

$$S_j^+(\omega) = \frac{j_{LO}}{\pi} + j_{LO}^2 \delta'(\omega) + j_{LO}\langle j_s \rangle \frac{2\Gamma/\pi}{(\omega - |\omega_s - \omega_{LO}|)^2 + \Gamma^2}. \quad (3.3.25)$$

The heterodyne light-beating spectrum is a Lorentzian of half width at half maximum $\Delta\omega_{1/2} = \Gamma$, centered at $\omega = \omega_s - \omega_{LO}$, with intensity proportional to j_{LO} and $\langle j_s \rangle$. It should be emphasized that Eqs. (3.3.23) to (3.3.25) are valid only if $I_{LO} \gg \langle I_s \rangle$. The self-beating term with a scaling factor of $\langle I_s \rangle^2$ is always present, as shown schematically in Fig. 3.2.1(c). Thus, it is usually quite difficult to achieve very high precisions on measuring the form of $g^{(1)}(\tau)$ by using the heterodyne technique. It should also be noted that the heterodyne technique with Eqs. (3.3.23) and (3.3.24) measures $g^{(1)}(\tau)$, while the self-beating technique with Eq. (3.3.17) measures $g^{(2)}(\tau)$.

For the *total* photocurrent autocorrelation function with imperfect optical mixing over an effective photocathode surface A, $R_i(\tau)$ has the form

$$R_i(\tau) \cong A[j_{LO} + \langle j_s \rangle]\delta(\tau) + j_{LO}^2 A^2$$

$$+ j_{LO}\langle j_s \rangle A^2 \begin{cases} 1, & A < A_{coh} \\ A_{coh}/A, & A > A_{coh} \end{cases}$$

$$\times \{\exp[i(\omega_{LO} - \omega_s)\tau]\exp(-\Gamma|\tau|) + \exp[-i(\omega_{LO} - \omega_s)\tau]\exp(-\Gamma|\tau|)\},$$

$$(3.3.26)$$

where we have taken $g^{(1)}(\tau) = \exp(-\omega_s\tau)\exp(-\Gamma|\tau|)$, and kept $\langle j_s \rangle$ in the shot-noise term $\langle i \rangle = \langle j \rangle A$. The quantity A_{coh} is a coherence area; $A_{coh} \propto \lambda^2/\Omega$ with Ω being the solid angle that the source subtends at the detector.

There are a variety of definitions for the coherence area. A more restrictive form is expressed in Eq. (6.9.1).

In Eq. (3.3.26), the ratio A_{coh}/A signifies the overall beating efficiency responsible for the net signal time correlation function. In practice, it is not a well-defined experimental quantity that can be computed from optical geometry. The main source of A_{coh}/A is discussed in Section 3.1: the finite macroscopic size of the scattering volume. In Section 3.1, we have idealized the photocathode as a point detector, while in Eq. (3.3.26), the ratio also includes effects due to wavefront mismatch over a finite effective photocathode surface area. Similarly, the corresponding power spectrum of the total photocurrent fluctuations under the assumption that the scattered field is a Lorentzian of width Γ centered around the optical frequency ω_s has the form

$$S_j^+(\omega) \simeq j_{LO}^2 A^2 \, \delta'(\omega) + \frac{(j_{LO} + \langle j_s \rangle)A}{\pi}$$

$$+ j_{LO}\langle j_s \rangle A^2 \begin{cases} 1, & A < A_{coh} \\ A_{coh}/A, & A > A_{coh} \end{cases} \frac{2\Gamma/\pi}{\Gamma^2 + (\omega - |\omega_s - \omega_{LO}|)^2}.$$

$$(3.3.27)$$

Finally, we introduce a heterodyne mixing efficiency ϵ representing the degree to which the wavefront of the scattered light and that of the local oscillator are matched over an area equal to a coherence area A_{coh} (Benedek, 1969):

$$S_i^+(\omega) \simeq i_{LO}^2 \delta'(\omega) + \frac{i_{LO} + \langle i_s \rangle}{\pi}$$

$$+ \epsilon i_{LO}\langle i_s \rangle \begin{cases} 1, & A < A_{coh} \\ A_{coh}/A, & A > A_{coh} \end{cases} \frac{2\Gamma/\pi}{\Gamma^2 + (\omega - |\omega_s - \omega_{LO}|)^2}.$$

$$(3.3.28)$$

ϵ is equal to unity if the two wavefronts are perfectly matched in phase, and becomes very small when the relative phases of the two wavefronts fluctuate many times by $\pm 2\pi$ over the coherence area.

The heterodyne technique offers no specific advantage over the self-beating method for simple Lorentzians. It has the disadvantage of requiring optical matching equivalent to that of the alignment of a Michelson interferometer, although in practice, ingenious ways have been devised to circumvent this problem. In the heterodyne scheme, the assumption of Gaussian statistics, which may not hold for certain biological systems, is not required. (Particles undergoing Brownian motions obey Gaussian statistics. Live sperm cells swimming around in a fluid do not obey Gaussian statistics.)

REFERENCES

Benedek, G. B. (1969). "Polarization Matière et Rayonnement, Livre de Jubile en l'Honneur du Professeur A. Kastler," pp. 49–84, Presses Universitaires de France, Paris.

Chu, B. (1970). *Ann. Rev. Phys. Chem.* **21**, 145.

Cummins, H. A. and Swinney, H. L. (1970). *Progr. Opt.* **8**, 135.

Forrester, A. T. (1961). *J. Opt. Soc. Amer.* **51**, 253.

Forrester, A. T., Gudnumdsen, R. A., and Johnson, P. O. (1955). *Phys. Rev.* **99**, 1691.

Glauber, R. J. (1963). *Phys. Rev.* **130**, 2529; **131**, 2766.

Glauber, R. J. (1964). *In* "Quantum Electronics III," Proc. 3rd Conf., Paris, 1963 (N. Bloembergen and P. Grivet, eds.), p. 111, Columbia Univ. Press, New York.

Glauber, R. J. (1965). *In* "Quantum Optics and Electronics," Les Houche Summer School of Theoretical Physics, Grenoble, 1964 (C. Dewitt, A. Blandin, and C. Cohen-Tannoudji, eds.), p. 63, Gordon and Breach, New York.

Glauber, R. J. (1969). *In* "Enrico Fermi XLII Course, Varenna, 1967" (R. Glauber, ed.), pp. 15–56, Academic Press, New York and London.

Hanbury Brown, R. and Twiss, R. Q. (1956). *Nature* **177**, 27.

Jakeman, E. and Pike, E. R. (1968). *J. Phys. A.* **1**, 128; **1**, 625.

Javan, A., Bennett, W. R., and Herriott, D. R. (1961). *Phys. Rev. Lett.* **6**, 106.

Lastovka, J. B. (1967). "Light Mixing Spectroscopy and the Spectrum of Light Scattered by Thermal Fluctuations in Liquids," Ph.D. thesis, Massachusetts Institute of Technology, Cambridge, Massachusetts.

Mandel, L. (1958). *Proc. Phys. Soc. London* **72**, 1037.

Mandel, L. (1959). *Proc. Phys. Soc. London* **74**, 233.

Mandel, L. (1963). *Prog. Opt.* **2**, 181.

Mandel, L. and Wolf, E. (1965). *Rev. Mod. Phys.* **37**, 231.

Mandel, L., Sudarshan, E. C. G., and Wolf, E. (1964). *Proc. Phys. Soc. London* **84**, 435.

Saleh, G. (1978). "Photoelectron Statistics," Springer-Verlag, Berlin.

Schulz-DuBois, E. O. (1983). *In* "Photon Correlation Techniques in Fluid Mechanics" (E. O. Schulz-DuBois, ed.), p. 15, Springer-Verlag, Berlin.

Siegert, A. J. F. (1943). MIT Radiation Lab. Report No. 465.

Sudarshan, E. C. G. (1963a). *Phys. Rev. Lett.* **10**, 277.

Sudarshan, E. C. G. (1963b). *In* "Proc. Symposium Optical Masers," p. 45, Wiley, New York.

Zinman, J. M. (1964). "Principles of the Theory of Solids," pp. 23–25, Cambridge Univ. Press, London and New York.

IV

PHOTON CORRELATION SPECTROSCOPY

In optical mixing spectroscopy, as we have discussed in the previous chapter, the most promising technique, which recognizes and takes advantage of the digital nature of photon statistics of scattered laser light (Pike, 1969) is that of photocount correlation measurements. Pike and his coworkers have done pioneering work in this field (Jakeman *et al.*, 1968; Jakeman and Pike, 1968, 1969a, b; Jakeman, 1970). However, there are several other schemes for computing the time correlation function from a train of photoelectron pulses.

In this chapter, we shall explore some of the basic principles governing photocount time correlation functions. In Section 4.1, we discuss the photocount autocorrelation function and establish the relationship between field and intensity time correlation functions. Section 4.2 describes a comparison of clipped and full correlators. In Sections 4.3–4.7, the schemes for complementary single-clipping, sampling, random-clipping, scaling, and add–subtract correlations are presented. The reader who is not concerned with these schemes may wish to skip those sections altogether. The statistical accuracy of estimating digital autocorrelations based on a single exponential decay analysis is summarized in Section 4.8. Finally, a time-of-arrival approach together with the concept of a structure function is described in Sections 4.9 and 4.10. It is clear that future correlators will be computer-based. Some early developments are sketched in Section 4.11.

4.1. PHOTOCOUNT AUTOCORRELATION

In digital intensity correlation, the autocorrelation function of photocounting fluctuations—which, for stationary processes where the starting time is unimportant, may be written as $\langle n(t)n(t + \tau)\rangle$ ($\equiv \langle n(0)n(\tau)\rangle$)—is related to the autocorrelation function of the short-time-averaged intensity fluctuations by the formula

$$\langle n(t)n(t + \tau)\rangle = \langle n(0)n(\tau)\rangle = (aT)^2\langle I(0)I(\tau)\rangle = (aT)^2\langle I(t)I(t + \tau)\rangle, \quad (3.3.5)$$

where $n(t + \Delta t)$ is the number of photocounts (i.e., photoelectron counts) during the time interval t to $t + \Delta t$ with Δt ($\equiv T$) being the sample time and $\Delta t \ll \tau_c$; $I(t)$ is the short-time-averaged (i.e., averaged over the time interval Δt) intensity of light incident on the unit (point) photocathode at time t; a is the quantum efficiency of the photodetector; and τ_c ($\equiv 1/\Gamma$) is the correlation (or coherence) time of the intensity fluctuations (not of the source). For an optical field that obeys Gaussian statistics, we have, according to Eq. (3.3.15),

$$\langle I(0)I(\tau)\rangle = \langle I\rangle^2(1 + |g^{(1)}(\tau)|^2) \quad (4.1.1)$$

for $\tau \neq 0$. By combining Eqs. (3.3.5) and (4.1.1), we get

$$\begin{aligned}
\langle n(0, T)n(\tau, T)\rangle &= (aT)^2\langle I\rangle^2(1 + |g^{(1)}(\tau)|^2) \\
&= \langle n\rangle^2(1 + |g^{(1)}(\tau)|^2)
\end{aligned} \quad (4.1.2)$$

with $\langle n\rangle = \bar{n} = \langle n(0)\rangle = \langle n(\tau)\rangle = \langle aTI\rangle$. In Eq. (4.1.2), we have explicitly shown $n(0)$ and $n(\tau)$ and $n(0, T)$ and $n(\tau, T)$, emphasizing that the photocounts were measured over a short time sampling interval T, and τ is the time interval (or time difference) between $n(0)$ and $n(\tau)$. The quantity $\langle n(\tau, T)\rangle$ is the average number of photocounts measured over a sample (or sampling) time interval T at delay time τ.

If $n(t, T)$ is a random process, the photocount autocorrelation function has the form

$$\langle n(t_1, T)n(t_2, T)\rangle = \sum_{n_1=0}^{\infty} \sum_{n_2=0}^{\infty} n_1 n_2 p(n_2, t_2, T; n_1, t_1, T), \quad (4.1.3)$$

where $p(n_2, t_2, T; n_1, t_1, T)$, or more simply $p(n_1, n_2)$, is the joint photocount distribution function: the probability of having n_1 photoelectron pulses over a sampling time interval T at time t_1 *and* n_2 photoelectron pulses over a sampling time interval T at time t_2 with $t_2 - t_1 = \tau$. In this more general approach we do not have to use the Gaussian property of $E(t)$.

In general, we want estimates of $G^{(1)}(\tau)$ from measurements of $G^{(2)}(\tau)$. For the optical fields of practical importance, we recall that for thermal light

(Gaussian statistics)

$$g^{(2)}(\tau) = 1 + |g^{(1)}(\tau)|^2, \qquad \tau \neq 0, \qquad (3.3.15)$$

provided that $T \ll \tau_c$ and $A < A_{coh}$. If the effective detector area A is greater than A_{coh}, then

$$g^{(2)}(\tau) = 1 + f(A)|g^{(1)}(\tau)|^2 = 1 + \beta|g^{(1)}(\tau)|^2 \qquad \tau \neq 0, \qquad (4.1.4)$$

where the coherence factor $\beta = f(A)$ $(\sim A_{coh}/A)$ is related to the number of spatial modes collected by the detector covering a photocathode surface A.

Equation (4.1.4) forms the basis of the self-beating technique. For a mixture of thermal ($\mathbf{E_s}$) and coherent ($\mathbf{E_{LO}}$) light and for $T \ll \tau_C$ and $A < A_{coh}$, the relation between $g^{(2)}(\tau)$ and $g^{(1)}(\tau)$ becomes (Saleh, 1978)

$$g^{(2)}(\tau) = 1 + c|g^{(1)}(\tau)|^2 + b|g^{(1)}(\tau)|\cos(\Delta\omega\,\tau), \qquad \tau \neq 0, \qquad (4.1.5)$$

where $g^{(1)}(\tau)$ is real and $\Delta\omega$ is the difference between the central frequencies of thermal and coherent light. $c = \langle n_s \rangle^2/\langle n \rangle^2$ and $b = 2\langle n_s \rangle\langle n_{LO}\rangle/\langle n \rangle^2$ with $\langle n_s \rangle$, $\langle n_{LO} \rangle$ $(\equiv n_{LO})$, and $\langle n \rangle$ being, respectively, the mean number of counts per sampling time T due to the scattered light, the mean number due to the coherent local oscillator, and the total number measured by the detector. If $\Delta\omega = 0$, we then have

$$g^{(2)}(\tau) = 1 + c|g^{(1)}(\tau)|^2 + b|g^{(1)}(\tau)|, \qquad (4.1.6)$$

where the contributions from the self-beating term $c|g^{(1)}(\tau)|^2$ and the heterodyne term $b|g^{(1)}(\tau)|$ are proportional to $c/b = \langle n_s \rangle/2n_{LO}$. Therefore, if $n_{LO} \gg \langle n_s \rangle$, or $b \gg c$, then Eq. (4.1.6) is reduced to

$$(3) \quad g^{(2)}(\tau) \cong 1 + b\,\mathrm{Re}\{g^{(1)}(\tau)\} \qquad (4.1.7)$$

which is equivalent to Eq. (3.3.23). In Eq. (4.1.7), we again emphasize the fact that we can determine the time correlation function of a nonthermal light field by using heterodyne detection.

There are several ways to measure the autocorrelation function $G^{(2)}(\tau)$. For a sequence of pulses with $n(t_j)$ as random variables, as shown in Fig. 4.1.1, the expectation value $\hat{S}^{(2)}$ for the sample correlation function is defined by

$$\hat{S}^{(2)}(\tau_l) = \frac{1}{N}\sum_{i=1}^{N} n(t_i)n(t_{i+l}), \qquad l = 1, 2, \ldots, L, \qquad (4.1.8)$$

with $\tau_l = lT$, $N = N_s - l$, and $L \ll N_s$. The accent $\hat{\ }$ (instead of $\langle\ \rangle$) is used for an expectation value in order to emphasize the fact that the summation is to N, which is finite. A symbol with $\hat{\ }$ does not denote a normalized quantity. In Eq. (4.1.8), N_s is the *number* of samples of counts measured over a total time interval $N_s T$ with T being the sampling time interval for each n, and L is the longest delay-time number with delay-time increment T. Thus, the

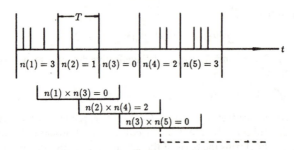

FIG. 4.1.1.   A schematic representation of photoelectron pulses and computation of the photoelectron-count correlation function. If T is the sampling time and N_s samples are measured, $N_s T$ is the total time interval $(0, N_s T)$ during which the measurements were performed. $n(t_i)$, denoted by $n(i)$, is the number of counts detected at time t_i over a sample (or sampling) time interval $T (\equiv \Delta t)$. For the illustration, $\hat{S}^{(2)}(\tau_2) = (1/N)\sum_{i=1}^{N} n(t_i)n(t_{i+2}) = (1/N)(0 + 2 + 0 + \cdots)$, where $\tau_2 = 2T$ and $N = N_s - 2$.

longest delay time $\tau_L = LT$. For example, in Fig. 4.1.1, the total time interval shown explicitly in the figure is $5T$, with $N_s = 5$. Equation (4.1.8) is an unbiased estimate for $G^{(2)}(\tau_l)$, where the factor $1/N$ is introduced to equalize the sums in different channels. If N_s is sufficiently large, we normally set $\hat{G}^{(2)}(\tau_l) = \sum_{i=1}^{N_s} n(t_i)n(t_{i+l}) \simeq G^{(2)}(\tau_l)$ without the factor $1/N$.

4.2. FULL AND CLIPPED HARDWARE DIGITAL CORRELATORS

A hardware correlator is needed in order to process Eq. (4.1.8) in real time, because the data rate is usually quite high (e.g. $T \sim 1$ μsec). Figure 4.2.1 shows a schematic drawing of an experiment using photon correlation spectroscopy (PCS). By mixing the scattered light intensity either with a portion of the unshifted (or shifted) incident radiation [referred to as homodyne (or heterodyne) detection], or with itself [referred to as the self-beating technique, as shown schematically in Fig. 3.2.1], the intensity spectrum or correlation can be measured using either a FFT spectrum analyzer or a digital correlator, respectively.

A digital correlator samples the incoming signal from a photomultiplier, after appropriate standardization of the photoelectron pulses, by counting the number of photon events occurring in a sample time $T (\equiv \Delta t)$, giving rise to a train of numbers $n(t)$, $n(t + T)$, $n(t + 2T), \ldots$, and computes the average

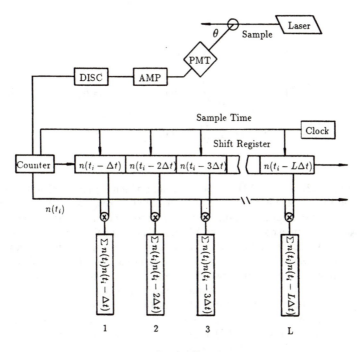

FIG. 4.2.1.   Schematic of the photon correlation spectroscopy (PCS) experiment. θ = scattering angle; $n(t_i)$ = number of photocounts measured during the sample (or sampling) time $\Delta t\, (\equiv T)$ at time t_i. For clipped correlation, a clipping circuit would precede the shift register. The contents of the shift register would become the binary signal $n_k(t)$, and the multiplications indicated by $\times$ would be replaced by AND gates. It should be noted that the resulting correlation function corresponding to the photocounts $n(t_i),\, n(t_i - \Delta t),\dots$ is the same as that predicted by Eq. (4.1.8) corresponding to the photocounts $n(t_i),\, n(t_i + \Delta t),\dots$ with $n(t_i + \Delta t) \equiv n(t_{i+1})$. The sampling time interval $T \equiv \Delta t$, and the subscript k denotes the clipping level. We have used the symbol Δt for the sampling time interval T to emphasize the short-time-interval nature of this process. PMT = photomultiplier tube; AMP = amplifier; DISC = discriminator; $1, 2, 3,\dots, L$ = storage counters.

products

$$\hat{S}^{(2)}(\tau_1 = T) = \frac{1}{N} \sum_{i=1}^{N} n(t_i)n(t_{i+1}), \qquad N = N_s - 1,$$

$$\hat{S}^{(2)}(\tau_2 = 2T) = \frac{1}{N} \sum_{i=1}^{N} n(t_i)n(t_{i+2}), \qquad N = N_s - 2, \qquad (4.2.1)$$

$$\hat{S}^{(2)}(\tau_3 = 3T) = \frac{1}{N} \sum_{i=1}^{N} n(t_i)n(t_{i+3}), \qquad N = N_s - 3,$$

$$\vdots$$

until $\tau_L = LT$. In Eq. (4.2.1), τ is the delay time, which is an integral multiple of T, and the average is performed over all sample times. This photoelectron-count autocorrelation function can be related to the Fourier transform of the power spectrum according to the Wiener–Khinchine theorem (Eqs. (2.4.13) and (2.4.14)); hence the two techniques of correlation and spectrum analysis, in principle, yield the same information. The input signal (the $n(t_i)$'s) is split in the digital correlator into two paths, and a delay is introduced in one of these (see Fig. 4.2.1), allowing the multiplications to be performed in parallel. In practice, these multiplications are sometimes approximated by *clipping* the signal in one path, as shown in Fig. 4.2.2, so that if $n(t_i)$ exceeds a particular value (the clipping level k), a one is propagated, and otherwise a zero. The products of Eq. (4.2.1) then become

$$\hat{S}_k^{(2)}(\tau_1 = T) = \frac{1}{N} \sum_{i=1}^{N} n(t_i)n_k(t_{i+1}),$$

$$\hat{S}_k^{(2)}(\tau_2 = 2T) = \frac{1}{N} \sum_{i=1}^{N} n(t_i)n_k(t_{i+2}) \qquad (4.2.2)$$

with

$$n_k(t_i) = \begin{cases} 1 & \text{if } n(t_i) > k, \\ 0 & \text{otherwise (if } n(t_i) \le k). \end{cases} \qquad (4.2.3)$$

Figure 4.2.3 shows a schematic representation of a single-clipped correlator.

In an ideal full correlator, the counting–delay–multiplication–addition operation at t_i is accomplished within one sampling time $T \, (= \Delta t)$ by

1. storing a string of L numbers $n(t_i - T)$, $n(t_i - 2T)$, $n(t_i - 3T),\ldots,$ $n(t_i - LT)$ in the shift registers,
2. simultaneously multiplying the L numbers in the shift registers by $n(t_i)$ using L multipliers, yielding $n(t_i)n(t_i - T)$, $n(t_i)n(t_i - 2T)$, $n(t_i)n(t_i - 3T),\ldots,n(t_i)n(t_i - LT)$ products,

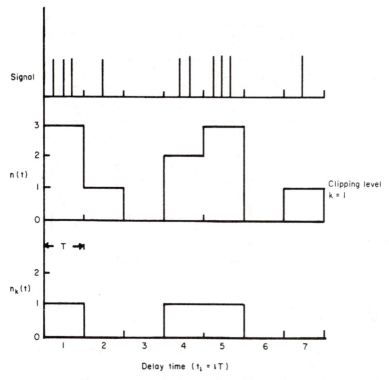

FIG. 4.2.2.   A schematic representation of the clipped photocount number $n_k(t)$ for a clipping level of $k = 1$. The sampling time interval T is the period of the shift clock in Fig. 4.2.1.

3. storing the products $n(t_i)n(t_i - T)$, $n(t_i)n(t_i - 2T)$, $n(t_i)n(t_i - 3T)$,..., $n(t_i)n(t_i - LT)$ in L separate adders (storage counters).

At the next time interval $t_i + T$ $(\equiv t_{i+1})$, the string of L numbers becomes $n(t_i)$, $n(t_i - T)$, $n(t_i - 2T)$,...,$n(t_i - (L - 1)T)$, and the multipliers produce $n(t_i + T)n(t_i)$, $n(t_i + T)n(t_i - T)$,...,$n(t_i + T)n(t_i - (L - 1)T)$, which are added to the previous set in the storage counters. After N_s samplings, the storage counters contain

$$\sum_{j=0}^{N_s-1} n(t_{i+j})n(t_{i+j-1}), \qquad \sum_{j=1}^{N_s-1} n(t_{i+j})n(t_{i+j-2}),...,$$

$$\sum_{j=1}^{N_s-1} n(t_{i+j})n(t_{i+j-L}),$$

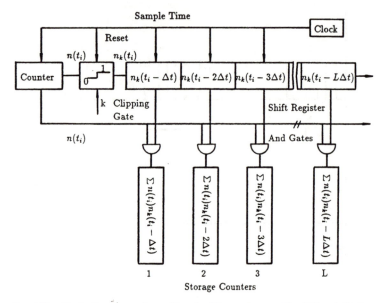

FIG. 4.2.3. Single-clipping correlator. Note the difference between Fig. 4.2.1 and this figure, where a clipping gate is introduced, and instead of the multipliers AND gates are used.

which are slightly different from Eq. (4.1.8) because the L storage shift registers have been filled with the batch of numbers $n(t_{i-1})$, $n(t_{i-2}), \ldots, n(t_{i-L})$ before we start the signal processing. As shown in Fig. 4.2.1, the corresponding storage counters are providing estimates of $\sum n(t_i)n(t_{i-l})$ with $l = 1, 2, \ldots, L$. A correlator similar to the above scheme has been constructed by Ash and Ford (1973) using three-bit shift registers as the L-memory.

Three problems exist for those older "full" correlators:

(1) The three-bit (or present-day four-bit) shift registers mean that the number of photoelectron pulses per sampling time, $n(t, T)$ (not $\langle n(t, T)\rangle$), cannot exceed 2^3 (or 2^4). Therefore, for long sampling times, the correlator can easily be saturated. Present-day commercial "full" correlators have at least a $4 \times n$ capability with provision for indicating when saturation has occurred. The notation $4 \times n$ means that only one of the two numbers used in formulating the intensity time correlation is limited to 2^4 counts per sampling time. $8 \times n$ correlators have been built. The limitation, however, is not serious even for long sampling times, since software correlators with an arbitrary number of bits and long delay times ($\gtrsim n$ sec) are easily constructed.

(2) The multiplication process forms the slowest link. Present-day commercial full correlators often use different schemes to realize Eq. (4.2.1), resulting in

sampling time intervals as short as 100 nsec. Special correlators with sampling time intervals as short as 5 nsec and 20 nsec have been constructed. At such short sampling time intervals, the correlators act more like clipped correlators (see next section), since most of the time $n(t, T)$ will be either 1 or 0.

(3) The correlators usually have a limited number of channels, e.g., $L = 64$ or 128 channels. With equal sampling time intervals, the correlator would have a bandwidth limit starting at T and ending at LT, i.e., the first channel of the time correlation function is $\langle n(t)n(t - T)\rangle$, and the last channel, $\langle n(t)n(t - LT)\rangle$. We shall discuss this problem in connection with data analysis (Chapter 7). It suffices to mention here that variable delay times τ can be useful for the characterization of systems with broad characteristic time (τ_c) distributions. For either multi-τ or log-delay-time correlators the limited channel number becomes less important.

In summary, modern correlators can accept high count rates (megahertz range) and cover a large range of delay times (say 6–8 orders of magnitude), making full correlation a standard process. Single-clipped correlation is used only for very short delay times ($\lesssim 50$ nsec) where $n(t, T)$ is usually 1 or 0 to start with.

The uninitiated reader may be confused by the notation for the number of samples (N_s), the sampling time (or sampling time interval T, which is also denoted by Δt), the number $n(t, T)$ of photocounts at time t over a sampling time interval T ($\equiv \Delta t$), the delay-time increment T, the delay time lT, etc. To review, Figs. 4.1.1 and 4.2.1 should be examined together. First, a string of signals (photocounts) as a function of time t is counted over a sampling time T ($=\Delta t$) by the counter, yielding N_s numbers, over a total time period of N_sT. In most correlators, the product $n(t_i)n(t_i - lT)$ is then formed, as shown in Fig. 4.2.1, and lT is the delay time, with l and T ($\equiv \Delta t$) being the delay channel number and the delay-channel (time-interval) increment, respectively.

If the incident laser beam has a tendency to drift slowly, the mean count rate $\langle n \rangle$ can also drift during the measurement time, resulting in a distorted $\hat{G}^{(2)}(\tau)$. Therefore, it becomes advantageous to make a number of shorter-time-interval measurements and to compute the average of estimates of the normalized correlation function based on data in each subinterval (Saleh, 1978):

$$\hat{s}^{(2)}(\tau_l) = \frac{1}{M}\sum_{m=0}^{M-1}\frac{\dfrac{1}{N_s}\displaystyle\sum_{j=mN_s+1}^{mN_s+N_s}n(t_j)n(t_{j+l})}{\dfrac{1}{N_s}\displaystyle\sum_{j=mN_s+1}^{mN_s+N_s}[n(t_j)]^2},\qquad(4.2.4)$$

where N_s is the *number* of sampling time intervals in each *subinterval* and N_sMT is the total duration of measurements with M being the *number* of subintervals. $\hat{s}(\tau_l)$, instead of $\hat{S}(\tau_l)$, is used to denote the estimate of the sample correlation function over M subintervals. The choice of M depends on the time constant of the drift of the mean count rate. Furthermore, N_s should be sufficiently large so that $\hat{s}^{(2)}(\tau)$ is an unbiased estimate of $G^{(2)}(\tau)$ (Oliver, 1974).

In practice, as computation of Eq. (4.1.8) requires an extensive amount of digital electronics, Jakeman and Pike (1969b) introduced the method of clipping a fluctuating signal before correlation, which permitted considerable simplification of the instrumentation. We have already introduced the concept of single clipping in Eqs. (4.2.2) and (4.2.3). The principle is further expanded as follows.

Let us recall the binary random variable $n_k(t)$ for the clipped photocount:

$$n_k(t) = \begin{cases} 1 & \text{if} \quad n(t) > k, \\ 0 & \text{if} \quad n(t) \le k, \end{cases} \qquad (4.2.3)$$

where k is a positive integer or zero. The functions of interest are $\langle n_k(0)n(\tau)\rangle$ and $\langle n_k(0)n_k(\tau)\rangle$, the single-clipped and double-clipped photocount autocorrelation functions. Experimentally $n_k(t)$ can be measured with simpler instrumentation than $n(t)$. With the delayed channel clipped, the past history of the system can be stored in shift registers as a linear chain of zeros and ones. Multiplication processes in computing the time correlation function are simplified when one (single clipping) or both (double clipping) of those numbers is always a one or a zero. We now need to relate correlations involving $n_k(t)$ and $g^{(2)}(\tau)$.

The two-dimensional generating function for the joint probability distribution involving counting over two discrete intervals is

$$Q(\lambda_1, \lambda_2) = \langle (1 - \lambda_1)^{n(0)}(1 - \lambda_2)^{n(\tau)}\rangle$$
$$= \sum_{n(0)=0}^{\infty} \sum_{n(\tau)=0}^{\infty} (1 - \lambda_1)^{n(0)}(1 - \lambda_2)^{n(\tau)}p[n(0), n(\tau)]. \qquad (4.2.5)$$

Thus, by definition, we get

$$\langle n(0)n(\tau)\rangle = \sum_{n(0)=0}^{\infty} \sum_{n(\tau)=0}^{\infty} n(0)n(\tau)p[n(0), n(\tau)]$$
$$= \frac{\partial^2}{\partial\lambda_1\,\partial\lambda_2}Q(\lambda_1, \lambda_2)\Big|_{\lambda_1 = \lambda_2 = 0}, \qquad (4.2.6)$$

$$\langle n_k(0)n(\tau)\rangle = \sum_{n(0)=k+1}^{\infty} \sum_{n(\tau)=0}^{\infty} n(\tau)p[n(0), n(\tau)]$$
$$= \sum_{n(0)=k+1}^{\infty} \frac{(-1)^{n(0)+1}}{n(0)!}\frac{\partial^{n(0)+1}}{\partial^{n(0)}\lambda_1\,\partial\lambda_2}Q(\lambda_1, \lambda_2)\Big|_{\lambda_1 = 1, \lambda_2 = 0}, \qquad (4.2.7)$$

and

$$
\begin{aligned}
\langle n_0(0)n_0(\tau)\rangle &= 1 - p[n(0) = 0] - p[n(\tau) = 0] \\
&\quad + p[n(0) = 0, n(\tau) = 0] \\
&= 1 - Q(\lambda_1, \lambda_2)|_{\lambda_1 = 1, \lambda_2 = 0} \\
&\quad - Q(\lambda_1, \lambda_2)|_{\lambda_1 = 0, \lambda_2 = 1} + Q(\lambda_1, \lambda_2)|_{\lambda_1 = \lambda_2 = 1},
\end{aligned}
\tag{4.2.8}
$$

where $\langle n_0(0)n_0(\tau)\rangle$ represents the correlation function for double clipping at zero photocount number. Double clipping at photocount levels other than zero produces more complex expressions.

Our next step is to evaluate the generating function $Q(\lambda_1, \lambda_2)$ for the joint probability distribution of intensity fluctuations. Equation (4.2.5) permits connection of the field to the photocount distribution by means of the relation (Arecchi *et al.*, 1966)

$$
p[n(0), n(\tau)] = \frac{(-1)^{n(0)}(-1)^{n(\tau)}}{n(0)!\,n(\tau)!}\left(\frac{d^{n(0)}}{d\lambda_1^{n(0)}}\frac{d^{n(\tau)}}{d\lambda_2^{n(\tau)}}Q(\lambda_1, \lambda_2)\right)_{\lambda_1 = \lambda_2 = 1}.
\tag{4.2.9}
$$

The result has been generalized by Bedard (1967) for N-fold joint photocount distributions of Gaussian (thermal) light of arbitrary spectral profile. The basic formula

$$
p(n, t, T) = \frac{1}{n!}\int_0^\infty e^{-aW}(aW)^n P(W)\,dW
\tag{3.2.8}
$$

relates the statistical photocount distribution $p(n, t, T)$ registered with a single detector in a sample time interval T to the probability density distribution $P(W)$ for the quantity

$$
W = \int_t^{t+T} I(T')\,dT'.
\tag{3.2.7}
$$

Alternatively, we can write Eq. (3.2.8) in a more compact form,

$$
p(n, t, T) = \left\langle \frac{(aW)^n}{n!}e^{-aW}\right\rangle,
\tag{4.2.10}
$$

with the angular brackets denoting the appropriate averaging process over the phase-space functional. The results can be extended to a twofold photocount distribution, whereby

$$
p(n_1, n_2) = \left\langle \prod_{i=1}^{2} \frac{(aW)_i^{n_i}}{n_i!}\exp[-(aW)_i]\right\rangle.
\tag{4.2.11}
$$

Therefore, we can introduce a generating function

$$
Q(\lambda_1, \lambda_2) = \langle \exp(-\lambda_1 U_1)\exp(-\lambda_2 U_2)\rangle,
\tag{4.2.12}
$$

where $U = aW$, or in terms of the joint probability density distribution, $p_2(E_1, E_2)$,

$$Q(\lambda_1, \lambda_2) = \iint p_2(E_1, E_2) \exp(-\lambda_1 |E_1|^2 T)$$
$$\times \exp[-\lambda_2 |E_2|^2 T] \, d^2E_1 \, d^2E_2. \tag{4.2.13}$$

For a stationary Gaussian–Markovian field, the result is

$$Q(\lambda_1, \lambda_2) = \{\langle n \rangle^2 [1 - |g^{(1)}(\tau)|^2]\lambda_1 \lambda_2 + \langle n \rangle (\lambda_1 + \lambda_2) + 1\}^{-1}, \tag{4.2.14}$$

which represents the degenerate form of a two-detector arrangement ($N = 2$) with (Bedard, 1967)

$$Q(\lambda_1, \lambda_2) = [1 + \langle n_1 \rangle \lambda_1 + \langle n_2 \rangle \lambda_2 + \langle n_1 \rangle \langle n_2 \rangle \lambda_1 \lambda_2 (1 - |\gamma_{12}|^2)]^{-1}.$$

Here γ_{12} is the second-order complex degree of coherence for the radiation field at detector 1 located at the space–time point R_1, t_1 and detector 2 located at R_2, t_2. In our case, we have $R_1 = R_2$, signifying light falling at a single space point on the photocathode, and $\tau = t_2 - t_1$. Thus, Eq. (4.2.14) is retrieved. By substituting Eq. (4.2.14) into Eqs. (4.2.6)–(4.2.8), we obtain for the thermal light

$$\boxed{G^{(2)}(\tau) = \langle n(0)n(\tau)\rangle = \langle n \rangle^2 [1 + |g^{(1)}(\tau)|^2],} \tag{4.2.15}$$

$$\boxed{\begin{aligned} G_k^{(2)}(\tau) &= \langle n_k(0)n(\tau)\rangle \\ &= \langle n \rangle \left(\frac{\langle n \rangle}{1 + \langle n \rangle}\right)^{k+1} \left(1 + \frac{1 + k}{1 + \langle n \rangle}|g^{(1)}(\tau)|^2\right), \end{aligned}} \tag{4.2.16}$$

and

$$\begin{aligned} G_{00}^{(2)}(\tau) &= \langle n_0(0)n_0(\tau)\rangle \\ &= \left(\frac{\langle n \rangle}{1 + \langle n \rangle}\right)^2 \left(1 + \frac{|g^{(1)}(\tau)|^2}{(1 + \langle n \rangle)^2 - \langle n \rangle^2 |g^{(1)}(\tau)|^2}\right). \end{aligned} \tag{4.2.17}$$

In deriving Eqs. (4.2.16) and (4.2.17) we have computed $\langle n_k(0)\rangle$ using Eqs. (3.2.8) and (3.2.9), and

$$p(n, t, T) = \int_0^\infty \frac{(aW)^n}{n!} \exp(-aW) \frac{1}{\langle W \rangle} \exp\left(-\frac{W}{\langle W \rangle}\right) dW. \tag{4.2.18}$$

The integral can be evaluated by a change of variable

$$y = aW[1 + (a\langle W \rangle)^{-1}], \qquad dy = a \, dW [1 + (a\langle W \rangle)^{-1}],$$

yielding (Chu, 1974)

$$p(n,t,T) = \frac{1}{[1 + \langle n(T)\rangle][1 + (\langle n(T)\rangle)^{-1}]^n} \tag{4.2.19}$$

and

$$\langle n_k(0)\rangle = \sum_{n(0)=k+1}^{\infty} p[n(0)] = \frac{1}{1+\langle n\rangle} \sum_{n=k+1}^{\infty} \left(\frac{\langle n\rangle}{1+\langle n\rangle}\right)^n$$

$$= \frac{1}{1+\langle n\rangle}\left[\sum_{n=0}^{\infty}\left(\frac{\langle n\rangle}{1+\langle n\rangle}\right)^n - \sum_{n=0}^{k}\left(\frac{\langle n\rangle}{1+\langle n\rangle}\right)^n\right]$$

$$= \frac{1}{1+\langle n\rangle}\left(\sum_{n=0}^{\infty} X^n - \sum_{n=0}^{k} X^n\right),$$

where $X = \langle n\rangle/(1 + \langle n\rangle)$. The second term is a geometrical progression. We then have

$$\boxed{\langle n_k(0)\rangle = \frac{1}{1+\langle n\rangle}\left(\frac{1}{1-X} - \frac{1-X^{k+1}}{1-X}\right) = \left(\frac{\langle n\rangle}{1+\langle n\rangle}\right)^{k+1}.} \tag{4.2.20}$$

The same information may be obtained easily by using the generating function for the single-photon probability distribution of intensity fluctuations, $(1 + \langle n\rangle\lambda)^{-1}$.

Equation (4.2.20) permits a direct comparison of the correlation functions Eqs. (4.2.16) and (4.2.17) with Eq. (4.2.15), which has the form

$$\langle n(0)n(\tau)\rangle/\langle n\rangle^2 = 1 + |g^{(1)}(\tau)|^2. \tag{4.2.15*}$$

Now the normalized single- and double-clipped photocount autocorrelation functions are given by

$$\frac{\langle n_k(0)n(\tau)\rangle}{\langle n_k(0)\rangle\langle n(\tau)\rangle} = 1 + \frac{1+k}{1+\langle n\rangle}|g^{(1)}(\tau)|^2 \tag{4.2.16*}$$

and

$$\frac{\langle n_0(0)n_0(\tau)\rangle}{\langle n_0\rangle^2} = 1 + \frac{|g^{(1)}(\tau)|^2}{(1+\langle n\rangle)^2 - \langle n\rangle^2|g^{(1)}(\tau)|^2} \tag{4.2.17*}$$

with the * symbol reminding us of the normalized correlation function. It should be interesting to note that the expression (Eq. (6) of Jakeman and Pike (1969b))

$$\frac{\langle n_0(0)n_0(\tau)\rangle}{\langle n_0\rangle^2} = \frac{1 + [(1 - \langle n\rangle)/(1 + \langle n\rangle)]|g^{(1)}(\tau)|^2}{1 - [\langle n\rangle/(1 + \langle n\rangle)]^2|g^{(1)}(\tau)|^2}$$

is identical to Eq. (4.2.17).

If $\langle n \rangle \ll 1$, i.e., the average (or mean) number of photocounts per sample time interval T is much less than one, Eq. (4.2.17) is reduced to Eq. (4.2.15), while if $\langle n \rangle \gg 1$, Eq. (4.2.17) approaches a constant value of unity, so that the spectral information is lost for this extreme form of clipping. As $\langle n \rangle \propto T$, $\langle n \rangle$ decreases with decreasing sample time interval T. On the other hand, in single clipping, little spectral information is lost. In fact, if $k = \langle n \rangle$, Eq. (4.2.16) reduces exactly to Eq. (4.2.15). A single-clipped correlator is just as effective as a full correlator, i.e., we can obtain the desired $|g^{(1)}(\tau)|^2$ from the measured $G^{(2)}(\tau)$ using either scheme with almost the same efficiency. Uninitiated readers may find it difficult to accept such a statement and should read Section 4.8. It is sufficient to emphasize that the statement is applicable to Gaussian signals that do not have long-time slow fluctuations, so that $\langle n \rangle$ remains nearly constant over many LT time intervals.

We should also note that in Eq. (4.2.15), we have set $\langle n(0) \rangle = \langle n(\tau) \rangle$. In fact, Eq. (4.2.15) should state that $\langle n(0)n(\tau) \rangle / [\langle n(0) \rangle \langle n(\tau) \rangle] = 1 + |g^{(1)}(\tau)|^2$, where $\langle n(0) \rangle$ is not necessarily equal to $\langle n(\tau) \rangle$. In other words, the baseline $\langle n(0) \rangle \langle n(\tau = lT) \rangle$ may vary with l due to long-time slow fluctuations and so may not be the same for different delay time ranges. Thus, the normalized intensity–intensity time correlation function with long-time delay-time increments should be $\langle n(0)n(\tau) \rangle / [\langle n(0) \rangle \langle n(\tau) \rangle]$, not $\langle n(0)n(\tau) \rangle / \langle n \rangle^2$.

In practice, it is important to realize that in the presence of temporal and spatial integration, $g_k^{(2)}(\tau = 0, T)$ is a more complicated function of the sampling time interval T, detector coherence area A_{coh}, and clipping level k (Koppel, 1971; Jakeman, 1974). Thus, deviation from linearity in a plot of $[g_k^{(2)}(\tau = 0, T) - 1][1 + \langle n \rangle]$ vs $1 + k$ is expected and does not necessarily indicate a defect in the clipping gate (Nieuwenhuysen and Clauwaert, 1977). Working with low light levels and/or small sample times, Nieuwenhuysen and Clauwaert suggested the use of a larger detector, close to the coherence area, in order to achieve a higher $\langle n \rangle$ and hence a higher statistical accuracy (Hughes et al., 1973).

4.3. COMPLEMENTARY SINGLE-CLIPPED AUTOCORRELATION FUNCTION

Complementary clipping is similar to normal clipping except that

$$n_{\text{ck}}(t_i) = \begin{cases} 1 & \text{if} \quad n(t_i) \leq k, \\ 0 & \text{if} \quad n(t_i) > k \end{cases} \tag{4.3.1}$$

instead of $n_k(t_i) = 1$ if $n(t_i) > k$ and 0 if $n(t_i) \leq k$ as defined by Eq. (4.2.3) for normal clipping. In Eq. (4.3.1), the subscript c denotes complementary clipping. The complementary single-clipped autocorrelation function $G_{ck}^{(2)}(\tau)$ is defined by $\langle n_{ck}(0)n(\tau)\rangle$.

Equations (4.2.7) and (4.2.16) are fundamental for the clipped-photon correlation technique. Rewriting Eq. (4.2.7), we have

$$
\begin{aligned}
\langle n_k(0)n(\tau)\rangle &= \sum_{n(0)=k+1}^{\infty} \sum_{n(\tau)=0}^{\infty} n(\tau)p[n(0),n(\tau)] \\
&= \sum_{n(\tau)=0}^{\infty} \left\{ \sum_{n(0)=0}^{\infty} n(\tau)p[n(0),n(\tau)] - \sum_{n(0)=0}^{k} n(\tau)p[n(0),n(\tau)] \right\} \\
&= \sum_{n(\tau)=0}^{\infty} n(\tau)p[n(\tau)] - \sum_{n(\tau)=0}^{\infty} \sum_{n(0)=0}^{k} n(\tau)p[n(0),n(\tau)] \\
&= \langle n\rangle - \langle n_{ck}(0)n(\tau)\rangle,
\end{aligned}
$$

or

$$\boxed{G_k^{(2)}(\tau) = \langle n\rangle - G_{ck}^{(2)}(\tau),} \qquad (4.3.2)$$

implying that the complementary single-clipped correlation function and the normal single-clipped correlation function are equivalent and carry the same information. Chen *et al.* (1972) first introduced the complementary single-clipped autocorrelation function $\langle n_{ck}(0)n(\tau)\rangle$.

By combining Eqs. (4.2.16) and (4.3.2), we get

$$
\begin{aligned}
\langle n_{ck}(0)n(\tau)\rangle &= \sum_{n(0)=0}^{k} \sum_{n(\tau)=0}^{\infty} n(\tau)p[n(0),n(\tau)] \\
&= \langle n\rangle - \langle n_k(0)n(\tau)\rangle \\
&= \langle n\rangle \left[1 - \left(\frac{\langle n\rangle}{1+\langle n\rangle}\right)^{k+1} \right] \\
&\quad - \langle n\rangle \left(\frac{\langle n\rangle}{1+\langle n\rangle}\right)^{k+1} \left(\frac{1+k}{1+\langle n\rangle}\right)|g^{(1)}(\tau)|^2,
\end{aligned}
\qquad (4.3.3)
$$

which gives an alternative approach for measuring $|g^{(1)}(\tau)|^2$. Figure 4.3.1 shows the two quantities $\langle n_k(0)n(\tau)\rangle$ and $\langle n_{ck}(0)n(\tau)\rangle$ as a function of time for $k=0$, $\langle n\rangle = 5$, $\tau_c = 150$ μsec.

The advantage of using $G_{ck}^{(2)}(\tau)$ instead of $G_k^{(2)}(\tau)$ is revealed only when we take into account the limitation of storage counters in real correlators. For example, according to Eqs. (4.2.16), (4.2.20), and (4.3.3),

$$G_k^{(2)}(\tau \approx 0) \approx \langle n_k\rangle\langle n\rangle \left(1 + \frac{1+k}{1+\langle n\rangle}\right) = \left(\frac{\langle n\rangle}{1+\langle n\rangle}\right)^{k+2}(2+\langle n\rangle + k)$$

$$(4.3.4)$$

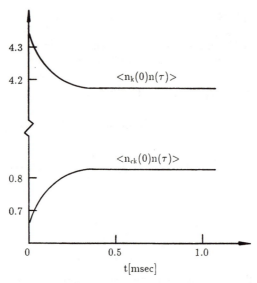

FIG. 4.3.1. $\langle n_k(0)n(\tau)\rangle$ and $\langle n_{ck}(0)n(\tau)\rangle$ as a function of delay time for $k = 0$, $\langle n\rangle = 5$, and $\tau_c = 150 \ \mu\text{sec}$ (after Chen *et al.*, 1972). $\langle n_{ck}(0)n(\tau)\rangle \equiv \langle n_k(0)n(\tau)\rangle_c$.

and

$$G_{ck}^{(2)}(\tau \approx 0) \approx \langle n\rangle - \left(\frac{\langle n\rangle}{1 + \langle n\rangle}\right)^{k+2}(2 + \langle n\rangle + k). \qquad (4.3.5)$$

It should be noted that Eqs. (4.3.4) and (4.3.5), like Eq. (4.1.1), are strictly valid only for $\tau \neq 0$. For $\langle n\rangle \gg 1$ and small $k \ (\ll \langle n\rangle)$, we have

$$G_k^{(2)}(\tau \approx 0) \approx \langle n\rangle \qquad (4.3.6)$$

and

$$G_{ck}^{(2)}(\tau \approx 0) \approx 0, \qquad (4.3.7)$$

implying that the number of counts to be stored is much smaller for $G_{ck}^{(2)}(\tau)$ than for $G_k^{(2)}(\tau)$. Thus, we can alleviate the overflow problem (to be discussed in greater detail in Section 4.4) using the complementary single-clipped autocorrelation function, which is particularly useful if an intense coherent beam is used as a local oscillator in the homodyne (or heterodyne) detection technique. We should recall that Eqs. (4.2.15), (4.2.16), and (4.3.3) are valid for Gaussian signals; the reader is referred to discussions by Saleh (1978) for more details concerning the mixing of coherent light, thermal light, and light with unspecified statistics.

If $\langle n \rangle \ll 1$, Eqs. (4.3.4) and (4.3.5) become

$$G_k^{(2)}(\tau \approx 0) \simeq \langle n \rangle^{k+2}(2 + k)$$

$$G_{ck}^{(2)}(\tau \approx 0) \simeq \langle n \rangle - \langle n \rangle^{k+2}(2 + k),$$

i.e.,

$$G_k^{(2)}(\tau \approx 0) \ll G_{ck}^{(2)}(\tau \approx 0). \tag{4.3.8}$$

Equation (4.3.8) implies that we should use normal clipping under the condition $\langle n \rangle \ll 1$.

4.4. SAMPLING SCHEME OF A SINGLE-CLIPPED DIGITAL CORRELATOR

The essence of the clipped sampling scheme used in digital autocorrelation of photon-counting fluctuations is shown schematically in Figs. 4.2.2 and 4.2.3. This method makes use of the fact that the zero crossings of a Gaussian signal contain most of the spectral information of the original unclipped correlation function. In the past, for practical reasons, emphasis has been on single-clipped digital correlators. This trend is, however, ending, since the cost of digital electronics has gone down, and full correlators that can accept large numbers of photocounts per sample time are beginning to appear (e.g., BI-8000 by Brookhaven Instruments). In addition, most of the $4 \times n$ correlators can work as full correlators so long as $n(t) \leq 2^4$ counts per sample time T.

In single clipping, the lth channel receives counts of $n_k(0)n(t_l = lT)$ for each sampling. After N_s samplings, the $G_k(t_l)$ content of the lth channel in storage counter l is, according to Eq. (4.2.8),

$$\hat{G}_k(t_l) = \sum_{r=1}^{N_s} n_k(t_r)n(t_{r+l}) \qquad \text{without } 1/N_s \text{ normalization,} \tag{4.4.1}$$

while the total photocounts during N_s samplings are

$$\hat{n} = \sum_{r=1}^{N_s} n(t_r) = N_s\langle n \rangle. \tag{4.4.2}$$

In Eq. (4.4.1), $\hat{G}_k(t_l)$ implies $\hat{G}_k^{(2)}(t_l)$.

In complementary single clipping, the lth channel receives counts of

$$(1 - n_k)n(t_l = lT)$$

for each sampling. After $'_s$ samplings, the content of the lth channel is

$$G_{ck}(t_l) = \sum_{r=1}^{N_s} [1 - n_k(t_r)]n(t_{r+l}) \qquad \text{without } 1/N \text{ normalization.} \tag{4.4.3}$$

Thus, $\hat{G}_k(t_l)$ and $\hat{G}_{ck}(t_l)$ contain the same information.

In practice, the digital correlator often has a buffer storage of a finite number of bits. Then, complementary clipping offers an advantage whenever $\langle n \rangle$ is large. For example, if we take $k = 0$, we obtain from Eq. (4.2.16) that, on the average, the maximum number of counts in a channel per sample time T for single clipping is about $\langle n \rangle \langle n_0 \rangle = \langle n \rangle \langle n \rangle / (1 + \langle n \rangle)$, which approaches $\langle n \rangle$ for large $\langle n \rangle$, while for complementary single clipping, the averaged maximum number of counts in a channel per sample time T is $\langle n \rangle [1 - \langle n \rangle / (1 + \langle n \rangle)] = \langle n \rangle / (1 + \langle n \rangle)$, which is always less than 1. Thus, complementary clipping alleviates the overflow problem, as we have already discussed in a slightly different way with regard to Eqs. (4.3.6) and (4.3.7).

In double clipping, we can also reformulate Eq. (4.2.8):

$$
\begin{aligned}
\langle n_0(0) n_0(\tau) \rangle &= \sum_{n(0)=0}^{\infty} \sum_{n(\tau)=0}^{\infty} p[n(0), n(\tau)] \\
&= 1 - p[n(0) = 0] - p[n(\tau) = 0] \\
&\quad + p[n(0) = 0, n(\tau) = 0] \\
&= 1 - 2p(0) + p(0, 0),
\end{aligned}
\tag{4.4.4}
$$

or, with the generating function (4.2.14),

$$
\begin{aligned}
\langle n_0(0) n_0(\tau) \rangle &= \frac{\langle n \rangle - 1}{\langle n \rangle + 1} + \frac{1}{(1 + \langle n \rangle)^2 - \langle n \rangle^2 |g^{(1)}(\tau)|^2} \\
&= \frac{\langle n \rangle - 1}{\langle n \rangle + 1} + p(0, 0),
\end{aligned}
\tag{4.4.5}
$$

where $p(0, 0) \equiv \langle n_0(0) n_0(\tau) \rangle_c$ represents the complementary double clipping at zero photocount number. The method of complementary double clipping is very useful whenever dust particles in the sample become a problem. At small scattering angles, the scattering by dust particles is usually very strong. As dust particles enter the effective scattering volume, the sudden increase in the scattered intensity essentially shuts off the correlator in the complementary double-clipping mode.

In an actual experiment, the content of the lth channel for N delays after M sweeps is

$$
M \hat{G}_k(t_l) = M \sum_{r=1}^{N} n_k(t_r) n(t_{r+l}) = MN \langle n_k(t_r) n(t_{r+l}) \rangle,
\tag{4.4.6}
$$

while the total photocounts during M sweeps with N channels are

$$
M \hat{n} = M \sum_{r=1}^{N} n(t_r) = MN \langle n \rangle.
\tag{4.4.7}
$$

Thus, MN gets canceled out between Eqs. (4.4.6) and (4.4.7). However,

Eqs. (4.4.6) and (4.4.7) do tell us that the counting capacity of the lth correlator channel must be more than $M\hat{G}(t_l)$ and that of the storage counter for total photocounts more than $MN\langle n \rangle$.

4.5. RANDOMLY CLIPPED AUTOCORRELATION FUNCTION (SALEH, 1978)

Random clipping is a scheme designed to handle non-Gaussian signals. If we let the clipping level k vary randomly from one sample to another with a probability distribution $q(k)$ that is statistically independent of photocount fluctuations, then the randomly clipped autocorrelation function is given by

$$\langle G_k^{(2)}(\tau) \rangle_k = \sum_{k=0}^{\infty} q(k) G_k^{(2)}(\tau). \tag{4.5.1}$$

Tartaglia *et al.* (1973) have pointed out that for $\langle n \rangle = 1$, a finite sum over all the clipping levels $k \leq 6$ will represent Eq. (4.5.1) as a full photocount autocorrelation function within an accuracy of 1% for Gaussian light, i.e.,

$$\langle G_k^{(2)}(\tau) \rangle_{k \leq 6} = \sum_{k=0}^{6} q(k) G_k^{(2)}(\tau) \propto G^{(2)}(\tau).$$

By substituting Eq. (4.2.7) into Eq. (4.5.1) and rearranging the summations, we get

$$\begin{aligned}
\langle G_k^{(2)}(\tau) \rangle_k &= \sum_{k=0}^{\infty} \sum_{n(0)=k+1}^{\infty} \sum_{n(\tau)=0}^{\infty} q(k) n(\tau) p[n(0), n(\tau)] \\
&= \sum_{n(\tau)=1}^{\infty} \sum_{n(0)=1}^{\infty} \left(\sum_{k=0}^{n(0)} q(k) \right) n(\tau) p[n(0), n(\tau)].
\end{aligned} \tag{4.5.2}$$

If k has a uniform distribution over a range 0 to S, i.e.,

$$q(k) = \begin{cases} 1/S, & k < S, \\ 0, & k \geq S, \end{cases} \tag{4.5.3}$$

then

$$\begin{aligned}
\langle G_k^{(2)}(\tau) \rangle_k &= \frac{1}{S} \sum_{n(\tau)=1}^{\infty} \sum_{n(0)=1}^{S-1} n(0) n(\tau) p[n(0), n(\tau)] \\
&+ \sum_{n(\tau)=1}^{\infty} \sum_{n(0)=S}^{\infty} n(\tau) p[n(0), n(\tau)],
\end{aligned} \tag{4.5.4}$$

where the second term on the right side is $[G_k^{(2)}(\tau)]_{k=S}$ and can be neglected for large S. Then,

$$\langle G_k^{(2)}(\tau) \rangle_k \simeq \frac{1}{S} G^{(2)}(\tau). \tag{4.5.5}$$

For thermal light, if $S > 10\langle n \rangle$, Eq. (4.5.5) has less than 1% error. The value of S necessary to justify this approximation depends on the statistical model (Saleh, 1978).

In practice, uniform random clipping can be achieved using a random-number generator to determine the clipping level. A simpler method is to use ramp clipping, whereby the finite summing is achieved by uniformly and sequentially clipping the incoming photocounts with $k = l$ during each of $l + 1$ sampling periods, and then starting over again at $k = l$, as shown in Fig. 4.5.1. In other words, Eq. (4.5.1) is reduced to

$$\langle G_k^{(2)}(\tau) \rangle_k = \frac{N_s}{l+1} \sum_{k=0}^{l} \langle n_k(0)n(\tau) \rangle \simeq \frac{N_s}{l+1} \langle n(0)n(\tau) \rangle \qquad (4.5.6)$$

where N_s is the total number of samplings, and we have taken

$$\sum_{k=l+1}^{\infty} \langle n_k(0)n(\tau) \rangle$$

to be negligible for sufficiently large l. For Gaussian light, we need

$$\left(\frac{\langle n \rangle}{1 + \langle n \rangle} \right)^{l+1} \ll 1 \qquad (4.5.7)$$

to set the level of l for uniform and sequential clipping.

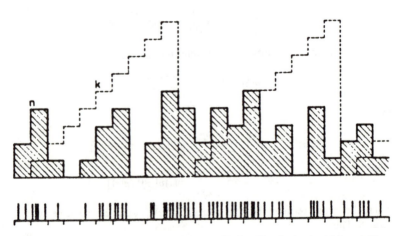

FIG. 4.5.1.   A schematic representation of ramp clipping. The clipping level k is varied in a sawtooth waveform, independent of the phase of the sampling-counting process. (Reprinted with permission from Springer-Verlag, After Saleh, 1978.)

4.6. SCALED AUTOCORRELATION FUNCTION

Correlation of *scaled* photon-counting fluctuations (Jakeman *et al.*, 1972; Koppel and Schaefer, 1973) provides a method for measuring the full photocount correlation function by means of a clipped correlator.

Scaling, like clipping, is a one-bit technique. It approximates uniform random clipping and provides an estimate of the time intensity correlation function, regardless of signal statistics. The electronic circuitry for scaling is considerably simpler than that required for uniform random clipping and full correlation. In scaling, we count one count and skip s counts instead of counting the $s + 1$ counts. Figure 4.6.1 shows a schematic representation of how a string of photoelectron pulses is counted, scaled and then clipped. For example, during the second sampling period between T and $2T$ and for $s = 6$, as shown in Fig. 4.6.1, we have $n = 6$, $r = 4$, $n^s = 1$, $n_0^s = 1$ where $n^s(t, T)$ is the number of scaled counts at time t and measured over a sampling time period T; $n_0^s(t, T)$ is the number of clipped-at-zero counts among scaled-by-s photocounts at time t and measured over a sampling time period T.

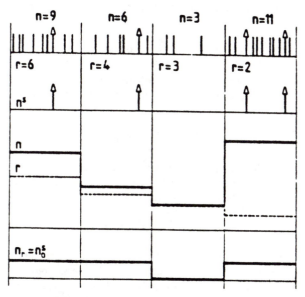

FIG. 4.6.1. A schematic representation of photoelectron pulses $n(t)$, the scaled-by-s photocount number $n^s(t)$, and the number of clipped-at-zero counts among scaled-by-s photocounts, $n_0^s(t)$. "Scaled-by-s" means that instead of counting all the photoelectron events, we count one event and skip s events. Figure 4.5.1 shows an example with the scaling factor $s = 6$. (Reprinted with permission from Springer-Verlag, After Saleh, 1978.)

The conditions are

$$n_0^s(t) = \begin{cases} 1 & \text{if} \quad n^s(t) > 0, \\ 0 & \text{if} \quad n^s(t) \le 0. \end{cases} \tag{4.6.1}$$

If r is the number of counts preceding the first scaled pulse at time t over the sampling period T, we have the number of zero-clipped scaled counts equal to the number of r-clipped counts:

$$n_0^s(t) = n_r(t) \tag{4.6.2}$$

with

$$n_r(t) = \begin{cases} 1 & \text{if} \quad n(t) > r, \\ 0 & \text{if} \quad n(t) \le r, \end{cases} \tag{4.6.3}$$

as, by definition, $n^s(t) > 0$ when $n > r$.

The full photocount autocorrelation function can be represented as an infinite sum over all separate single-clipped photocount autocorrelation functions:

$$G^{(2)}(\tau) = \langle n(0)n(\tau) \rangle = \sum_{k=0}^{\infty} \langle n_k(0)n(\tau) \rangle = \sum_{k=0}^{\infty} G_k^{(2)}(\tau). \tag{4.6.4}$$

If $q(r)$ is the probability distribution of the remainder, the probability of recording one or more counts during the sampling time interval T in the scaled channel is

$$\sum_{r=0}^{s-1} q(r) \sum_{n=s-r}^{\infty} p(n, T), \tag{4.6.5}$$

where $p(n, T)$ is the probability of having n photoelectron pulses in the time interval T. The joint probability of having one or more counts in the scaled channel and $n(\tau)$ counts in the other channel is

$$\sum_{r=0}^{s-1} q(r) \sum_{n(0)=s-r}^{\infty} p[n(0), n(\tau)], \tag{4.6.6}$$

where $p[n(0), n(\tau)][\equiv p(n_2, T, t_2; n_1, T, t_1)$ is the joint probability of having $n_1(\equiv n(0))$ photoelectron pulses at time $t_1(\equiv 0)$ in the sampling time interval T and $n_2(\equiv n(\tau))$ photoelectron pulses at time $t_2(\equiv \tau)$ in the sampling time interval T. The correlation function of photocounts scaled in one channel has the form

$$G_s^{(2)}(\tau) = \sum_{k=0}^{s-1} q(s - k - 1)G_k^{(2)}(\tau), \tag{4.6.7}$$

where

$$G_k^{(2)}(\tau) = \langle n_k(0)n(\tau) \rangle = \sum_{n(0)=k+1}^{\infty} \sum_{n(\tau)=0}^{\infty} n(\tau)p[n(0), n(\tau)]. \qquad (4.2.7)$$

Equation (4.6.7) reveals that scaling in one channel averages the single-clipped correlation function over a finite distribution of clipping levels. If $q(r)$ is uniform, scaling represents uniform random clipping over the same distribution of clipping levels. Then, Eq. (4.6.7) is reduced to

$$G_s^{(2)}(\tau) = \frac{1}{s} \sum_{n(0)=0}^{s} \sum_{n(\tau)=0}^{\infty} n(0)n(\tau)p[n(0), n(\tau)] + [G_k^{(2)}(\tau)]_{k=s}. \qquad (4.6.8)$$

If s is chosen (large enough) so that $G_s^{(2)}(\tau) \gg [G_k^{(2)}(\tau)]_{k=s}$, then a measure of $G_s^{(2)}(\tau)$ becomes a measure of the full photocount autocorrelation function. Jakeman *et al.* (1972) have shown that $q(r)$ is uniform. For Gaussian light, Eq. (4.6.8) becomes

$$\boxed{\begin{aligned} G_s^{(2)}(\tau) = \frac{\langle n \rangle^2}{s} \Bigg\{ & 1 - \left(\frac{\langle n \rangle}{1 + \langle n \rangle} \right)^s \\ & + |g^{(1)}(\tau)|^2 \left[1 + \frac{1 + \langle n \rangle + s}{1 + \langle n \rangle} \left(\frac{\langle n \rangle}{1 + \langle n \rangle} \right)^s \right] \Bigg\}, \end{aligned}} \qquad (4.6.9)$$

which differs from the true autocorrelation function by $<1\%$, owing to the cutoff at $s - 1$ in the sum of Eq. (4.6.6), if we choose $\langle n \rangle \gtrsim 1$, $s > 10\langle n \rangle$, and $\tau \simeq 25\langle \bar{n} \rangle T$.

Jakeman *et al.* (1972) found that at least for Gaussian light, scaling at small s is essentially as accurate as clipping, which is comparable in accuracy to full correlation. They also suggested a method for quick selection of the optimum scaling level in any experiment by picking an s such that the correlation counting rate is much greater for scaled clipped-at-zero correlation than for scaled clipped-at-one correlation. The scaled clipped-at-zero correlation function is then a good approximation to the full correlation function.

4.7. ADD–SUBTRACT AUTOCORRELATION FUNCTION

An add–subtract correlation function (Jen *et al.*, 1977) is similar to a single-clipped correlation function except that

$$a_k(t_i) = \begin{cases} 1 & \text{if} \quad n(t_i) > k, \\ -1 & \text{if} \quad n(t_i) \leq k \end{cases} \qquad (4.7.1)$$

instead of $n_k(t_i) = 1$ if $n(t_i) > k$ and 0 if $n(t_i) \leq k$ as defined by Eq. (4.2.3) for normal clipping. The add–subtract autocorrelation function $G_{as}^{(2)}(\tau)$ is defined by $\langle a_k(0)n(\tau)\rangle$. By rewriting Eq. (4.2.7),

$$\langle n_k(0)n(\tau)\rangle = \sum_{n(0)=k+1}^{\infty} \sum_{n(\tau)=0}^{\infty} n(\tau)p[n(0),n(\tau)],$$

we have for $G_{as}^{(2)}(\tau)$

$$\langle a_k(0)n(\tau)\rangle = \sum_{n(0)=k+1}^{\infty} \sum_{n(\tau)=0}^{\infty} n(\tau)p[n(0),n(\tau)]$$

$$- \sum_{n(0)=0}^{k} \sum_{n(\tau)=0}^{\infty} n(\tau)p[n(0),n(\tau)], \qquad (4.7.2)$$

where $p[n(0),n(\tau)]$ is the joint probability distribution for having $n(0)$ counts at time $t=0$ and $n(\tau)$ counts at time $t=\tau$; k is the clipping level. As $\langle n\rangle = \sum_{n(0)=0}^{\infty} \sum_{n(\tau)=0}^{\infty} n(\tau)p[n(0),n(\tau)]$,

$$G_{as}^{(2)}(\tau) = 2\sum_{n(0)=k+1}^{\infty} \sum_{n(\tau)=0}^{\infty} n(\tau)p[n(0),n(\tau)] - \sum_{n(0)=0}^{\infty} \sum_{n(\tau)=0}^{\infty} n(\tau)p[n(0),n(\tau)]$$

$$= 2G_k^{(2)}(\tau) - \langle n\rangle. \qquad (4.7.3)$$

Thus, the add–subtract correlation function yields essentially the same temporal information as the single-clipped correlation function except that a different value for the baseline is expected. In a normal single-clipped correlator,

$$\hat{G}_k(\tau_l) = \sum_{i=1}^{N_s} n_k(t_i)n(t_{i+l}), \qquad l = 1, 2, \ldots, L,$$

In an add–subtract correlator,

$$\hat{G}_{as}(\tau_l) = \sum_{i=1}^{N_s} a_k(t_i)n(t_{i+l}), \qquad l = 1, 2, \ldots, L.$$

It is worth noticing that within a time period of N_sT, a correlator with cleared shift counters performs $(N_s - L)L + L(L+1)/2 = [N - (L-1)/2]L$ multiplications (or additions) provided that N/L is an integer and T is the sampling time interval. Thus, the utilization of data is highly efficient. The multiplication (or summation) schemes, as shown in Figs. 4.2.1 and 4.2.3, are known as multistart multistop techniques, because each number is used as $n(t)$ and as $n(t - \tau)$ in forming the correlation function. If a single-start multi-stop technique is used, the correlation function is estimated by

$$\hat{G}_{as}(\tau_l) = \sum_{l=1, L+1, 2L+1}^{N_s} a_k(t_i)n(t_{i+l}), \qquad (4.7.4)$$

where only $(N_s/L)L = N_s$ multiplications (or additions) are performed within the time period $N_s T$. Thus, it takes a single-start multistop correlator $\sim L$ times longer to accumulate the same amount of correlation data as in a multi-start multistop correlator. As L is usually quite large, it is always advisable to use only multistart multistop correlators.

4.8. STATISTICAL ACCURACY OF ESTIMATING DIGITAL AUTOCORRELATION OF PHOTON-COUNTING FLUCTUATIONS (JAKEMAN *ET AL.*, 1971a–c; PIKE 1972)

Following the introduction of light-beating spectroscopy by Benedek and clipped photocount autocorrelation by Jakeman and Pike (1969b), several extensive studies on the statistical accuracy of the spectral linewidth in optical mixing spectroscopy have been reported. Special cases of the problem, corresponding experimentally to using a detector area much larger than the coherence area of the light, have been discussed by Benedek (1968), Haus (1969), and Cummins and Swinney (1970). Degiorgio and Lastovka (1971) have presented the statistical errors inherent in intensity correlation spectroscopy owing to the stochastic nature of both the scattering and the photo-emission processes. Independently, Jakeman *et al.* (1971b,c) have investigated the statistical errors (caused by the finite duration of experiments) in the intensity autocorrelation function and in the spectral linewidth of Gaussian–Lorentzian light. Analysis of Gaussian light by clipped photocount auto-correlation has also been extended to include the effects of finite-duration sampling intervals and incomplete spatial coherence by Koppel (1971). The details of these developments can best be obtained by reading the original papers, especially the work of Degiorgio and Lastovka (1971) on intensity correlation spectroscopy and of Jakeman *et al.* (1971b) on statistical accuracy in the digital autocorrelation of photon-counting fluctuations. While the analysis of data using three variables generally leads to larger errors, and the one-parameter fit of Kelly (1971) is more restrictive, the two-parameter approach of Pike and his coworkers for single exponential decays represents a first realistic attempt to solve this complex problem. It should be recognized that the time correlation functions from macromolecular solutions or colloidal suspensions are seldom single-exponential ones. Nevertheless, the analysis should provide us useful insight into how the experiments should be performed. Pike (1972) has summarized some of the pertinent guidelines for single-clipped autocorrelation functions as follows.

(1) Given a fixed number of channels L for the correlator, the ratio of the sample time T to the coherence (or characteristic) time $\tau_c (\equiv 1/\Gamma)$ has an optimum value corresponding to $LT/\tau_c \simeq 2$–3.

(2) Increase in the effective scattering area A will increase the light flux but decrease the correlation coefficients (Scarl, 1968; Jakeman et al., 1970a). The effects compensate asymptotically (Jakeman et al., 1971a), so that there is no advantage in using more than one coherence area if we are concerned *only* with time-correlation-function measurements. With high light flux the area can be reduced to well below a coherence area. Figure 4.8.1 (Pike, 1972) shows the percentage error in linewidth as a function of A/A_{coh} for fixed values of the number of channels ($L = 20$), sample-time/coherence-time ratio ($\Gamma T = 0.1$), experiment duration (10^4 coherence times), and single clipping at zero ($k = 0$). A decrease in the ratio A/A_{coh} decreases the averaging effects of the signal statistics without losing accuracy from photon statistics; but the count rate also decreases.

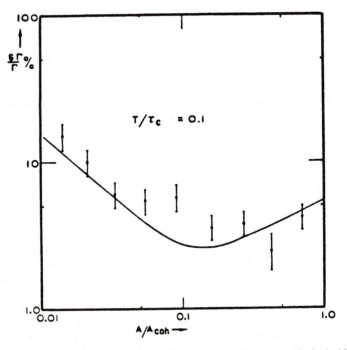

FIG. 4.8.1.   The linewidth accuracy for 20 channels single-clipped at zero, obtained with an experimental duration of 10^4 coherence times, as a function of detector-area/coherence-area ratio. The solid line shows the theory for a point detector with equivalent count rates. The points are experimental values using a sample time 0.1 times the coherence time. The number of photodetections per coherence time ranged from 0.5 to 23. The accuracy increases with area until the count rate becomes too high for clipping at zero, when it decreases again. (After Pike, 1972.)

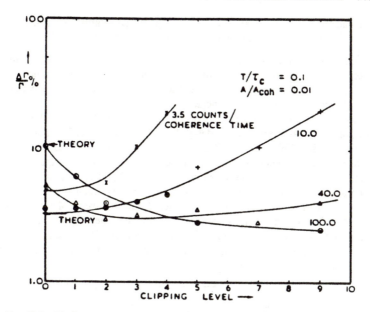

FIG. 4.8.2.   The linewidth accuracy for 20 channels obtained with an experiment duration of 10^4 coherence times, as a function of single-clipping level and number of counts per coherence time. The points are experimental values using a sample time 0.1 times the coherence time and a detector area 0.01 times the coherence area. The maximum accuracy obtainable does not improve significantly with count rate above about 10 counts per coherence time. Theoretical values are shown at zero clipping level only; the behavior at higher values has been confirmed by computer simulations. (After Pike, 1972.)

(3) Clipping levels should be set between one-half and two times the mean counting rate $\langle n \rangle$. Figure 4.8.2 (Pike, 1972) shows the percentage error in linewidth as a function of single-clipping level and photocounts per coherence time for fixed number of channels ($L = 20$), sample-time/coherence–time ratio ($\Gamma T = 0.1$), detector area ($A/A_{coh} = 0.01$), and experiment duration (10^4 coherence times). Using an appropriate clipping level, the accuracy improves with increasing number of photocounts per coherence time. However, the improvements quickly saturate, depending on the number of channels, as the number of photocounts per coherence time (τ_c) exceeds about 10.

In summary, we learn that for *single*-exponential decay curves there is no special advantage in having a correlator with an excessive number of channels, such as 128. Rather, it is essential

1. to set the clipping level at the mean counting rate per delay time,
2. to have $A/A_{coh} < 1$ with 10–100 photoelectrons per coherence time, and

3. to adjust the total delay (LT) to ~ 2 optical correlation (or coherence) times, which for a 20-channel correlator corresponds to $\Gamma T \sim 0.1$.

In addition, the condition $A/A_{coh} = 1$ is near optimum for a very wide range of scattering cross sections if the number of photocounts per coherence time $r < 10$. If $r > 10$, choose A such that the average number of photocounts per coherence time, $\langle n \rangle \tau_c / T$, is increased to ~ 10.

As we increase the number of delay channels in a correlator to 400 from 20 (say), the mean counting rate per delay channel will be reduced by a factor of 20. At very low light flux levels, the efficiency at even single zero clipping decreases. The dependence of the percentage error in linewidth on the number of channels is more complex, as shown in Fig. 4.8.3 (Jakeman *et al.*, 1971b). Generally, however, *there is no net gain of accuracy when the number of channels exceeds a certain limit.* The reader should be aware of this fact in practice. Many experimentalists tend to use more channels than necessary in order to achieve the desired precision and accuracy. The excess channels mainly add time to data analysis and management. In contrast, for correlators

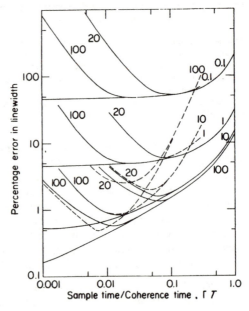

FIG. 4.8.3.   The percentage error in linewidth as a function of sample time for the photon flux and number of channels shown on the right-hand and the left-hand side of the graph, respectively: solid line, unclipped; dashed line, clipped at zero. When $r = 0.1$ and 1 the clipped and unclipped results coincide for $\Gamma T < 0.1$. Here $N_s \Gamma T = 10^4$. (After Jakeman *et al.*, 1971b.)

with equal delay-time increments, additional channels were often necessary just to cover the bandwidth required. However, within a fixed bandwidth having the first channel of the time correlation function represented by $\langle n(0)n(T)\rangle$ and the last channel by $\langle n(0)n(LT)\rangle$ (where LT is the maximum delay time) and with a fixed experiment duration time, the error in linewidth cannot be reduced significantly by improving the resolution of the instrument, i.e., by using more channels. A more detailed discussion is presented in Chapter VII.

In Fig. 4.8.3 a comparison is also made between the unclipped and the single-clipped-at-zero correlations. When $r = 0.1$ and 1, the clipped and unclipped results coincide for $\Gamma T \leq 0.1$. Unexpectedly, for $r = 100$ the clipped results actually provide less error in linewidth than the unclipped ones at small values of ΓT.

Figure 4.8.4 (Jakeman *et al.*, 1971b) shows the percentage error in linewidth as a function of number of photons per coherence time, r, at various values of ΓT, using $N_s \Gamma T = 10^4$ and an infinite number of channels in the unclipped case. It should be noted that the argument breaks down for more complex correlation functions that are not single exponentials. Again, the error in linewidth cannot be reduced indefinitely by increasing r.

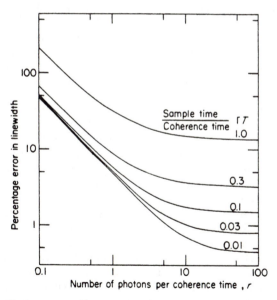

FIG. 4.8.4.   The percentage error in linewidth as a function of photon flux for the values of ΓT shown and an infinite number of channels in the unclipped case. $N_s \Gamma T = 10^4$. (After Jakeman *et al.*, 1971b.)

The important points here are as follows:

1. We should try to get $r \geq 10$.
2. Each run should be normalized independently against its own total counts, since this gives greater accuracy than subsequent normalization against a long-term mean value by assuming $\langle n(0) \rangle = \langle n(\tau) \rangle$.

The best approach is to normalize a large number of short runs independently. The length of the runs can be reduced until the bias, of order $1/N_s$, measurably affects the results, i.e., for $N_s \simeq 10^3$, where N_s is the total number of samples. A more detailed discussion has been presented elsewhere (Hughes et al., 1973).

The single-clipped correlator has played an important role since its inception. It remains competitive with a full correlator because of the relative ease of its construction and its efficiency, as has been discussed in this section. Future correlators are likely to measure $G^{(2)}(\tau)$, instead of $G_k^{(2)}(\tau)$, with less restriction on the overflow problem. The counting process for full autocorrelation has been discussed in some detail (Saleh, 1978). However, the quantitative formulations are beyond the level of the present monograph.

4.9. TIME-OF-ARRIVAL SCHEMES (SALEH, 1978)

Time correlation functions can be estimated by photoelectron statistical methods other than the autocorrelation-function technique. These methods are based on measurements of probability distributions, including the probability of photoelectronic coincidence, the probability distribution of inherent times, and single and joint photoelectron counting probabilities. The probability-distribution measurements can be used for very short delay-time increments, not accessible by present-day digital autocorrelators. We shall be concerned with a few selected schemes that are complementary to the photon correlation technique because of the accessible short delay-time increments, which can be used to bridge the frequency ranges of photon correlation spectroscopy and Fabry–Perot interferometry.

At low count rates, the probability of registering a count during the sample time interval $T(\ll \tau_c)$ is small. Then, the probability of detecting two photons separated by a sample time T can be approximated by the joint count rate at time intervals centered at $t = 0$ and $t = T$ (Glauber, 1968; Blake and Barakat, 1973). Scarl (1968) used the distribution of separation times between individual photons to measure the width of a mercury-vapor line. For very fast time correlations, Scarl used a double discriminator system in order to eliminate the time slewing by allowing only pulses well above the threshold of the timing discriminator to be analyzed. The measurement of pulse separations for photoelectrons uses a time-to-amplitude converter (TAC) and pulse-height analyzer (MCA)—the same approach as in nanosecond fluorescence decay studies (Lewis *et al.*, 1973)—for the determination of rotational diffusion coefficients of macromolecules (Tao, 1969) and reaction rates (DeLuca *et al.*, 1971). Unfortunately, such an approach permits only measurements of time arrivals for pairs of individual photons because of the limitations of the time-to-amplitude converter. Nevertheless, correlation times outside the range of present-day digital photon correlators, i.e., with sample times less than ~ 20 nsec per channel, can be measured using the TAC–MCA approach. It is also expected that for intermediately fast delay-time increments (50–100 nsec) the double discriminator–coincidence circuit (Scarl, 1968) can be avoided, since the rise times of the photoelectron pulses after appropriate amplification and discrimination are in the nanosecond range. Thus, measurements of separation times between individual photons provide us with an alternative route to measurements of the intensity time correlation function for very fast correlation-time studies with resolutions down to about 2 nsec. In the nanosecond region, a multiple time-to-amplitude converter in the form of a time-to-digital converter (TDC) can be used to measure time correlations.

Chopra and Mandel (1972) developed a correlator based on digital storage of the arrival times of up to six photoelectric pulses following a start pulse. They also adapted the same technique to the development of a third-order intensity correlator (Chopra and Mandel, 1973). In fast-decay-time studies the finite pulsewidth and instrumental dead time have to be taken into account.

Kelly and Blake (1971) and Kelly (1972) measured the width of a Lorentz spectral line using a digital photoelectron correlation device as well as a pulse-separation detector (TAC–MAC) and showed that the pulse-separation measurements can yield information similar to autocorrelation-function measurements over a wide range of count rates if the results are correctly interpreted. For macromolecules in solution and phase transition studies, the best approach is to use digital photon counting for the time-averaged intensity and digital photon correlation (clipped or unclipped) for decay-time studies, as we have described in detail in previous sections.

4.9.1. PROBABILITY OF COINCIDENCE

The joint probability that two photoelectrons occur simultaneously—one at position r_1 and time instant t_1 within δt_1, and another at position r_2 and time instant t_2 within δt_2—is proportional to the intensity autocorrelation function $\langle I(r_1,t_1)I(r_2,t_2)\rangle$, where $I(r,t)$ is the instantaneous intensity at position r and time t. If we assume that light falls on a single photomultiplier tube, the joint probability that the first start pulse arrives at time $t + T$ within δT and another (single) stop pulse arrives at time $t + T + \tau$ within

$\delta\tau(\tau = m\,\delta\tau, m = 0, 1, 2, \ldots)$ is expressible in the form

$$p_2(t + T, t + T + \tau)\delta T\delta\tau$$

$$= \langle\exp[-a\int_t^{t+T} I(t')\,dt']\,a^2 I(t + T)I(t + T + \tau)\,\delta T\delta\tau\rangle. \quad (4.9.1)$$

It should be noted that t is the time when the correlator is enabled after a counting sequence, i.e., when it is ready to accept a start pulse and T is not the sample time here. In Eq. (4.9.1), we have included an exponential factor that represents the absence of pulses during the time interval t to $t + T$. When the average rate at which start pulses are arriving is sufficiently small, i.e., if $\langle I(t)\rangle$ is to be observed in the time range $[0, t_{\max}]$, then, under the condition

$$a\int_0^{t_{\max}} \langle I(t')\rangle\,dt' \ll 1, \quad (4.9.2)$$

the exponentially decaying multiplicative factor in Eq. (4.9.1) can be neglected and we have for a stationary field, using the single-photoelectron-decay technique,

$$\boxed{p_2(1, 1, \tau) = a^2\langle I(0)I(\tau)\rangle.} \quad (4.9.3)$$

Equation (4.9.2) states that the mean number of counts in the time interval of interest (e.g., the characteristic decay time) is much less than 1. At higher intensities, the exponential factor cannot be ignored and the single-photoelectron-decay technique suffers from the so-called pile-up distortion. On the other hand, a time-of-arrival correlator in which the arrival time of six photoelectron pulses can be measured following a single start pulse (Chopra and Mandel, 1972) permits a much higher counting rate. The finite number of stop channels imposes a limitation on the maximum usable signal count rate, since the measured correlation function will be biased if any photoelectron pulses are lost in the measurement interval.

The probability of registering a photoelectron pulse in the Nth stop channel is

$$\boxed{p(>N) = \sum_{r=N}^{\infty} \frac{(\bar{n}_s T)^r}{r!} e^{-\bar{n}_s T},} \quad (4.9.4)$$

where $\bar{n}_s$ is the mean signal count rate. For example, if $N = 6$ and $\bar{n}_s T \approx 1$, the probability that the photoelectron pulses are lost in the measurement interval (T) if up to six pulses (events) per measurement interval can be recorded is less than 10^{-3}. Thus, the count rate of $1/T$ is a considerable improvement over the requirement $\bar{n}_s \ll 1/T$. The Chopra–Mandel correlator has a maximum count-rate capacity of 10,000 counts per second and a delay-time resolution (or increment) of 1 μsec.

Another version of the TOA correlator was described by Friberg *et al.* (1982). With modern electronics, a fast TOA correlator could be constructed. We were able to use a delay-time increment of 2.5 nsec. The probability distribution in the arrival time of eight pulses following a start pulse was obtained by means of an eight-channel time-to-digital converter (TDC). For a stationary field, both the probability and the correlation function depend only on the time interval $t_2 - t_1$ ($= \tau$) between the two time arguments. Generally, $I(t_1)$ and $I(t_2)$ can come from two distinct detectors, and hence they are also position-dependent. Here we assume that light falls normally on a single photomultiplier tube with an effective surface area A and quantum efficiency a. So the joint probability $p_2(t + T, t + T + \tau)\delta T\delta\tau$ obeys Eq. (4.9.1).

The digital time-of-arrival photoelectron correlator (TOAC), as shown in Fig. 4.9.1, has τ_L ($= L\delta\tau \simeq 5 \,\mu$sec) with $L = 2048$ and delay-time increment $\Delta\tau$ (or sampling time interval $\delta\tau$) = 2.5 nsec. The system has an overhead time of

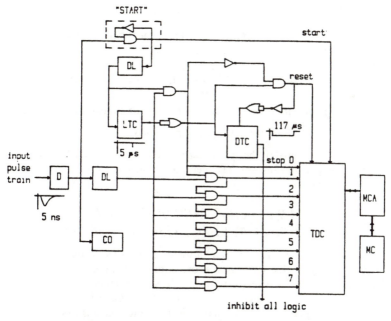

FIG. 4.9.1.   Schematic of the time-of-arrival correlator (TOAC). D, four-channel discriminator; DL, delay line; CO, photon counter; DTC, dead-time clock; LTC, live-time clock; TDC, time-to-digital converter; MCA, multichannel analyzer; MC, LeCroy System 3500 microcomputer. (After Dhadwal *et al.*, 1987.)

112 μsec, due mainly to the conversion of time-interval data in the TDC. The probability distribution of the arrival time of eight photoelectron pulses following a start pulse is obtained by coupling an eight-channel TDC to a multichannel analyzer (MCA), which has 2048 delay intervals for *each* of the eight independent channels. If all the photoelectron pulses are counted during the time interval $\tau_{L=2048} \simeq 5$ μsec, i.e., $N \leq 8$, then the total number of counts recorded in the MCA has the form (after Dhadwal *et al.*, 1987, p. 1446)

$$\langle n(m\,\Delta\tau) \rangle = M\bar{n}_s\,\Delta\tau\,[1 + \beta|g^{(1)}(m\,\Delta\tau)|^2], \tag{4.9.5}$$

where $m = 1$ to 2048, M is the number of times that the measurement sequence is initiated, $\Delta\tau$ is the delay-time increment of the correlator, $\bar{n}_s\,\Delta\tau$ is the average number of counts per delay-time increment, β ($= f(A)$ of Eq. (4.1.4)) is the spatial coherence factor, and $|g^{(1)}(m\,\Delta\tau)|$ is the normalized first-order electric field correlation function. The first term of Eq. (4.9.5) represents the contribution due to "accidental coincidences," and the second term is due to the photon bunching effect. In the absence of intensity correlation and for $m \to \infty$, $|g^{(1)}(m\,\Delta\tau)| = 0$ and

$$\langle n(\infty) \rangle = M\bar{n}_s\,\Delta\tau. \tag{4.9.6}$$

In the case that $\langle n(m\,\Delta\tau)\rangle$ and $\langle n(\infty)\rangle$ can be measured separately, the normalized intensity–intensity correlation function is

$$\beta|g^{(1)}(m\,\Delta\tau)|^2 = \frac{\langle n(m\,\Delta\tau)\rangle}{\langle n(\infty)\rangle} - 1, \qquad m = 1,\ldots,2048, \tag{4.9.7}$$

which is the net unnormalized intensity correlation function in the relation

$$G^{(2)}(m\,\Delta\tau) = A(1 + \beta|g^{(1)}(m\,\Delta\tau)|^2). \tag{4.9.8}$$

In Eq. (4.9.8), $G^{(2)}(m\,\Delta\tau)$ is the unnormalized intensity correlation function, A is the baseline, β is the same spatial coherence factor, and $m\,\Delta\tau$ is the delay time of the mth channel.

4.9.2. CORRELATOR DESCRIPTION
(AFTER DHADWAL ET AL., 1987, PP. 1446–1447)

We present a detailed description of the time-of-arrival correlator (TOAC) as shown in Fig. 4.9.1, because such a correlator is not available commercially, but can be assembled using existing off-the-shelf components. High-speed logic gates with a rise time of 2.0 nsec (LeCroy Research Systems Corp.) are used for signal routing. A pulse train of photoelectrons from a photomultiplier tube, after amplification and appropriate discrimination, is fed into the input of the TOAC, which has an eight-channel TDC (LeCroy model

2228A) for measuring the arrival time of eight stop pulses following a single start pulse in a burst of photoelectrons. The joint probability distributions in the arrival times of each of the eight photoelectron pulses following a single start pulse are measured using a multichannel analyzer, which has 2048 delay times in each of its eight separate channels. The measurements are controlled by the LeCroy 3500 microcomputer system. This arrangement greatly improves the analysis over an earlier version (Patkowski *et al.*, 1978). A measurement time of 5 μsec gives a delay-time resolution (or increment) of 2.5 nsec, which far exceeds the 1-μsec resolution (or delay-time increment) described elsewhere (Davidson, 1969; Chopra and Mandel, 1972; Friberg *et al.*, 1982). The MCA requires 117 μsec for transferring the data from the TDC to its memory. As a result, we have a maximum cycle time of 122 μsec, which together with the short measurement time interval and eight stop channels means that the accumulation time can be reduced considerably from typical values of 5–10 hours per correlation-function curve.

Another important feature of the TOAC is the absence of any periodic timing signals, which are normally used for defining the measurement time interval followed by a data-transfer time interval. In our particular implementation, two independent clocks, which are triggered by the incoming photoelectron pulse train, are used to define a live time interval of 5 μsec, and a dead time interval of 117 μsec in the presence of one or more stop pulses and 2 μsec in the absence of stop pulses. In this fashion, the efficiency of the system is very high, that is, very few events are lost for a single-start multi-stop scheme.

The discriminator (LeCroy model 821) provides four balanced outputs, which prevents loading effects from becoming significant. The signal pulse train is fed into the start AND gate (LeCroy model 622), which detects the first pulse in a given photon burst and uses it to trigger the live-time clock. Any subsequent pulses received, within the measurement interval of 5 μsec, are ignored by the start AND gate in order to prevent the TDC from being restarted before the end of the measurement interval. The arrival time of all subsequent pulses, up to eight, is recorded by the TDC. At the end of the live time, a separate dead-time pulse of 117 μsec is generated using an oscillator (LeCroy model 222). This pulse, derived from an independent clock, is used to veto the discriminator and disable all logic gates while the data from the TDC is being transferred to the MCA. In the event that no stop pulses are detected in the measurement interval, the dead-time clock is disabled after a 2-μsec delay, which is required to reset the TDC. In this manner we are able to reduce the system overhead considerably. Note that the start of the next measurement sequence is initiated by the first pulse in the next photoelectron burst. Therefore, there is a random delay associated with the end of a measurement sequence and a start of the next one. If P_1 is the width of the measurement cycle when one or more stops are detected within the live time, P_2 is the width of the measurement cycle in the absence of stop pulses, and w is the width of the live-time clock, then the mean cycle time for a single measurement, Δt, is

$$\Delta t = P_1 + \frac{P_2}{w\bar{n}_s}. \tag{4.9.9}$$

In our particular setup $P_1 = 122$ μsec, $P_2 = 7$ μsec, and $w = 5$ μsec. In the derivation of Eq. (4.9.9) we have assumed that the random separation between measurement sequences is negligible. This is plausible, since we are measuring effects due to photon bunching.

The MCA has a dead time of 105 nsec, which means that any photoelectron pulses detected within this interval will be lost. To overcome this dead-time effect we introduced an appropriate delay between the start of TDC ramp and the timing logic gates. In other words, the start AND gate, on detecting the first pulse, sends a flag to the TDC to start the ramp. If there were no dead time in the MCA, the pulse train would be fed into the logic unit simultaneously. Thus, in this particular implementation, the input pulse train is delayed before being sent to the timing logic gates. The appropriate delay was obtained by using a 70-foot cable.

The histogram of the number of counts in a particular channel of the MCA memory is therefore a measure of the joint probability distribution of the arrival times of, say, the second pulse, following a single start pulse. The combined expected number of counts, $\langle n(m\,\Delta\tau)\rangle$, is then the sum of all eight arrival distributions. In practice we can arrange the signal count rate so that the probability of detecting pulses in, say, the Nth stop is very small, as given by Eq. (4.9.4). This, of course, restricts the signal count rate. For example, if $P(>6) = 10^{-3}$, then the maximum signal count rate is 200,000 counts/sec ($\bar{n}_s T \approx 1$); this compares favorably with the value 10,000 counts/sec given by Chopra and Mandel. Thus with eight stop channels the maximum signal count rate can be increased still further.

In summary, a correlator based on the measurement of the joint probability distribution function in the arrival times of eight stop pulses following a single start pulse can be assembled using commercially available components. This correlator has a measurement interval of 5 μsec and a delay-time resolution of 2.5 nsec. With an eight-stop capability, much higher count rates can be used without fear of distorting the intensity–intensity time correlation function by saturation. Thus the accumulation time becomes tolerable for acceptable signal-to-noise ratios in time-correlation measurements with a short delay-time increment.

4.10. PHOTON STRUCTURE FUNCTION

According to Eq. (2.4.5), the intensity correlation function $G^{(2)}(\tau)$ has the form

$$
\begin{aligned}
G^{(2)}(\tau) &= \lim_{T\to\infty} \frac{1}{2T} \int_{-T}^{T} I(t)I(t+\tau)\,dt \\
&= \lim_{T\to\infty} \frac{1}{2T} \int_{0}^{2T} I(t)I(t+\tau)\,dt,
\end{aligned}
\tag{4.10.1}
$$

in which we have taken the periodogram of length $2T$. It should be noted that T is not the sample time here. An alternative approach is to examine how interferometry is performed. All interferometric methods involve the superposition of an instantaneous and a delayed signal. The superposition is then squared and averaged. For example, in conventional amplitude ($A(t)$) interferometry,

$$
S^{(1)}(\tau) = \lim_{T\to\infty} \frac{1}{2T} \int_{0}^{2T} |A(t) - A(t+\tau)|^2 \, dt.
\tag{4.10.2}
$$

Thus, in intensity ($I(t)$) interferometry, one can define a structure function

$$
S^{(2)}(\tau) = \lim_{T\to\infty} \frac{1}{2T} \int_{0}^{2T} [I(t) - I(t+\tau)]^2 \, dt.
\tag{4.10.3}
$$

The photon structure function (Schatzel, 1983; Schulz-DuBois and Rehberg, 1981; Oliver and Pike, 1981) is an old concept dating back to Einstein's formula for the mean square of the displacement X of Brownian particles,

$$
\langle [X(t) - X(t+\tau)]^2 \rangle = 2D\tau,
\tag{4.10.4}
$$

where the angular bracket indicates an ensemble average and τ is the lag time. Following Eq. (4.10.4), a photon structure function can be defined as

$$
\boxed{D(\tau_l) = \langle (n_i - n_{i+l})^2 \rangle,}
\tag{4.10.5}
$$

where n_i ($\equiv n(t_i)$) is the number of photons detected during a sample time interval i of sampling time duration T. Equation (4.10.5) requires that n_i be a stochastic process with stationary increments.

The photon structure function differs from the photon correlation function

$$
G^{(2)}(\tau_l) = \langle n_i n_{i+l} \rangle,
\tag{4.10.6}
$$

which merely expresses Eq. (4.2.6) in a slightly different form. If a linear drift is present in n_i, Eq. (4.10.5) remains well defined, whereas Eq. (4.10.6) becomes a function of i as well as l. The structure function can tolerate stronger low-frequency noise, while the correlation function is less sensitive to high-frequency photon shot noise, especially for weak signals. For a stationary photon signal n_i,

$$D(\tau_l) = \langle n_i^2 - 2n_i n_{i+l} + n_{i+l}^2 \rangle$$
$$= \langle n_i^2 \rangle - 2\langle n_i n_{i+l} \rangle + \langle n_{i+l}^2 \rangle,$$

$$\boxed{D(\tau_l) = 2G^{(2)}(0) - 2G^{(2)}(\tau_l).} \tag{4.10.7}$$

Figure 4.10.1 shows a schematic representation of the photon correlation and structure functions. For a fair comparison, the estimators

$$\hat{D}(\tau_l) = \frac{1}{N} \sum_{i=1}^{N} [n(t_i) - n(t_{i+l})]^2 \tag{4.10.8}$$

and

$$\hat{S}^{(2)}(\tau_l) = \frac{1}{N} \sum_{i=1}^{N} n(t_i)n(t_{i+l}), \tag{4.1.8}$$

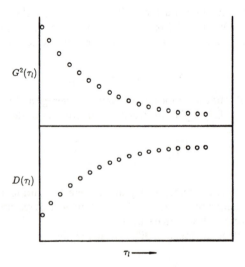

FIG. 4.10.1.   Schematic representation of photon correlation (upper) and structure (lower) functions. Here we have set $G^{(2)}(\tau_l) \simeq \hat{S}^{(2)}(\tau_l)$, neglecting the normalization factor $1/N$.

with $\tau_l = lT$, must be modified by subtracting the constant photon noise component and the background, respectively. The modified estimators are

$$\hat{D}_2(\tau_l) = \hat{D}(\tau_l) - 2\langle n_i \rangle_i \qquad (4.10.9)$$

$$\hat{S}_2^{(2)}(\tau_l) = \hat{S}^{(2)}(\tau_l) - \langle n_i \rangle^2, \qquad (4.10.10)$$

where $\langle \; \rangle$ denotes a finite-time average. For T much less than the coherence time τ_c and spatially coherent detection, the photon signals n_i are Poisson-distributed about a mean $\mu_i \propto aTI$. The mixed moments of the photon signal are (Saleh, 1978)

$$\langle n_i \rangle = \mu_i, \qquad (4.10.11)$$

$$\langle n_i n_j \rangle = \mu_i \mu_j + \delta_{ij} \mu_i, \qquad (4.10.12)$$

$$\langle n_i n_j n_k \rangle = \mu_i \mu_j \mu_k + \delta_{ij} \mu_i \mu_k + \delta_{ik} \mu_i \mu_j + \delta_{jk} \mu_i \mu_j + \delta_{ijk} \mu_i, \quad (4.10.13)$$

By assuming a nonzero τ_l, the expectations and variances of the modified estimators can be computed for arbitrary stationary processes μ_i. Interested readers should consult the literature directly. Schatzel and his coworkers have also extended the structure function concept to laser Doppler velocimetry (Schatzel and Merz, 1984; Schatzel *et al.*, 1985; Schatzel, 1987).

4.11. COMPUTER-BASED CORRELATORS

We have discussed a variety of schemes for computing the time correlation function. The earlier attempts invariably try to simplify the computation hardware by some form of quantization of signals, such as single clipping. However, the ultimate norm is to compare those correlation measurements with values from a full ideal correlator, as characterized by Eq. (4.1.3). On the other hand, full correlators still have limitations in that they can accept only a small number of photocounts per sample time (typically $n(t, T) \le 2^4$), which often becomes insufficient for large delay-time increments (e.g., $\Delta\tau \equiv T \ge 10$ msec). With $T = 10$ msec and $\bar{n}_s \sim 10^5$ counts/sec, we have $\bar{n}(t, T) = 10^5 \times 10^{-2} = 10^3$ counts/(sample time) $\gg 2^4$. Thus, such a $4 \times n$ correlator will be overloaded.

Historically, after Foord *et al.* (1970) introduced the single-clipped photon-counting technique, there was a flurry of activity in

1. hardware correlator construction (Ohbayashi and Igarashi, 1973; Mole and Geissler, 1975), including a 1024-channel digital correlator for spectral analysis of radio-astronomy signals with a maximum sampling rate of 20 MHz (Ables *et al.*, 1975), a 3-bit full correlator (Asch and Ford, 1973), a VHF autocorrelator capable of accepting input signal frequencies up to 75 MHz by batch processing (Wooding and Pearl, 1974), a simple

low-cost digital events analyzer (Yangos and Chen, 1980), and a high-product-rate 10-nsec 256-point correlator (Norsworthy, 1979);

2. specific schemes for measuring autocorrelation functions (Kam *et al.*, 1975); and

3. special-purpose correlators, such as a real-time correlator with a peak detector (Tai *et al.*, 1975) or a wide-band correlator (Padin and Davis, 1986).

Although new construction, especially with microprocessor-based operations, continues (Bisgaard *et al.*, 1984; Devanand and Sekhar, 1985; Noullez and Boon, 1986; Subrahmanyam *et al.*, 1987), these correlators often do not exceed the specifications of commercial correlators.

From the outset, one recognizes that the correlator is a special-purpose computer, hard-wired to calculate a single function. It is therefore always tempting to use a general-purpose computer to calculate the correlation function. All on-line correlation measurements of microprocessor-based correlators employ hard-wired front ends to perform the appropriate countings per sample time. The information is then transferred, sometimes with the aid of a fast buffer memory, and the correlation function computed by a general-purpose computer (e.g., see Wijnaendts van Resandt, 1974; Ogata and Matsuura, 1974; Gray *et al.*, 1975; Knox and King, 1975; Matsumoto *et al.*, 1976; Nemoto *et al.*, 1981; Abbey *et al.*, 1983). As all present-day commercial correlators are microprocessor-based, it will suffice to mention here that the details in trying to achieve the full-time correlation (Eq. (4.1.3)) as shown in Fig. 4.1.1. may vary because there are different approaches to implementing the multiplier function as shown in Fig. 4.2.1. The more popular correlators construct $\hat{G}^{(2)}(\tau_l) = \sum_{i=1}^{N_s} n(t_i)n(t_{i+l})$. A more efficient approach in computing the time correlation function is to try to exploit the time-interval information on the pulse sequence in a photoelectron pulse train as we have discussed in Section 4.9. The scheme used in the TOA correlator is a single-start multi-stop one which does not maximize the statistical accuracy of the desired information.

Matsumoto *et al.* (1976) first developed the computer-based photoelectron counting system, which was further refined by Nemoto *et al.* (1981), with dead time down to 310 nsec. The main idea is to be able to record the time intervals of an arbitrarily long sequence of photoelectron pulses and then to construct the probability distribution function $p(n, t)$ (Eq. (3.2.8)) and the conditional probability $p_2(\tau)\,d\tau$ (Eqs. (4.9.1) and (4.9.3)) by sampling from the pulse train as many datum points as one needs. The measured $p_2(\tau)$ is equal to $G^{(2)}(\tau)$ up to a proportionality constant. The scheme developed by Matsumoto *et al.* (1976) is as follows:

The time intervals between neighboring pulses in a photoelectron pulse train are stored in a memory along with their sequence in the train. From the stored data on a photoelectron pulse train, the probabilities $p(n, T)$ and $p_2(\tau)$ are constructed so as to maximize the statistical accuracy.

To find $p(n, T)$, an appropriately chosen time window T is shifted throughout the pulse train and the number of possible locations of the window covering n and only n successive pulses is counted. More specifically, $T_s(i, i + n)$, the shortest time covering n successive pulses starting from the $(i + 1)$th pulse, as shown in [Fig. 4.11.1(a)], is determined. Likewise, $T_L(i, i + n)$, the longest one, also shown in [Fig. 4.11.1(a)], is determined. The former is given by the sum of $(n - 1)$ successive data in the memory starting from the $(i + 1)$th data. The latter is likewise given by the sum of $(n + 1)$ successive data starting from the ith data. In the next step, $T_M(i, i + n)$, the shorter of the ith and $(i + n)$th pulse intervals, and then $f_i(n, T)$, the number of possible locations of the window T covering n successive pulses starting from the ith pulse, as illustrated in [Fig. 4.11.1(b)], are determined. $p(n, T)$ is obtained by summing $f_i(n, T)$ over all i.

The conditional probability $p_2(\tau)$ is readily obtained from the distribution of time intervals between all possible pair of pulses in the photoelectron pulse train. First, compute $S_i(n)$, which is the sum of $(n + 1)$ successive data [on the] pulse train, for all the possible value of i and n. The frequency distribution for $S_i(n)$ is equivalent to $p_2(\tau)$, which is equal to $\langle I(t)I(t + \tau)\rangle/\langle I(t)\rangle^2 \equiv g^{(2)}(\tau)$ up to a proportional constant.

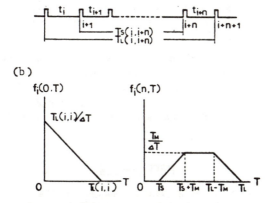

FIG. 4.11.1. Scheme for finding $p(n, T)$. In (a), the definitions are illustrated for $T_s(i, i + n)$, the shortest time covering n successive pulses starting from the $(i + 1)$th pulse, and for $T_L(i, i + n)$, the longest one. In (b), the definition is given for $f_i(n, T)$, the number of possible locations of the time window T covering n successive pulses starting from the ith pulse, where ΔT is a unit of window shifting, and $T_M(i, i + n)$ is the shorter of the ith and $(i + n)$th pulse intervals. $p(n, T)$ is found by summing $f_i(n, T)$ over all i. (After Matsumoto *et al.*, 1976).

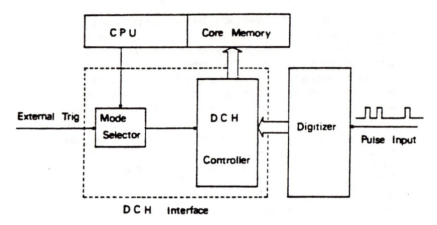

Fig. 4.11.2. Scheme of the computer-based photoelectron counting system. An incoming pulse train is fed into the digitizer, where the pulse time intervals are converted into digital data. The data are transferred through a data-channel (DCH) controller to the computer core memory. (After Matsumoto *et al.*, 1976.)

The system consists of a pulse interval digitizer, a data channel interface, and a minicomputer system, which are assembled as shown in [Fig. 4.11.2]. The pulse interval digitizer digitizes the time intervals between neighboring pulses in a pulse train. The data channel interface transfers the data from the digitizer directly to the magnetic core memory of the computer system. The computer system is used to process the stored data on pulse intervals into the statistical quantities such as $p(n, T)$ and $p_2(\tau)$.

In the improved version (Nemoto *et al.*, 1981), the correlator has 512 channels with an improved procedure to obtain the true statistical average for photon correlation, in addition to reducing the dead time from 800 nsec (Matsumoto *et al.*, 1976) to 310 nsec. Further exploration of this time-interval scheme with smaller delay-time increments and dead times should be of interest.

REFERENCES

Abbey, K. M., Shook, J., and Chu, B. (1983). *In* "The Application of Laser Light Scattering to the Study of Biological Motion" (J.C. Earnshaw and M. W. Steer, eds.), NATO ASI Series, Life Sciences, **59**, pp. 77–87, Plenum, New York.

Ables, J. G., Cooper, B. F. C., Hunt, A. J., Moorey, G. G., and Brooks, J. W. (1975). *Rev. Sci. Instrum.* **46**, 284.

Arecchi, F. T., Berne, A., and Sona, A. (1966). *Phys. Rev. Lett.* **17**, 260.

Asch, R. and Ford, N. C. (1973). *Rev. Sci. Instrum.* **44**, 506.

Bedard, G. (1967). *Phys. Rev.* **161**, 1304.

Bisgaard, C., Johnsen, B., and Hassager, O. (1984). *Rev. Sci. Instrum.* **55**, 737.

Benedek, G. B. (1968). "Polarisation, Matière et Rayonnement," pp. 49–84, Presses Universitaires de France, Paris.

Blake, J. and Barakat, R. (1973). *J. Phys. A* **6**, 1196.

Chen, S. H., Tartaglia, P., and Polonsky-Ostrowsky, N. (1972). *J. Phys. A* **5**, 1619.

Chopra, S. and Mandel, L. (1972). *Rev. Sci. Instrum.* **43**, 1489.

Chopra, S. and Mandel, L. (1973). *Rev. Sci. Instrum.* **44**, 466.

Chu, B. (1974). "Laser Light Scattering," Academic Press, New York.

Cummins, H. Z. and Swinney, H. L. (1970). *Progr. Opt.* (E. Wolf, ed.) **8**, 135, North-Holland, Amsterdam.

Davidson, F. (1969). *Phys. Rev.* **185**, 446.

Degiorgio, V. and Lastovka, J. B. (1971). *Phys. Rev. A.* **4**, 2033.

DeLuca, M., Brand, L., Cebula, T. A., Seliger, H. H., and Makula, A. F. (1971). *J. Biol. Chem.* **246**, 6702.

Devanand, K. and Sekhar, P. (1985). *J. Phys. E.* **18**, 904.

Dhadwal, H. S., Chu, B., and Xu, R.-L. (1987). *Rev. Sci. Instrum.* **58**, 1445.

Foord, R., Jakeman, E., Oliver, C. J., Pike, E. R., Blagrove, R. J., Wood, E., and Peacocke, A. R. (1970). *Nature, Lond.* **227**, 242.

Friberg, S., Anderson, J., and Mandel, L. (1982). *Rev. Sci. Instrum.* **53**, 205.

Glauber, R. J. (1968). *In* "Fundamental Problems in Statistical Mechanics II" (E. G. D. Cohen, ed.), pp. 140–187, North-Holland, Amsterdam.

Gray, A. L., Hallett, F. R., and Rae, A. (1975). *J. Phys. E* **8**, 501.

Haus, H. A. (1969). *In* "Proc. Int. School of Physics Enrico Fermi, Course XLII," p. 111, Academic Press, New York.

Hughes, A. J., Jakeman, E., Oliver, J., and Pike, E. R. (1973). *J. Phys. A* **6**, 1327.

Jakeman, E. (1970). *J. Phys. A* **3**, 201.

Jakeman, E. (1974). *In* "Photon Correlation and Light Beating Spectroscopy" (H. Z. Cummins and E. R. Pike, eds.), p. 75, Plenum, New York.

Jakeman, E. and Pike, E. R. (1968). *J. Phys. A* **1**, 128.

Jakeman, E. and Pike, E. R. (1969a). *J. Phys. A* **2**, 115.

Jakeman, E. and Pike, E. R. (1969b). *J. Phys. A* **2**, 411.

Jakeman, E., Oliver, C. J., and Pike, E. R. (1968). *J. Phys. A* **1**, 406.

Jakeman, E., Oliver, C. J., and Pike, E. R. (1970a). *J. Phys. A* **3**, L45.

Jakeman, E., Oliver, C. J., and Pike, E. R. (1971a). *J. Phys. A* **4**, 827.

Jakeman, E., Pike, E. R., and Swain, S. (1971b). *J. Phys. A* **4**, 517.

Jakeman, E., Pike, E. R., and Swain, S. (1971c). *J. Phys. A* **3**, L55.

Jakeman, E., Oliver, C. J., Pike, E. R., and Pusey, P. N. (1972). *J. Phys. A* **5**, L93.

Jen, S., Shook, J., and Chu, B. (1977). *Rev. Sci. Instrum.* **48**, 414.

Kam, Z., Shore, H. B., and Feher, G. (1975). *Rev. Sci. Instrum.* **46**, 269.

Kelly, H. C., (1971). *IEEE J. Quantum Electron.* **QE-7**, 541.

Kelly, H. C., (1972). *J. Phys. A* **5**, 104.

Kelly, H. C. and Blake, J. G. (1971). *J. Phys. A* **4**, L103.

Knox, A. and King, T. A. (1975). *Rev. Sci. Instrum.* **46**, 464.

Koppel, D. E. (1971). *J. Appl. Phys.* **42**, 3216.

Koppel, D. E. and Schaefer, D. W. (1973). *Appl. Phys. Lett.* **22**, 36.

Lewis, C., Ware, W. R., Doemey, L. J., and Nemzek, T. L. (1973). *Rev. Sci. Instrum.* **44**, 107.

Matsumoto, G., Shimizu, H., and Shimada, J. (1976). *Rev. Sci. Instrum.* **47**, 861.

Mole, A. and Geissler, E. (1975). *J. Phys. E* **8**, 417.

Nemoto, N., Tsunashima, Y., and Kurata, M. (1981). *Polymer J.* **13**, 827.

Nieuwenhuysen, P. and Clauwaert, J. (1977). *Rev. Sci. Instrum.* **48**, 699.

Norsworthy, K. H. (1979). *Phys. Scripta* **19**, 369.

Noullez, A. and Boon, J. P. (1986). *Rev. Sci. Instrum.* **57**, 2523.

Ogata, A. and Matsuura, K. (1974). *Rev. Sci. Instrum.* **45**, 1077.

Ohbayashi, K. and Igarashi, T. (1973). *Japan J. Appl. Phys.* **12**, 1606.

Oliver, C. J. (1974). *in* "Photon-Correlation and Light-Beating Spectroscopy," edited by H. Z. Cummins and E. R. Pike, Plenum, New York, 151.

Oliver, C. J. and Pike, E. R. (1981). *Optica Acta* **29**, 1345.

Padin, S. and Davis, R. J. (1986). *Radio Sci.* **21**, 437.

Pike, E. R. (1969). *Riv. Nuovo Cimento, Ser. 1* **1**, Numero Speciale, 277–314.

Pike, E. R. (1972). *J. Phys. (Paris) Suppl.* **33**, C1–177.

Saleh, B. (1978). "Photoelectron Statistics," Springer-Verlag, Berlin, Heidelberg, New York.

Scarl, D. B. (1968). *Phys. Rev.* **175**, 1661.

Schatzel, K. (1983). *Optica Acta* **30**, 155.

Schatzel, K. (1987). *Appl. Phys. B* **42**, 193.

Schatzel, K. and Merz, J. (1984). *J. Chem. Phys.* **81**, 2482.

Schatzel, K., Drewel, M., Merz, J., and Schroder, S. (1985). *Inst. Phys. Conf.* **77**, 185.

Schulz-DuBois, E. O. and Rehberg, I. (1981). *Appl. Phys.* **24**, 323.

Subrahmanyam, V. R., Devraj, B., and Chopra, S. (1987). *J. Phys. E* **20**, 341.

Tai, I., Hasegawa, K., and Sekiguchi, A. (1975). *J. Phys. E* **8**, 207.

Tao, T. (1969). *Biopolymers* **8**, 669.

Tartaglia, P., Postol, T. A., and Chen, S. H. (1973). *J. Phys. A* **6**, L35.

Wijnaendts van Resandt, R. W. (1974). *Rev. Sci. Instrum.* **45**, 1507.

Wooding, E. R. and Pearl, P. R. (1974). *J. Phys. E* **7**, 514.

Yangos, J. P. and Chen, S.-H. (1980). *Rev. Sci. Instrum.* **51**, 344.

V

INTERFEROMETRY

5.1. GENERAL CONSIDERATION

5.1.1. INTRODUCTION

Although optical mixing spectroscopy is capable of extremely high resolutions with linewidths down to 1 Hz, the spectral range is limited to perhaps a few megahertz with commercial digital correlators. Typical delay-time increments Δt can be as short as 10 nsec; standard correlators have the Δt down to 100 nsec, and time-of-arrival correlators (Chapter 4) can have $\Delta t \sim 2$ nsec. In quasielastic light scattering (or Brillouin scattering), whenever the linewidths (or the Doppler shifts) pass beyond the megahertz range, it is advantageous to consider the more familiar form of interference spectroscopy, even though few experiments thus far have required its use.

There are several types of interferometers, with the Fabry–Perot interferometer being the most suitable one for our purposes. The theory and description of the Fabry–Perot etalon have been discussed in standard optics textbooks (Rossi, 1957; Born and Wolf, 1965; Klein, 1970). However, we shall review the essential aspects of the theory, since most chemists and biochemists are not familiar with the principles of optics. The purpose of this chapter is to provide sufficient background information so that the reader may feel reasonably comfortable when he or she is actually required to use the Fabry–Perot interferometer. It is advisable to consult the pertinent references for details.

5.1.2. REVIEW ON SINUSOIDAL AND EXPONENTIAL WAVES

A plane sinusoidal wave traveling with velocity c in the positive x-direction has the form

$$E(x, t) = E^0 \cos[\omega(t - x/c)] = E^0 \cos(\omega t - kx), \qquad (5.1.1)$$

where $\omega \, (= 2\pi\nu)$ is the angular frequency. The wave number $k \, (= 2\pi/\lambda)$ obeys the relation $ck = \omega$, while the period $\tau = 2\pi/\omega$.

A wave AB in a medium with a refractive index n_1 incident upon the boundary surface between the two media with indices of refraction n_1 and n_2 splits into a reflected wave BD and a refracted wave BC, as shown in Fig. 5.1.1a. If we disregard phase changes at this interface, the incident, re-

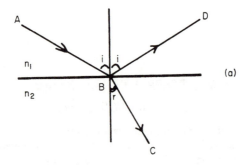

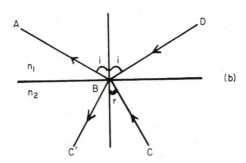

Fig. 5.1.1.   Refraction and reflection at an interface.

flected, and transmitted (refracted) waves are, respectively,

$$E_i = E^0 \cos[\omega(t + X_A/c_1)],$$
$$E_r = \rho E^0 \cos[\omega(t + X_D/c_1)],$$
$$E_t = \tau E^0 \cos[\omega(t - X_C/c_2)],$$

where X_A, X_D, and X_C are the distances from B measured along the incident, reflected, and refracted rays, respectively, while c_1 and c_2 are the velocities of light in media 1 and 2, respectively. ρ is the ratio of the reflected to the incident amplitude at the boundary from medium 1 to medium 2, and τ is the ratio of the transmitted to the incident amplitude at the boundary from 1 to 2.* Both ρ and τ are positive if there is no phase change, and become negative if the phase is inverted.

We shall now apply the reversibility principle, which implies that the wave equations remain valid, in the absence of dissipation of energy, when we change the sign of the time, i.e., from t to $-t$. The electric field for the reversed reflected ray (DB), as shown in Fig. 5.1.1b, has the form

$$E'_r = \rho E^0 \cos[\omega(-t - X_D/c_1)] = \rho E^0 \cos[\omega(t + X_D/c_1)],$$

which gives rise to a reflected wave (BA) and a refracted (transmitted) wave (BC'). Similarly, for the reversed refracted wave (CB)

$$E'_t = \tau E^0 \cos[\omega(t + X_C/c_2)]$$

we have a refracted wave (BA) and a reflected wave (BC').

The principle of optical reversibility assures us that the reversed beam in the direction of X_A must reproduce the *reversed* incident wave, and the two waves in the $X_{C'}$ direction must cancel. Thus, the equations (Rossi, 1967)

$$\rho^2 E^0[\omega(t - X_A/c_1)] + \tau\tau' E^0 \cos[\omega(t - X_A/c_1)] = E^0 \cos[\omega(t - X_A/c_1)]$$

and

$$\rho\tau E^0 \cos[\omega(t - X_{C'}/c_2)] + \tau\rho' E^0 \cos[\omega(t - X_{C'}/c_2)] = 0$$

give

$$\rho^2 + \tau\tau' = 1 \tag{5.1.2}$$

and

$$\rho = -\rho', \tag{5.1.3}$$

---

* The use of boldface for ρ and τ does not indicate vector quantities, but is only meant to distinguish them from, e.g., the density ρ and the period τ.

where ρ' and τ' are the ratios of the reflected and transmitted amplitudes to the incident amplitude at the interface from medium 2 to medium 1. $\rho^2\,[=(-\rho')^2]$ is the ratio of the reflected to the incident intensity and is known as the reflectance. It should be noted that the principle of reversibility does not hold for half-silvered mirrors.

Equation (5.1.1) is a solution to the wave equation

$$\nabla^2 E - \frac{1}{c^2}\frac{\partial^2 E}{\partial t^2} = 0, \tag{5.1.4}$$

which has a complex solution of the form[†]

$$E = E^0 e^{i(\omega t - kx)} = E^0 e^{i\phi} e^{i(\omega t - kx)}. \tag{5.1.5}$$

Since the electric field E is a real function of real variables, we retrieve Eq. (5.1.1) by taking the real part of Eq. (5.1.5):

$$E = \mathrm{Re}(E^0 e^{i(\omega t - kx)}) = E^0 \cos(\omega t - kx). \tag{5.1.1'}$$

We shall now review the rules governing the addition of sinusoidal waves by considering the combination of two wave displacements $E_1(x,t)$ and $E_2(x,t)$:

$$E_1(x,t) = E_1^0 \cos(\omega t - kx + \phi_1) = \mathrm{Re}\{E_1^0 \exp[i(\omega t - kx + \phi_1)]\},$$

$$E_2(x,t) = E_2^0 \cos(\omega t - kx + \phi_2) = \mathrm{Re}\{E_2^0 \exp[i(\omega t - kx + \phi_2)]\},$$

where ϕ is the phase of the wave at $x = 0$. The resultant electric field has the form

$$
\begin{aligned}
E(x,t) &= E_1(x,t) + E_2(x,t) \\
&= \mathrm{Re}\{[E_1^0 \exp(i\phi_1) + E_2^0 \exp(i\phi_2)]\exp[i(\omega t - kx)]\} \quad (5.1.6) \\
&= \mathrm{Re}\{E^0 \exp(i\phi)\exp[i(\omega t - kx)]\},
\end{aligned}
$$

where the relations of $E^0 \exp(i\phi)$, $E_1^0 \exp(i\phi_1)$, and $E_2^0 \exp(i\phi_2)$ are shown in Fig. 5.1.2.

The magnitude E of the vectors $\mathbf{E}_1$ and $\mathbf{E}_2$ obeys the equation

$$
\begin{aligned}
E^{02} &= [E_1^0 \exp(i\phi_1) + E_2^0 \exp(i\phi_2)][E_1^0 \exp(-i\phi_1) + E_2^0 \exp(-i\phi_2)] \\
&= E_1^{02} + E_2^{02} + E_1^0 E_2^0\{\exp[i(\phi_1 - \phi_2)] + \exp[-i(\phi_1 - \phi_2)]\} \quad (5.1.7) \\
&= E_1^{02} + E_2^{02} + 2E_1^0 E_2^0 \cos(\phi_2 - \phi_1).
\end{aligned}
$$

When $E_1^0 = E_2^0$, we have

$$E^{02} = 2E_1^{02}[1 + \cos(\phi_2 - \phi_1)].$$

[†] Strictly speaking, the symbol **E** should be used.

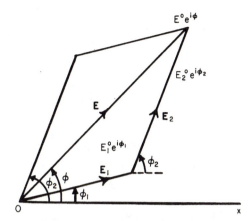

FIG. 5.1.2.   Addition of two sinusoidal functions.

To get ϕ, we take

$$E^0 e^{i\phi} = E_1^0 \exp(i\phi_1) + E_2^0 \exp(i\phi_2)$$

with

$$E^0 \cos\phi = E_1^0 \cos\phi_1 + E_2^0 \cos\phi_2$$

and

$$E^0 \sin\phi = E_1^0 \sin\phi_1 + E_2^0 \sin\phi_2.$$

Thus,

$$\tan\phi = \frac{E_1^0 \sin\phi_1 + E_2^0 \sin\phi_2}{E_1^0 \cos\phi_1 + E_2^0 \cos\phi_2}. \tag{5.1.8}$$

5.1.3. INTERFERENCE IN A DIELECTRIC SLAB

Consider a plane-parallel dielectric slab of refractive index n_2 and thickness d imbedded in a medium of refractive index n_1 as shown in Fig. 5.1.3.

If we take the incident wave (AB) to be $E = E^0 \cos \omega t$, then the reflected (BD) and transmitted (BC) waves are respectively, $\rho E^0 e^{i\omega t}$ and $\tau E^0 e^{i\omega t}$. The rays DE and $D'E'$ are parallel. We first compute the optical path lengths of the rays BD and BCD', which are denoted by l_1 and l_2, respectively. With $l_1 = n_1 \overline{BD}$ and $l_2 = n_2 (\overline{BC} + \overline{CD'})$, we get

$$l_2 - l_1 = n_2(\overline{BC} + \overline{CD'}) - n_1 \overline{BD}.$$

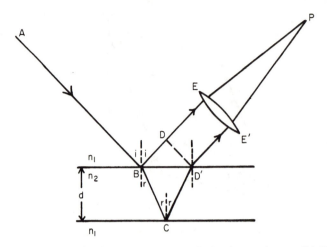

Fig. 5.1.3. Two-beam interference produced by reflection from a plane-parallel dielectric slab.

From the geometry in Fig. 5.1.3 and Snell's law, we also get

$$n_1 \overline{BD} = n_1 \overline{BD'} \sin i = n_1 (2d \tan r) \sin i$$

$$= \frac{n_2 \sin r}{\sin i} 2d \frac{\sin r}{\cos r} \sin i = \frac{2n_2 d \sin^2 r}{\cos r} \qquad (5.1.9)$$

and

$$n_2 (\overline{BC} + \overline{CD'}) = 2n_2 d / \cos r.$$

Thus,

$$l_2 - l_1 = 2n_2 d \left(\frac{1}{\cos r} - \frac{\sin^2 r}{\cos r} \right) = 2n_2 d \cos r. \qquad (5.1.10)$$

The phase difference due to the optical-path-length difference is $2\pi(l_2 - l_1)\lambda_0$, where λ_0 is the wavelength in vacuum. However, reflection at B for the ray ABD has a reflection coefficient ρ, while reflection at C for the ray $ABCD'$ has a reflection coefficient ρ', which is equal to $-\rho$. Thus, the phase difference due to both optical-path-length difference and reflection at upper and lower boundaries is

$$\phi = 2\pi \frac{l_2 - l_1}{\lambda_0} - \pi = 2\pi \left(\frac{2n_2 d \cos r}{\lambda_0} - \frac{1}{2} \right). \qquad (5.1.11)$$

Interference of the two rays at P will produce a maximum intensity at $\phi = 2\pi m$ and a minimum intensity at $\phi = \pi m$ with $m = 0, 1, 2, \ldots$. If $\cos r \approx 1$, we then have

$$d = (2m + 1)\tfrac{1}{4}\lambda \qquad \text{(interference maxima)}, \qquad (5.1.12)$$

$$d = m(\tfrac{1}{2}\lambda) \qquad\quad \text{(interference minima)}, \qquad (5.1.13)$$

where $\lambda = \lambda_0/n_2$.

The first (DE) and second $(D'E')$ reflected waves can be represented in terms of the complex exponential notation by

$$\rho E^0 e^{i\omega t}$$

and

$$\tau\tau'\rho' E^0 e^{i\omega t} e^{-i\phi} = \tau\tau'\rho E^0 e^{i\omega t} e^{-i\alpha} e^{\pi i}(-1)$$
$$= \tau\tau'\rho E^0 e^{i\omega t} e^{-i\alpha},$$

respectively, where $e^{\pi i} = -1$ and $\alpha = 2\pi(l_2 - l_1)/\lambda_0$.

5.1.4. INTERFERENCE BY MULTIPLE REFLECTION

We shall consider the case where Eqs. (5.1.2) and (5.1.3) hold. The pertinent geometrical and physical relationships can be derived with the aid of Fig. 5.1.4. For the transmitted waves, we first compute the optical path length from B to F and from B to C' which are, respectively.

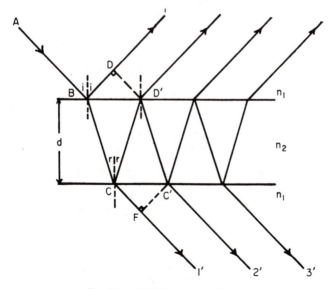

FIG. 5.1.4. Multiple wave interference.

$$l'_1 = n_1\overline{CF} + n_2\overline{BC} = n_1\overline{CC'}\sin i + n_2\overline{BC}$$
$$= n_1 2d\tan r\sin i + n_2\overline{BC}$$
$$l'_2 = n_2\overline{BC} + n_2(\overline{CD'} + \overline{D'C'}) = n_2\overline{BC} + n_2(2d/\cos r).$$

Thus,

$$l'_2 - l'_1 = 2n_2(d/\cos r) - n_1 2d\tan r\sin i = 2dn_2\cos r, \qquad (5.1.10')$$

and the phase difference due to the optical-path-length difference between neighboring waves is

$$\alpha = 2\pi\frac{l_2 - l_1}{\lambda_0} = 2\pi\frac{2d\cos r}{\lambda} = k(2d\cos r), \qquad (5.1.14)$$

where $\lambda = \lambda_0/n_2$ and $k = 2\pi/\lambda$.

The optical disturbance of the various transmitted waves at the point of interference can be listed as follows:

$$E_{t1'} = \tau\tau'E^0 e^{i\omega t},$$
$$E_{t2'} = \tau\tau'\rho^2 E^0 e^{i(\omega t - \alpha)},$$
$$\vdots$$
$$E_{tm'} = \tau\tau'\rho^{2(m-1)}E^0 e^{i[\omega t - (m-1)\alpha]}$$
$$= \tau\tau'E^0 e^{i\omega t}(\rho^2 e^{-i\alpha})^{m-1}.$$

The total transmitted electric field E_t is

$$E_t = E_{t1'} + E_{t2'} + \cdots + E_{tm'} + \cdots$$
$$= \tau\tau'E^0 e^{i\omega t}\sum_{m=1}^{\infty}(\rho^2 e^{-i\alpha})^{m-1} \qquad (5.1.15)$$
$$= \frac{\tau\tau'E^0 e^{i\omega t}}{1 - \rho^2 e^{-i\alpha}}.$$

The energy transmission coefficient T will be given by

$$T = \frac{|E_t|^2}{|E^0|^2} = \frac{|\tau\tau'|^2}{(1 - \rho^2 e^{-i\alpha})(1 - \rho^2 e^{+i\alpha})},$$

or

$$T = \frac{|\tau\tau'|^2}{1 + \rho^4 - 2\rho^2\cos\alpha}. \qquad (5.1.16)$$

Similarly, we can compute the reflected intensity. Various reflected waves at the point of interference can be represented by

$$E_{r0} = \rho^2 E^0 e^{i\omega t},$$

$$E_{r1} = \tau\tau'\rho E^0 e^{i(\omega t - \alpha)},$$

$$E_{r2} = \tau\tau'\rho^3 E^0 e^{i(\omega t - 2\alpha)},$$

$$\vdots$$

$$E_{rm} = \tau\tau'\rho^{2m-1} E^0 e^{i(\omega t - m\alpha)},$$

$$\vdots$$

The total reflected electric field E_r is

$$
\begin{aligned}
E_r &= E_{r0} + E_{r1} + E_{r2} + \cdots + E_{rm} + \cdots \\
&= E^0 e^{i\omega t}\left[\rho' + \tau\tau'\rho e^{-i\alpha} \sum_{m=1}^{\infty} (\rho^2 e^{-i\alpha})^{m-1}\right] \\
&= \left(\rho' + \frac{\tau\tau'\rho e^{-i\alpha}}{1 - \rho^2 e^{-i\alpha}}\right) E^0 e^{i\omega t}
\end{aligned}
\tag{5.1.17}
$$

or

$$E_r = \frac{\rho(e^{-i\alpha} - 1)E^0 e^{i\omega t}}{1 - \rho^2 e^{-i\alpha}}.$$

Thus, the energy reflection coefficient R is

$$R\left[= \frac{|E_r|^2}{|E^0|^2}\right] = \frac{2\rho^2(1 - \cos\alpha)}{1 + \rho^4 - 2\rho^2\cos\alpha}. \tag{5.1.18}$$

Equations (5.1.16) and (5.1.18) are the fundamental equations for the Fabry–Perot interferometer. With $\cos\alpha = 1 - 2\sin^2(\tfrac{1}{2}\alpha)$, we can also express the two equations in somewhat different but equivalent forms:

$$T = \frac{(1 - \rho^2)^2}{(1 - \rho^2)^2 + 4\rho^2 \sin^2(\tfrac{1}{2}\alpha)}, \tag{5.1.16a}$$

$$R = \frac{4\rho^2 \sin^2(\tfrac{1}{2}\alpha)}{(1 - \rho^2)^2 + 4\rho^2 \sin^2(\tfrac{1}{2}\alpha)}, \tag{5.1.18a}$$

where $T + R = 1$.

It should be noted that in deriving Eqs. (5.1.16) and (5.1.18), we have assumed that Eqs. (5.1.2) and (5.1.3) are valid. With metallic films, such as half-silvered mirrors, owing to absorption, Eqs. (5.1.16) and (5.1.18) may no longer hold. In other words, $\rho^2 + \tau\tau' \neq 1$ and $\rho \neq -\rho'$.

For mirrors with absorption we denote the change of phase on reflection by ϵ (Ditchburn, 1963). The phase difference between two neighboring waves is

$$\Delta = \frac{2\pi n_2 2d \cos r}{\lambda_0} + 2\epsilon, \tag{5.1.19}$$

which differs from Eq. (5.1.14) by 2ϵ. The resultant total transmitted electric field is

$$\begin{aligned}
E_t &= E^0 e^{i\omega t} \tau\tau'[1 + \rho^2 e^{-i\Delta} + \cdots + \rho^{2(m-1)} e^{-i(m-1)\Delta} + \cdots] \\
&= \tau\tau' E^0 e^{i\omega t}/(1 - \rho^2 e^{-i\Delta}).
\end{aligned} \tag{5.1.20}$$

Thus, the corresponding energy transmission and reflection coefficients are the same as Eqs. (5.1.16) and (5.1.18) except that α is replaced by Δ:

$$T = \frac{(\tau\tau')^2}{(1 - \rho^2)^2 + 4\rho^2 \sin^2(\frac{1}{2}\Delta)}, \tag{5.1.21}$$

$$R = \frac{4\rho^2 \sin^2(\frac{1}{2}\Delta)}{(1 - \rho^2)^2 + 4\rho^2 \sin^2(\frac{1}{2}\Delta)}. \tag{5.1.22}$$

Maxima occur when $\cos\Delta = 1$ (i.e., when $\Delta = 2\pi m$), and the minima are halfway between the maxima (i.e., when $\Delta = m\pi$), as we have discussed in Eqs. (5.1.12) and (5.1.13).

5.2. FABRY–PEROT INTERFEROMETER: GENERAL CHARACTERISTICS

The Fabry–Perot interferometer utilizes interference by multiple reflection. Figure 5.2.1 shows the basic elements for such an etalon or interferometer, where the slab is really an air gap between two partially silvered mirrors, M_1 and M_2, spaced at a distance d apart. For convenience, the Fabry–Perot etalon mirrors usually come in pairs with $\rho_1^2 = \rho_2^2 = \rho^2$ and $\tau_1^2 = \tau_2^2 = \tau^2$. In addition, the approximation $\rho^2 + \tau\tau' = 1$ holds.

The transmission coefficient T can be characterized by α with

$$T = [1 + F \sin^2(\frac{1}{2}\alpha)]^{-1}, \tag{5.2.1}$$

where $F \ (= 4\rho^2/(1 - \rho^2)^2)$ is called the contrast, and

$$\alpha = (2\pi/\lambda)(2d \cos r) = k_2(2d \cos r).$$

$T_{max} = 1$ when $\sin(\frac{1}{2}\alpha) = 0$ or $\alpha = 2\pi m$ with $m = 0, 1, 2, \ldots$; T_{min} occurs when $\sin(\frac{1}{2}\alpha) = 1$ or $\alpha = m\pi$ with $m = 0, 1, 2, \ldots$. Thus,

$$T_{min} = \left(\frac{1 - \rho^2}{1 + \rho^2}\right)^2. \tag{5.2.2}$$

Glass flats

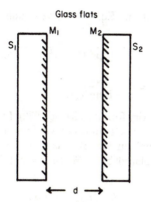

FIG. 5.2.1.   Basic elements of a Fabry–Perot interferometer. M_1, M_2: high-reflection mirrors; d: spacing between the two flats; glass flats with wedges ($S_1 \parallel\!\!\!/ M_1$; $S_2 \parallel\!\!\!/ M_2$) so that $S_1 \parallel\!\!\!/ S_2$ when $M_1 \parallel M_2$.

Figure 5.2.2 shows a plot of T vs α for (a) $\rho^2 \ll 1$ and (b) $\rho^2 \approx 1$. Large values of ρ^2 are obtained either with metallic layers or dielectric coatings.

In real situations, T_{max} is not likely to be equal to one. Then, by means of Eq. (5.1.21), we have

$$T_{\text{max}} = |\tau\tau'|^2/(1 - \rho^2)^2 \qquad (5.2.3)$$

and

$$T_{\text{min}} = T_{\text{max}}\left(\frac{1 - \rho^2}{1 + \rho^2}\right)^2. \qquad (5.2.4)$$

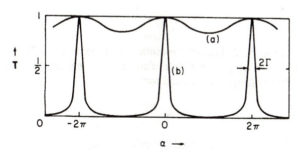

FIG. 5.2.2.   Intensity transmission coefficient T in the interference fringes using two plane-parallel plates: (a) $\rho^2 \ll 1$ and (b) $\rho^2 \simeq 1$, with negligible absorption effects and with 2Γ given by Eq. (5.2.8).

By combining Eq. (5.2.3) with Eq. (5.1.21), we obtain

$$T = \frac{T_{\max}}{1 + [4\rho^2/(1 - \rho^2)^2] \sin^2(\frac{1}{2}\Delta)}$$

$$= \frac{T_{\max}}{1 + F \sin^2(\frac{1}{2}\Delta)}, \tag{5.2.5}$$

which is known as the Airy formula (Klein, 1970). In Fig. 5.2.2, we note that $T_{\min}$ is very small when $\rho^2 \approx 1$. In other words, when the mirrors are good reflectors, T remains small unless Δ is close to $\pm 2\pi m$. If we neglect the effects of absorption, Δ is replaced by α. When $\alpha = \pi + 2m\pi$, $\sin(\frac{1}{2}\alpha) = \pm 1$ and $T = T_{\min}$. Then,

$$T_{\max}/T_{\min} = 1 + F. \tag{5.2.6}$$

If we take $\rho^2 = 0.8$, then $F = 3.2/(1 - 0.8)^2 = 80$. Thus, when $\rho^2 \approx 1$, F can be a very large number. If F is very large, Eq. (5.2.5) without the absorption effects, where we simply replace Δ by α, is a Lorentzian in α for α close to $2\pi m$. Then,

$$T = \frac{T_{\max}}{1 + F \sin^2(\frac{1}{2}\alpha)} = \frac{T_{\max}}{1 + F(\frac{1}{2}\alpha - m\pi)^2}$$

$$= T_{\max} \frac{1/F}{1/F + (\frac{1}{2}\alpha - m\pi)^2}, \tag{5.2.7}$$

where $\sin^2(\frac{1}{2}\alpha) \approx \sin^2(\frac{1}{2}\alpha - m\pi) \simeq (\frac{1}{2}\alpha - m\pi)^2$ and $|\frac{1}{2}\alpha - m\pi| \ll 1$. The half width at half maximum Γ is the value of $\frac{1}{2}\alpha - m\pi$ at which $T = \frac{1}{2}T_{\max}$. Thus,

$$\tfrac{1}{2}T_{\max} = T_{\max}[1 + F(\tfrac{1}{2}\alpha - m\pi)^2]^{-1},$$

which gives

$$\Gamma = \frac{2}{\sqrt{F}} = \frac{1 - \rho^2}{\rho}, \tag{5.2.8}$$

as shown in Fig. 5.2.2. The full linewidth at half maximum is 2Γ.

The order of a fringe in a Fabry–Perot interferometer obeys Eq. (5.1.14), where the bright fringes correspond to

$$m = \frac{2n_2 d \cos r}{\lambda_0} \tag{5.2.9}$$

with $m_{\max} = 2n_2 d/\lambda_0$, which need not be an integer.

The free spectral range is the spectral separation of adjacent transmission maxima (bright fringes) that can be observed without overlap, as shown in Fig. 5.2.3. For two wavelengths λ_a and λ, if λ_a exceeds λ by $\Delta\lambda$, defined as

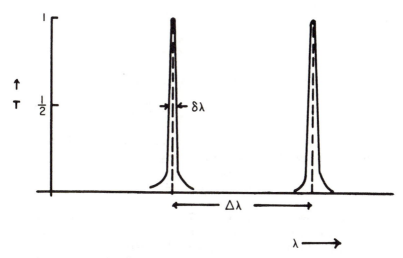

FIG. 5.2.3.   A schematic plot of T vs wavelength λ. $\Delta\lambda$: free spectral range in wavelength; $\delta\lambda$: instrumental bandwidth in wavelength.

the free spectral range, the mth-order fringe for λ_a will coincide with the $(m + 1)$th-order fringe for λ:

$$(m + 1)\lambda = m\lambda_a = m(\lambda + \Delta\lambda).$$

With Eq. (5.2.9), we then have

$$\Delta\lambda = \lambda/m = \lambda^2/2d \cos r \tag{5.2.10}$$

If $\cos r = 1$, or $r \ll \frac{1}{2}\pi$,

$$\Delta\lambda \simeq \lambda^2/2d, \tag{5.2.11}$$

which is the usual free spectral range expressed in terms of the wavelength. Since the wave number $\tilde{\nu} = 1/\lambda$ and $d\tilde{\nu} = -d\lambda/\lambda^2$, the corresponding free spectral range expressed in terms of $\tilde{\nu}$ is

$$\Delta\tilde{\nu} = 1/2d. \tag{5.2.11a}$$

Similarly, with $\lambda\nu = c$ and $d\nu = -(c/\lambda^2)\,d\lambda$, we get

$$\Delta\nu = c/2d. \tag{5.2.11b}$$

The free spectral range is better expressed in frequency units (Eq. 5.2.11b), so it has the same value going to the next higher or lower peak. Equation (5.2.8) gives us the instrumental bandwidth 2Γ when T is plotted against α. In terms of λ, we have

$$\delta\lambda = 2\Gamma\frac{\lambda}{2\pi m} = \frac{4}{\sqrt{F}}\frac{\lambda}{2\pi m}. \tag{5.2.12}$$

The resolving power $\mathscr{R}^\dagger$, defined as $\lambda/\delta\lambda$, is thus given by

$$\mathscr{R}^\dagger = \frac{\lambda}{\delta\lambda} = \frac{\sqrt{F}}{2}\pi m = \frac{\pi m\rho}{1-\rho^2}. \tag{5.2.13}$$

The ratio of the free spectral range to the instrumental bandwidth is called the *finesse* $\mathscr{F}$. By Eqs. (5.2.10) and (5.2.12), we get

$$\mathscr{F} = \frac{\Delta\lambda}{\delta\lambda} = \frac{\lambda}{m}\frac{\sqrt{F}}{4}\frac{2\pi m}{\lambda} = \frac{\pi\sqrt{F}}{2} = \frac{\pi\rho}{1-\rho^2}. \tag{5.2.14}$$

The finesse can also be defined as $\Delta\tilde{v}/\delta\tilde{v}$ or $\Delta v/\delta v$ and is a very important characteristic of the Fabry–Perot interferometer. With modern-day multi-layer dielectric coatings, a finesse of greater than 100 can be achieved.

For example, if we take $\rho^2 = 0.99$, then $F = 3.96/(10^{-2})^2 = 3.96 \times 10^4$ and $\mathscr{F} = (3.96 \times 10^4)^{1/2}(3.14)/2 = 3.13 \times 10^2$. In practice, the finesse is usually degraded by other factors, such as irregularities in the surfaces of the mirrors. Thus, Eq. (5.2.14) represents the instrumental finesse as limited by reflectance alone but in the absence of other losses.

For $\rho \simeq 1$, Eq. (5.2.14) can be reduced to

$$\mathscr{F}_R = \pi/(1-\rho^2), \tag{5.2.14a}$$

which is the reflectivity-limited finesse, valid for most of the present-day plane-parallel Fabry–Perot etalons. By substituting Eq. (5.2.14a) into Eq. (5.1.21), we can express the transmission coefficient T in terms of finesse $\mathscr{F}$ and the parameter Δ:

$$T = \left(\frac{\tau\tau'\mathscr{F}}{\pi}\right)^2\left\{1 + \left[\frac{2\mathscr{F}}{\pi}\sin(\tfrac{1}{2}\Delta)\right]^2\right\}^{-1}, \tag{5.1.21a}$$

where we have taken $\mathscr{F}$ as the instrumental finesse. When other loss mechanisms are present, the total net finesse is related to the individual contributions $\mathscr{F}_i$, by

$$\mathscr{F}^{-1} = \sum_i \mathscr{F}_i^{-1}. \tag{5.2.15}$$

In addition to the reflectivity-limited finesse, the major factors that tend to degrade the instrumental finesse are irregularities in the mirror surface, mirror misalignment, diffraction at the mirror aperture, and absorption.

We shall first consider the effect of irregularities in the mirror surfaces on the finesse. If the irregularity is smooth, and is of the order of λ/m across the effective   aperture,   then   the   plane-figure-limited   finesse $\mathscr{F}_f$ is approximately

$$\mathscr{F}_f \approx \lambda/2f \approx \tfrac{1}{2}m, \qquad (5.2.16)$$

where $f\,(=\lambda/m)$ is the mean surface deformation. It should be noted that an angular misalignment of the plates is equivalent to a corresponding plate deformation

$$\mathscr{F}_\beta \approx \lambda/2d\beta, \qquad (5.2.17)$$

where β is the tilt angle in radians and d is the plate separation distance. The instrumental finesse due to plate irregularities may be reduced by decreasing the effective aperture. However, in practice the reduction of etalon aperture significantly increases diffraction losses. On the other hand, a spherical Fabry–Perot etalon does not have this defect, nor does it require such careful angular alignment, since angular misalignment in a spherical Fabry–Perot etalon merely redefines the optical axis of the system.

For a plane-parallel Fabry–Perot etalon, the diffraction-limited finesse is approximately equal to

$$\mathscr{F}_D \approx D^2/2\lambda d, \qquad (5.2.18)$$

where D is the aperture diameter.

If a small fraction L of the radiation incident on the mirror is lost in making a single transit, then by analogy to Eq. (5.2.14a), the resultant contribution to the finesse is

$$\mathscr{F}_L \approx \pi/L. \qquad (5.2.19)$$

Table 5.2.1 shows a comparison of $\mathscr{F}_i$ for the plane-parallel and the spherical Fabry–Perot etalons. The listings clearly indicate that we require high reflectivity and smoothness of the mirrors.

Table 5.2.1

A Comparison of Finesse in Plane-Parallel and Spherical-Mirror Fabry–Perot Etalons

Finesse	Factor	Plane parallel	Spherical[a]
$\mathscr{F}_R$	Reflectivity	$\pi/(1-\rho^2)$	$\pi\rho^2/[1-(\rho^2)^2] \approx \pi/2(1-\rho^2)$
$\mathscr{F}_f$	Surface deformation	$\tfrac{1}{2}m$	$\tfrac{1}{2}m$
$\mathscr{F}_\beta$	Plate tilt	$\lambda/(2d\beta)$	Only redefines optical axis
$\mathscr{F}_D$	Mirror aperture	$D^2/2\lambda d$	Negligible
$\mathscr{F}_L$	Absorption	π/L	$\pi/2L$

(a) Hercher (1968).

For plane-parallel Fabry–Perot etalons, we also need more careful alignment of the mirrors and cannot try to increase $\mathscr{F}$ by reducing the etalon aperture because of $\mathscr{F}_D$. It appears that the spherical-mirror Fabry–Perot etalon has unique advantages. However, the plane-parallel Fabry–Perot etalon requires relatively low tolerance in determining the mirror separation distance, and its free spectral range can be varied according to Eq. (5.2.11), while in the confocal arrangement, a precise mirror separation distance fixes the free spectral range for each set of mirrors of a spherical Fabry–Perot etalon.

The spherical-mirror Fabry–Perot interferometer may be mode-matched to the laser for cavity resonance at frequencies satisfying the condition

$$v_0 = \frac{c}{2d}\left[q + \pi^{-1}(1 + m + n)\cos^{-1}\left(1 - \frac{d}{R}\right)\right]. \qquad (5.2.20)$$

Mode matching can double the free spectral range as well as the instrumental transmission. However, the angular tolerance on this alignment using a TEM_{00}-mode laser is extremely critical. The confocal spherical-mirror interferometer, being one of the best known mode-degenerate interferometers, is more useful for practical applications. The theory, design, and use of this interferometer has been described in detail (Hercher, 1968).

Table 5.2.2

Pertinent Formulas for Plane-Parallel and Spherical-Mirror
Fabry–Perot Etalons with High Reflectivity

	Plane parallel	Spherical[a] (mode-degenerate)
Resonance condition v_0	$mc/2d$	$(c/2dl)(lq + 1 + m + n)$
Airy formula $I/I_0(\rho^2 \simeq 1)$	$\left[\left(1 + \dfrac{A}{\tau^2}\right)^2\right]^{-1}$ $\times \left[1 + \left(\dfrac{4\pi d}{c(1 - \rho^2)}\right)^2\right.$ $\left. \times (v - v_0)^2\right]^{-1}$	$\left[l\left(1 + \dfrac{A}{\tau^2}\right)^2\right]^{-1}$ $\times \left[1 + \left(\dfrac{4\pi d}{c(1 - \rho^2)}\right)^2\right.$ $\left. \times (v - v_0)^2\right]^{-1}$
Free spectral range Δv	$c/2d$	$c/2ld$
Instrumental band-width δv	$c(1 - \rho^2)/2\pi d$	$c(1 - \rho^2)/2\pi d$
Finesse $\mathscr{F}_R$	$\pi/(1 - \rho^2)$	$\pi/l(1 - \rho^2)$
Resolving power $\mathscr{R}$[†] (or quality factor Q)	$2\pi vd/c(1 - \rho^2)$	$2\pi vd/c(1 - \rho^2)$

(a) Notation: A is the dissipative loss of the mirrors; q and m, n are integers denoting the longitudinal-and transverse-mode numbers; l is an integer satisfying the condition $\cos^{-1}(1 - d/R) = \pi/l$, where R is the radius of curvature. In the confocal arrangement, $l = 2$ (and $d = R$).

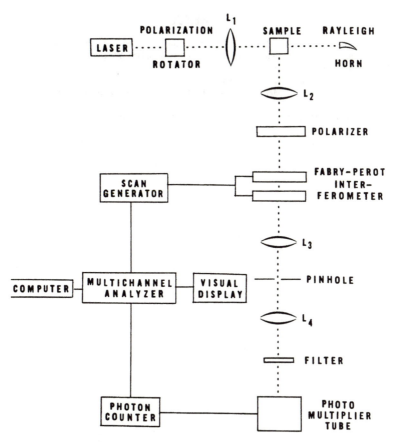

FIG. 5.2.4. Schematic representation of a typical scanning Fabry–Perot interferometric setup (courtesy of G. D. Patterson, 1980). The interested reader should consult the experiments of Patterson, who has published some of the best Fabry–Perot measurements.

Table 5.2.2 lists a comparison of the pertinent formulas for plane-parallel and spherical-mirror Fabry–Perot interferometers with high reflectivity ($\rho^2 \simeq 1$). A typical scanning Fabry–Perot interferometric setup is shown in Fig. 5.2.4. Super Invar Fabry–Perot interferometers manufactured by Burleigh Instruments, Inc.* have been found to be quite stable.

* Not an endorsement.

REFERENCES

Born, M. and Wolf, E. (1965). "Principles of Optics," 3rd revised ed., Pergamon, Oxford.

Ditchburn, R. W. (1963). "Light," p. 143, Wiley (Interscience), New York.

Hercher, M. (1968). *Appl. Opt.* **7**, 951.

Klein, M. (1970). "Optics," Wiley, New York.

Patterson, G. D. (1980). *In* "Methods of Experimental Physics," Vol. 16A, pp. 170–204, Academic Press, New York.

Rossi, B. (1967). " Optics," Addison-Wesley, Reading, Massachusetts.

VI

EXPERIMENTAL METHODS

6.1. INTRODUCTION

This chapter describes the pertinent features of the experimental aspects of laser light scattering. Particular attention is paid to the photon-counting and photon-correlation techniques, as well as optical setups using a single illuminating laser beam. Experiments with two interfering laser beams, either by means of a reference-beam scheme or a real-fringe technique, such as those often performed in electrophoretic light scattering (Ware and Flygare, 1971) or laser Doppler velocimetry (Durst, Melling, and Whitelaw, 1976), are beyond the scope of this book. This is not to say that these schemes are not important. Interested readers may want to proceed with an excellant review by Schatzel (1987), in which he has provided a proper perspective on the various techniques used in dynamic light scattering.

The discussion of the experimental methods in this chapter can be divided into three major parts. Section 6.2 deals with the physical and operational characteristics of the laser, which is the light source and also acts as the optical local oscillator in heterodyne spectroscopy. Sections 6.3 and 6.4 analyze the several optical setups that have been used in laser light scattering for studying phase transitions, liquids, macromolecules in solution, and colloidal particles in suspension. Section 6.5 presents the photon-counting (for time-averaged intensity) and photon-correlation (for intensity–intensity time correlation) measurements; power spectral analysis by means of an FFT spectrum analyzer is also discussed briefly.

For the uninitiated reader, more recent developments on light scattering optics are discussed in 6.4, with special emphasis on design considerations (6.4.7). The photon counting and current detection schemes are presented in 6.6 and 6.7, respectively. Finally, we keep the discussions on Fabry-Perot interferometry from the first edition for completeness (6.8) and introduce the use of fiber optics (6.9), which should play a very important role in future laser light scattering (LLS) development. A comparison of current correlators available commercially is presented in 6.10. The interested reader should examine the detailed specifications before any purchases.

The main purpose of this chapter is to try to provide sufficient background information so that a person who wants to utilize optical mixing spectroscopy will learn how to proceed in setting up the experiments using a single illuminating laser beam near an optimal signal-to-noise ratio. It is advisable to consult the references for details. This is especially true for biologists and chemists who are not familiar with lasers, optics, and electronics. In teaching chemistry graduate students, the author is aware of some of the difficulties that one encounters in trying to understand a technique of this type. Many experimental difficulties that the students have experienced will be discussed. It is hoped that some of their obstacles can be avoided by reading this chapter. Thus, for the more experienced reader, the discussions may, at times, appear obvious or tedious. The best advice is to skip those sections that look trivial.

6.2. THE LASER

6.2.1. GENERAL FEATURES

Helium–neon and argon ion lasers have been so well developed that most good commercial cw lasers are suitable for use as light sources in optical mixing spectroscopy. Our experience has been such that we always use a TEM_{00q} laser, where TEM means that the normal-mode disturbance is a transverse electromagnetic wave. The subscripts 00 stand for $m = 0$, $n = 0$, which correspond to the uniphase wavefront, while the integer q tells us the number of half wavelengths (or axial modes) that are contained in the resonator. Figure 6.2.1 shows the spectral width of a uniphase-wavefront He–Ne laser with several longitudinal modes. Each spectral line can act as a local oscillator. Superposition of line broadening centered at the zero frequency in light-beating spectroscopy permits the presence of many axial modes. However, when the linewidth of interest becomes very broad and is of the order of the spacings between the axial modes, then it is more appropriate for us to use optical interferometry and a single-mode single-frequency laser. These

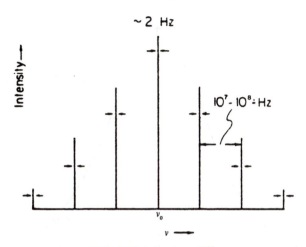

FIG. 6.2.1. Spectral width of a uniphase-wavefront helium–neon laser with several longitudinal modes $\lambda_0 = 632.8$ nm. (After Chu, 1968.)

types of lasers are also available on the market. Most such lasers are not frequency-stabilized, but use a thermally compensated intracavity Fabry–Perot etalon in order to isolate the axial mode of interest. Mode hopping is sometimes a problem if long-term stability is required. The experimentalist should be familiar with the detailed specifications of the laser.

The uniphase modes have several advantages that we can utilize in quasi-elastic light scattering experiments.

The intensity has a Gaussian profile given by

$$I = I_0 \exp(-2r^2/r_0^2), \tag{6.2.1}$$

where I is the intensity at a radial distance r from the cavity axis, I_0 is the peak intensity on the cavity axis ($r = 0$), and r_0 is the beam radius where the intensity falls to $1/e^2$ or 0.135 of the peak on-axis value, as shown schematically in Fig. 6.2.2. Integrating Eq. (6.2.1) from the cavity axis ($r = 0$) to r determines the power $P(r)$ in a Gaussian beam passing through an aperture of radius r:

$$P(r) = \int_0^r 2\pi r I(r)\, dr = \tfrac{1}{2}\pi I_0 r_0^2 [1 - \exp(-2r^2/r_0^2)], \tag{6.2.2}$$

or,

$$P(r)/P_{\text{tot}}(r = \infty) = 1 - \exp(-2r^2/r_0^2), \tag{6.2.3}$$

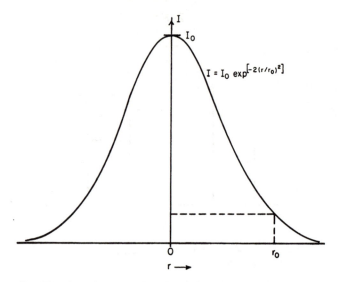

FIG. 6.2.2.   Intensity cross section of a uniphase-wavefront Gaussian beam.

where the total power $P_{tot}(r = \infty) = \frac{1}{2}\pi I_0 r_0^2$. Thus, 86.5% of the power is contained within the spot diameter $2r_0$. Figure 6.2.3 shows a collimated laser beam of diameter D, focused by a lens of focal length f to a spot diameter $d_s (= 2r)$. The equation

$$d_s \cong 2.44\lambda f/D \qquad (6.2.4)$$

is a useful linear far-field relationship, which shows that the focused spot size d_s is proportional to the wavelength λ of the laser and the focal length f of the

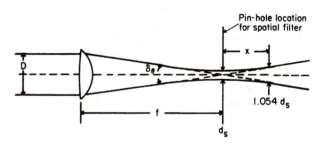

FIG. 6.2.3.   Collimated beam focused by a lens. D, laser-beam diameter; f, focal length; d_s, spot diameter. $d_s = 2.44\,(\lambda f/D)$; $x = (\pi/12)d_s^2/\lambda$ if $d/d_s = 1.054$.

focusing lens and is *inversely* proportional to the diameter D of the laser beam. Equation (6.2.4) is the relation between the lens aperture diameter $D_f(=D)$ and the diameter of the first dark ring of the Airy disk of diameter d in the image formed by a thin lens of focal length f (Klein, 1970).

In order to obtain a minimum on-axis spot size, we must take into account the aberration *angular radius* θ_a in addition to the diffraction angular radius $\theta_d(=d_s/2f) \approx 1.22\lambda/D$. The reader should note that the angular radius θ is not the scattering angle θ. If we assume $\theta = \theta_a + \theta_d$ and use a third-order approximation of abberation theory for a perfectly centered thin lens, we then have for the total on-axis spot angular radius

$$\theta \simeq 1.22\frac{\lambda}{D} + \frac{1}{2}\left(\frac{D}{f}\right)^3 K(n), \tag{6.2.5}$$

when n is the refractive index of the lens; and for a planopositive lens with infinite conjugate ratio and spherical curved surface facing the infinite conjugate, the lens-shape-dependent index function $K(n) \approx 4(n^2 - 2n + 2/n)/[32(n-1)^2]$. The optimum aperture for the lens, D_{opt}, is obtained by minimizing θ, or by setting $\partial\theta/\partial D = 0$, which yields

$$D_{opt} = \left[\frac{2(1.22)}{3}\frac{\lambda f^3}{K(n)}\right]^{1/4} \tag{6.2.6}$$

and a corresponding focal ratio

$$\frac{f}{D_{opt}} = \left[\frac{3}{2(1.22)}\frac{fK(n)}{\lambda}\right]^{1/4} \tag{6.2.7}$$

Equation (6.2.6) was derived from a minimization of the total on-axis spot angular radius. It represents the optimum lens aperture diameter that one should use for a given thin lens of focal length f. However, as the laser-beam diameter D should be close to D_f in order to let most of the laser power pass through the lens aperture, D_f is determined by the type of laser we use and by Eqs. (6.2.1)–(6.2.3). For a given lens, n is determined by the lens material. Thus, with Eq. (6.2.6), we have $D_{opt} \propto f^{3/4}$, and the optimal focal ratio $(f/D)_{opt}$ is fixed.

In the design of a laser light-scattering spectrometer, there are many conflicting optimization requirements. Equation (6.2.6) is one such condition. It will become evident later on that we cannot optimize all the requirements for a universal laser light-scattering spectrometer, since the optimal conditions for small-angle vs large-angle scattering and static vs dynamic light scattering are different.

The intensity profile along the (cavity) optical axis of the beam can best be described by defining a distance x such that the power density is reduced to (say) 90% of the value at d_s where $x = 0$, or $d_x = (1/0.9)^{1/2} d_s = 1.054 d_s$ with d_x being the beam diameter at x. If we take $d_x = d_s \{1 + [4\lambda x/(\pi d_s^2)]^2\}^{1/2}$ (Kogelnik and Li, 1966), then

$$\frac{d_x}{d_s} = 1.054 = \left[1 + \left(\frac{4\lambda x}{\pi d_s^2} \right)^2 \right]^{1/2},$$

or

$$x = \frac{0.333 \pi d_s^2}{4\lambda} = \frac{\pi}{12} \frac{d_s^2}{\lambda}. \tag{6.2.8}$$

Now if we redefine the depth of focus as the distance from the focus over which the intensity is reduced to 50% of the value at d_s, then $d_x = 2^{1/2} d_s$ and x becomes three times the 90%-intensity depth of focus (Marshall, 1971).

The uncertainty in the divergence angle $\delta\theta$ due to the incident laser beam divergence has the form

$$\delta\theta_{\text{INC}} \sim d_s/f, \tag{6.2.9}$$

where the subscript INC denotes the incident beam and $\delta\theta_{\text{INC}}$ is the angular divergence subtended by the incident beam at the scattering center. It is very important to remember $\delta\theta_{\text{INC}}$ when we design instruments for small-angle light-scattering work, as $\delta\theta_{\text{INC}}$ represents one of the uncertainties in momentum transfer, which often influences indirectly the spectral distribution of scattered laser light. $\delta\theta$ also plays an important role in estimating the effective coherence area, e.g., Eq. (3.3.26).

Equation (6.2.6) determines the pinhole size of a spatial filter for diffraction-limited operation by inserting the pinhole at f in Fig. 6.2.3. Usually, a short-focal-length lens with pinholes of the order of $10–50 \ \mu\text{m}$ in diameter but computed according to the relation $d > 2.44 \ \lambda f/D$ is used. The inequality is introduced to take account of imperfections and eccentricity of the pinhole. In a truncated Gaussian beam, it is worthwhile to remember Eqs. (6.2.1) and (6.2.3) for choosing beam apertures, as we have already mentioned. Figure 6.2.4 shows the relative time-averaged power $P(r)/P_{\text{tot}}$ as a function of relative radius r/r_0. For example $P(r)/P_{\text{tot}} = 0.956$ if $r/r_0 = 1.25$. Spatial filtering of the laser light is very helpful for small-angle light scattering experiments, where stray light due to laser plasma-tube window and cavity-mirror imperfections cannot be easily kept out by means of baffles, pinholes, and slits.

6.2.2. HELIUM–NEON CONTINUOUS-WAVE GAS LASER

The design of optical and detection systems should be such that the efficiency of the instruments has been maximized for the particular problem of interest. Then, a laser of the lowest appropriate power may be utilized. For systems that scatter light strongly, internal-mirror He–Ne plasma tubes with output powers of about $3–5 \ \text{mW}$ in the TEM_{00} mode have worked satis-

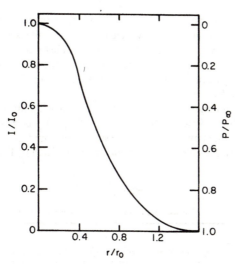

FIG. 6.2.4. Relative integrated power as a function of relative radius (r/r_0) in a Gaussian beam $I/I_0 = \exp[-2(r/r_0)^2]$, $P/P_{tot} = 1 - \exp[-2(r/r_0)^2]$.

factorily. By mounting the He–Ne laser together with the goniometer on a small, mechanically stable platform, we can usually ignore vibration isolation for self-beating experiments. There are several advantages to using a He–Ne laser; (1) it is reasonably priced; (2) it needs little or no adjustment in mirror alignment; (3) it is usually very short, so that $(\Delta v)_{axial}$ spacings between the axial modes are large. For example, if the mirror spacing l in a hemispherical system is 33 cm, then $\Delta v_{axial} = c/2l = 450$ MHz. Thus, it is possible to use such a laser for line-broadening studies of the central Rayleigh component close to the 100-MHz region by means of optical interferometry. Finally, (4) it has an average lifetime of over 10,000 hours.

For high-powered He–Ne lasers with adjustable mirror spacings, it is essential that the lasers themselves be checked for amplitude-modulation effects. Although mode locking tends to take place naturally in sufficiently long cavities, operation in the locked condition reduces the noise modulation. The lasers work best when they are tuned to the maximum output power. Long-term stability is usually not needed in photon correlation.

6.2.3. ARGON ION LASER

The argon ion laser has a higher ouput power and, in the blue–green region, offers an increased detection efficiency than the He–Ne laser because of the frequency dependence of the sensitivity of the photomultiplier tubes. Furthermore, it offers a higher scattering efficiency, since $I_s \propto 1/\lambda^4$. However, in dynamic light scattering experiments, the more stringent requirements for wavefront matching at shorter wavelengths with finite scattering volume reduce the efficiency to $\propto 1/\lambda^2$, as shown for example by combining Eq. (3.1.7) with $1/\lambda^4$. Argon ion lasers are bulkier and more expensive. For systems that fluoresce, we may at times have to take advantage of the longer wavelengths available from a He–Ne laser or a krypton laser.

The main requirement is that the laser must have a sufficiently noise-free low-frequency homodyne spectrum. The low-frequency components of laser light can be examined by monitoring the scattering from a ground-glass plate. Most commercial He–Ne and argon ion lasers, as listed in Table 6.2.1, show

Table 6.2.1

Typical He–Ne and Argon Ion Laser Specifications[a]

He–Ne lasers (632.8 nm)				
Manufacturer	Spectra Physics	Uniphase Corp.	Spectra Physics	Spectra Physics
Model number	105	1305	124B	127-25
TEM_{00} power (mW)	5.0	5.0	15.0	25.0
Beam diameter[b] (mm at e^{-2} points)	0.81	0.81	1.10	1.25
Beam divergence (mrad, full angle)	1.00	1.00	0.8	0.7
Beam amplitude noise (% RMS, 0–100 kHz)	0.5	0.2	0.3	1.0
Long-term amplitude stability (%/day)	5.0	2.5	2.0	2.0
Polarization (L = linear, R = random)	R or L	R or 500[c]	L	L

Argon ion lasers (457–530-nm blue–green lines)					
Manufacturer	Spectra Physics		Coherent Inc.		
Model number	2020-03	2020-05	Innova 70-2	70-3	90-3
TEM_{00} power (W)	3	5	2	3	3
Beam diameter[b] (mm at e^2 point)	1.5	1.5	1.5	1.5	1.5
Beam divergence (mrad, full angle)	0.5	0.5	0.5	0.5	0.5
Beam amplitude noise (% MRS, 0–100 kHz)	0.2	0.2	0.5	0.5	0.2
Long-term amplitude stability (%/day)	0.5 (30 min)	0.5 (30 min)	0.5	0.5	0.5 (30 min)
Polarization (L = linear)	L	L	L	L	L

Table 6.2.1 (*continued*)

Manufacturer	Lexel		
Model number	3500-3	85-1	95-2
TEM_{00} power (W)	3	1	2
Beam diameter[b]			
(mm at e^2 points)	≤ 1.3	1.1	1.3
Beam divergence			
(mrad, full angle)	≤ 0.5	0.7	0.6
Beam amplitude noise			
(% RMS, 10 Hz–2 MHz)	0.2	0.5	0.2
Long-term amplitude			
stability (%/day)	$\leq \pm 1.0$ (8 hr)	$\leq \pm 0.2$ (1 hr)	$\leq \pm 0.2$ (1 hr)
Polarization			
(L = linear)	L	L	L

(a) Generally, we search for a laser with low beam amplitude noise and long-term amplitude stability. If a polarizer is used to determine the vertical (or horizontal) polarization of the incident beam, random orientation of polarization of the laser source is acceptable. For lasers with random orientation of polarization, we only need to rotate the polarizer (without the use of a polarization rotator) in order to achieve an incident beam with either a vertical or a horizontal polarization. All the listed lasers can be used satisfactorily for photon correlation spectroscopy. Although a 1–2-mW He–Ne laser can be used to perform experiments in photon correlation spectroscopy (PCS), it is advisable to use at least a 5-mW He–Ne laser. A 2-W argon ion laser is more than adequate for most experiments in PCS. Listing of particular manufacturers does not mean a preference for their products, but merely reflects personal experience of the author.

(b) According to Eq. (6.2.4), $d_s \propto D^{-1}$, i.e., the spot diameter which defines the cross section of the scattering volume is *inversely* proportional to the laser beam diameter. As focusing of the laser beam is essential in order to reduce the scattering volume for efficient optical mixing, a proper choice of the focusing lens depends on the laser beam diameter as well as the uncertainty $\delta\theta$ that we can tolerate.

(c) Linear polarization with an extinction ratio of at least 500:1.

no detectable noise spectrum. The reader should also note that near-visible cw diode lasers, with the advantage of very small packaging, can be another suitable laser source for dynamic light scattering (Brown and Grant, 1987).

Many commercial argon ion lasers are capable of single-frequency operation, which greatly increases the coherence length and suppresses the noise spectrum arising from multimode instabilities. Jackson and Paul (1969) also suggested that the intracavity etalon for single-frequency operation should not be aligned normal to the laser beam, as the unstable multicavity configuration could produce a noise spectrum due to mode hopping.

In summary, commercial lasers are quite suitable for optical mixing spectroscopy, though a careful assessment of the light source is advised. The use of an intracavity etalon for single-mode output could improve the performance of an argon ion laser, even though it is often not essential, i.e., lasers without an intracavity etalon provide adequate performance in photon correlation spectroscopy.

6.2.4. LASER DIODES

Laser diodes have been used extensively for communication purposes. They emit electromagnetic radiation mainly beyond the visible range ($\lambda_0 > 700$ nm). However, development has progressed rapidly, and visible-light laser diodes have now become readily available. Furthermore, the device lifetime is usually very long, the cost compared with cw gas lasers is at worst competitive, and the sizes are much smaller. Thus, a brief mention of laser diodes is included in this book, even though applications to laser light scattering have so far been few. Laser light scattering experiments using near-infrared wavelengths could become useful for examining highly absorbing colored conducting polymer solutions.

Laser diodes can normally be categorized into gain-guided and index-guided structures. For laser light scattering, index-guided cw diode lasers tend to be more suitable. Index-guided lasers tend to be more expensive, but have a clean far-field beam profile suitable for dynamic light scattering. Most diode lasers have wavelengths in the range of 0.78–1.6 μm. It is anticipated that laser diodes will move towards the shorter wavelengths in the visible range, for beyond the 632.8 nm of a He–Ne laser. The device wavelength is determined by the band gap of the material used in constructing the actual chip containing the laser facet. The cost of the laser (without power supply) ranges from tens of dollars to thousands, depending on power and wavelength.

Aside from basic handling precautions, such as sensitivity to static electric discharge and proper grounding, an external temperature-controlled ($\pm 0.01°$C) thermoelectric (TE) heat sink is recommended for the operation of a cw laser diode for laser light scattering experiments. The temperature stabilization will also provide a more stable peak output wavelength and longer life. The dc power supply should incorporate features to eliminate current transients. Stability of 50 ppm up to 200 mA is available. High-power laser diodes can be purchased from Spectra Diode Labs (San Jose, California), and many other manufacturers exist. Power supplies are available from vendors such as ILX Lightwave Corporation (Bozeman, Montana) or Melles Griot (Irvine, California); no endorsement is intended.

6.3. THE OPTICAL SYSTEM

Optical systems in laser light scattering have been designed mainly for either the angular distribution of scattered light or its spectral distribution.

Optimization in photon correlation is often achieved at the expense of other needs. Ideally, it would be wonderful to have an optical system capable of (1) measuring the intensity as well as the spectral distribution of scattered light over the largest possible angular range ($0° \lesssim \theta \lesssim 180°$), (2) studying polarized and depolarized scattering, and (3) using cells of different designs suitable for studies of macromolecules in solution or phase transitions. In practice, as we have already mentioned, it is difficult to achieve optimum conditions satisfying all the requirements in one system.

6.3.1. CYLINDRICAL-TYPE CELLS AND DETECTION OPTICS

Figure 6.3.1 shows several cylindrical-type cells, which require virtually no angular corrections due to refraction of scattered light from the scattering medium to the detector, since the entrance and emergent light beams are always at right angles to the faces of the scattering cell. The absorbing glass in Fig. 6.3.1(a) and (b) at the exit walls reduces the back reflection. Simple cylindrical cells with diameters as small as 6 mm have been used successfully in light scattering experiments where refraction and reflection difficulties are alleviated by immersing the scattering cell in an immersion fluid that matches the refractive index of the scattering cell.

Figure 6.3.2 shows a schematic representation of the detection optics. According to Fig. 6.3.2, L_z is limited to the diameter of the cell; $d_1/\sin \theta$ determines the effective scattering volume, while d_2 is mainly responsible for the angular uncertainty $\delta\theta$ of the collecting system. Such a setup is very good for measurements of the angular distribution of time-averaged scattered intensity and is relatively insensitive to the distance between the scattering volume and the lens L_1. However, location of d_2 at the focal length of the lens can be difficult to achieve. Furthermore, for small scattering volumes, d_1 has to be so small that an alternative approach, as shown in Fig. 6.3.3, is preferred. In Fig. 6.3.3, we can select a lens with a variety of focal lengths to provide different magnification ratios between the object O and the image I using a fixed

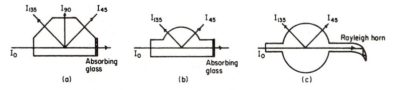

FIG. 6.3.1. Cylindrical-type cells: (a) hemioctagonal cell; (b) cell with flat entrance and exit windows; (c) cell with flat entrance window and Rayleigh horn at the exit.

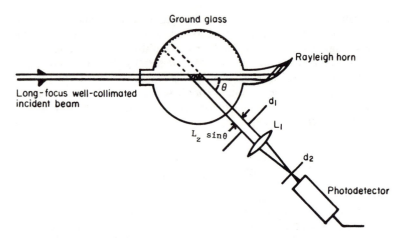

FIG. 6.3.2. A schematic representation of the detection optics. The cylindrical-type cell is shown only for illustration purposes. The same detection optics can be used for light-scattering cells of other geometry. d_1 is usually located as close to the scattering cell as is convenient, while d_2 is close to the photodetector in order to avoid another lens for transferring the scattered light from d_2 to the photocathode.

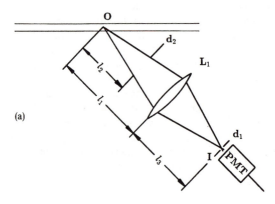

FIG. 6.3.3. Schematic representations of the alternative detection optics. (a) Same as Fig. 6.3.2 except that the locations of d_1 and d_2 have been interchanged. Furthermore, L_1 projects the object at O to the image plane at I with magnification ratio of $I/O = l_3/l_1$. (b) Same as (a) except that d_2 has been moved closer to the photodetector. This is the preferred detection optics because (1) the scattering volume can easily be defined by d_1; (2) the magnification ratio can be designed into the optics for a predetermined total length for the detector arm; (3) the angular aperture is controlled mainly by $(d_1 + d_2)/l_2$; and (4) slit diffraction from d_2 (in (a)) has been eliminated. It should be noted that d_1 not only acts as a field aperture but also influences the angular aperture $(d_1 + d_2)/l_2$. Instead of having d_2 (between d_1 and the PMT) immediately in front of the PMT, it is also appropriate to have d_2 immediately after lens L_1 (between L_1 and d_1). The same variation is applicable for the configuration in Fig. 6.3.2.

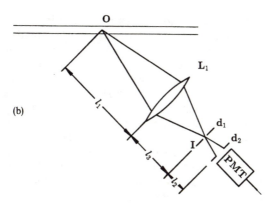

(b)

FIG. 6.3.3. (*continued*)

IO length. Figure 6.3.3(b) shows a more straightforward scheme for the detection optics. We can define the scattering volume v to be proportional to $d_1(l_3/l_1)\sin\theta \cdot \pi(d_s/2)^2$ with d_s being the diameter of the incident beam cross section, provided that $d_1 > d_s$. When one has $d_1 > d_s$ for the detection optics, the field aperture cuts through the incident-beam cross section $\pi(d_s/2)^2$. Thus, the scattering volume v is proportional to $d_1 d_s^2$. However, if $d_1 < d_s$, then $v \propto d_1^2 d_s$, indicating that the effects of d_1 on the total scattered intensity are closely coupled with the optical geometry. The angular aperture of the scattered light is determined by $(d_1 + d_2)/l_2$. Furthermore, by transferring the angular aperture from d_2, a position in front of the lens L_1, as shown in Fig. 6.3.3(a), to a position in front of the photodetector, as shown in Fig. 6.3.3(b), we have essentially eliminated the slit diffraction from d_2.

In Fig. 6.3.3(b), it can be seen that d_1 does not only play the role of a field aperture; it also influences the angular aperture of the detection optics. By increasing d_1 to $d_1' = 2d_1$, and for $d_1 < d_s$, the scattering volume is increased by a factor of 4; but the 2-D angular aperture is increased by a factor of only $(\frac{3}{2})^2 = 2.25$ with $d_1 = d_2$. Then, the total scattered intensity increase is ~ 9 times that of the initial scattered intensity provided that the incident beam has a uniform intensity profile. In reality, the incident beam has a Gaussian profile (Eq. (6.2.2)). The net increase in the scattered intensity will not be ~ 9 times; but less than 9 times the initial scattered intensity after correction of the laser-beam intensity profile in the vertical direction by means of Eq. (6.2.3). The detection schemes of Figs. 6.3.2 and 6.3.3 are valid for all optical-cell designs. As shown, we have not taken into account refraction effects due to the cell geometry, which could alter the effective distance between the object (scattering volume) and the lens L_1. Cylindrical-type cells, especially cylindrical cells, are easier to use than light-scattering cells of other optical geometry.

We have found little incentive to consider light-scattering cells with optical geometry different from the cylindrical ones, except when specific needs arise.

Another important point in our discussion of the detection optics is that optical mixing has taken place at d_2 of Fig. 6.3.3(b) or d_1 of Fig. 6.3.3(a). So long as the cone of light striking the photocathode is smaller than the photocathode area, we can move the photomultiplier further back, away from the last diaphragm (d_2 of Fig. 6.3.3(b) or d_1 of Fig. 6.3.3(a)) without appreciably reducing the optical-beating efficiency. In fact, we can attach a fiber-optic cable, for example, at d_1 of Fig. 6.3.3(a), as shown in Fig. 6.3.4. The fiber bundle (FB), which does not preserve coherence, permits us to move the PMT to a remote position. In Fig. 6.3.4, we have used a commercially available fiber-optic eyepiece, made by Gamma Scientific (California). The eyepiece has an optical fiber of diameter d_1, positioned on the optic axis, to collect the scattered light, and a fiber-optic bundle (FB) to transmit the light to a photodetector. The experimental arrangement used, as shown in Fig. 6.3.4, has an angular aperture $\delta\theta \sim \tan^{-1}(d_2/l_2)$, where d_2 is the diameter of the aperture diaphragm. A system of latex particles was used to make dynamic light scattering measurements at various scattering angles for two different eyepiece, each having a different fiber diameter (d_1), and for aperture diameters of 0.508 and 0.178 mm. The spatial coherence factor β was extracted from the

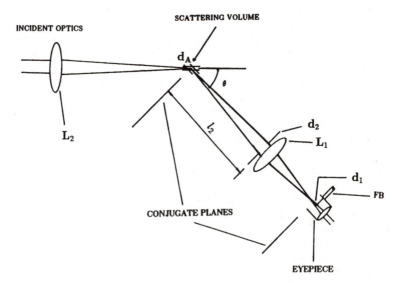

FIG. 6.3.4. A schematic of the fiber-optic eyepiece detection system. FB, fiber bundle; d_1, fixed optical fiber of diameter d_1; L_1, output lens; L_2, input lens. (After Dhadwal and Chu, 1987.)

Table 6.3.1

Spatial Coherence Effects of the Fiber-Optic Detection System in Fig. 6.3.4

| | $d_2 = 0.508$ mm, $\delta\theta \approx 0.08°$ | | | | $d_2 = 0.178$ mm, $\delta\theta \approx 0.03°$ | | | |
| | $d_1 = 0.700$ mm | | $d_1 = 0.450$ mm | | $d_1 = 0.700$ mm | | $d_1 = 0.450$ mm | |
θ (deg)	$(\Delta\theta)_{coh}$ (deg)	β	$(\Delta\theta)_{coh}$ (deg)	β	$(\Delta\theta)_{coh}$ (deg)	β	$(\Delta\theta)_{coh}$ (deg)	β
30	0.006	0.022	0.008	0.044	0.006	0.11	0.008	0.15
45	0.006	0.024	0.009	0.046	0.006	0.12	0.009	0.14
60	0.006	0.026	0.009	0.045	0.006	0.11	0.009	0.14
90	0.006	0.028	0.009	0.044	0.006	0.14	0.009	0.15

Values of $(\Delta\theta)_{coh}$, are based on Eq. (3.1.13) with $\lambda = 0.366$ μm, magnification ratio $d_A/d_1 = 2.5$, and $L_z = d_A/\sin\theta$. (Dhadwal and Chu, 1987.)

measured correlation data, where the intensity–intensity correlation function has the form $G^{(2)}(\tau) = A(1 + \beta|g^{(1)}(\tau)|^2)$ with A, τ, and $|g^{(1)}(\tau)|$ being the background, delay time, and electric field correlation function, respectively. The results are tabulated in Table 6.3.1.

Also included in Table 6.3.1 are the coherence-angle requirements for the various configurations. Ideally, the angular aperture of the detection system should be as closely matched as possible to the coherence angle. This was not the case, and we considered only one aspect of $(\Delta\theta)_{coh}$. We were, however, able to show, as expected, that decreasing the aperture diameter d_2 reduces the total signal strength and has a dramatic effect on the value of β. We also observe that β increases with decreasing fiber diameter d_1. Instead of using a fiber bundle that has a high attenuation factor ($\approx 50\%$/meter), a single multimode fiber with very low losses can be used for this purpose (Haller et al., 1983).

The fiber-optic detection optics (Section 6.9) is also different from optical-fiber probes used in optical mixing spectroscopy (Section 6.3.2).

6.3.2. OPTICAL-FIBER PROBE (AFTER DHADWAL AND CHU, 1987)

The optical fiber, of the type designed for communications, is now near the theoretical limit of fabrication and is rapidly finding applications outside the communications area. Optical fibers offer compactness, remote-sensing, and miniaturization capabilities to the experimenter in the field of light scattering. Indeed, the first experimental use of an optical fiber was described by Tanaka and Benedek (1975). This was followed by the fiber-optic Doppler anemometer (FODA), designed by Dyott (1978), and more recently by Auweter and

Horn (1985). The experimental setups of Dyott and of Auweter and Horn, representing a backscattering homodyne (using a local oscillator with unshifted frequency) detection system, make use of a single multimode fiber, which is used for both transmitting the laser light to the scattering medium and detecting the scattered laser light.

Optical fibers, whether multimode or in bundles, are basically incoherent, that is, the coherence properties of the optical field entering one end of the fiber cannot be related to the coherence properties of the optical field emanating from the other end of the fiber. This can be readily understood in terms of the numerical aperture, NA, of the optical fiber, which defines a cone of light guided by the optical fiber without significant attenuation, as shown in Fig. 6.3.5. The consequence of this incoherence is that light entering one end of the fiber, at *any* numerical aperture, always exits at the other end of the fiber at the *full* numerical aperture. This effect necessarily dictates that, in order to take account of the spatial coherence of the scattered electric field, the angular aperture of the detection system should be controlled *before* the scattered light is allowed to enter the fiber, and preferably not after exiting from the other end of the optical fiber, as discussed by Auweter and Horn (1985).

A typical multimode optical fiber with a step-graded refractive-index profile (Dyott, 1978) has a numerical aperture of 0.15 in air and a corresponding critical acceptance angle of 6.4° in water. We know that a detection system with an uncertainty in the scattering angle of 6.4° is generally useless as far as dynamic light scattering is concerned. However, the required coherence angle

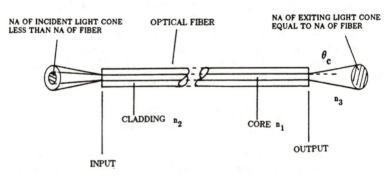

FIG. 6.3.5.   Optical-fiber waveguide. NA, numerical aperture; θ_c, critical acceptance angle; n_1, n_2, n_3, refractive indices of the core, cladding, and scattering medium. (After Dhadwal and Chu, 1987.)

$$NA = \sin^{-1}(\theta_c) = \frac{\sqrt{n_1^2 - n_2^2}}{n_3}$$

$(\Delta\theta)_{\text{coh}}$ is a function of the scattering volume (L_x, L_y, L_z) and the scattering angle θ. If the incident laser beam is propagating in the z-direction with polarization in the x-direction, as shown in Figs. 3.1.3 and 6.3.6(a), then the half angle of coherence in the scattering plane $(y-z)$ obeys Eq. (3.1.13). If the detection system has an aperture d_A, then $L_z = d_A/\sin\theta$ and Eq. (3.1.13) is

- a -

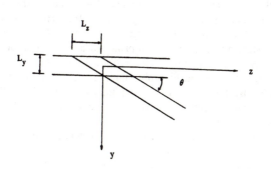

- b -

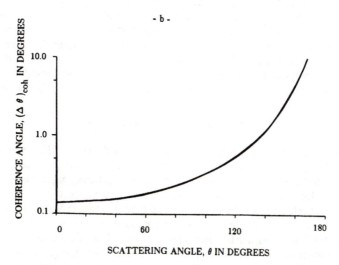

FIG. 6.3.6. Calculation of the coherence angle $(\Delta\theta)_{\text{coh}}$: (a) coordinate system—incident beam is propagating in the z-direction with polarization in the x-direction; (b) plot of $(\Delta\theta)_{\text{coh}}$ based on Eq. (6.3.1) with $L_x = L_y = D_A = 50\ \mu\text{m}$ and $\lambda = 0.475\ \mu\text{m}$. (After Dhadwal and Chu, 1987.)

reduced to

$$(\Delta\theta)_{\text{coh}} = \frac{\lambda}{2(d_A + L_y \cos\theta)}. \tag{6.3.1}$$

Figure 6.3.6(b) shows a plot of $(\Delta\theta)_{\text{coh}}$ as a function of θ for $d_A = L_y = 50$ μm, as is the case for the experimental arrangements described by Dyott and by Auweter and Horn. From the plot we see that they were fortunate when using the optical fiber in the backscattering configuration because the spatial coherence requirements on $(\Delta\theta)_{\text{coh}}$ are very favorable in this direction. For example, the critical acceptance angle of 6.4° is smaller than the required coherence angle of 7.9° at a scattering angle of 165°. However, it should be noted that for real fiber-optic backscattering probes L_z does not increase indefinitely with increasing θ, but approaches a limiting value of about a hundred times the core diameter of the optical fiber. Thus, in practice, the coherence angle may be considerably smaller in the backscattering direction than indicated by Fig. 6.3.6(b). This explains why Dyott's FODA and the optical arrangement reported by Auweter and Horn worked very well for dynamic light scattering.

It should be noted that the aperture arrangement used by Auweter and Horn (1985) does not really restrict the acceptance angle of the optical fiber in water—it serves mainly to reduce the intensity of the detected light. Thus we can conclude that the optical fiber, without additional modifications, is not a suitable probe for dynamic light scattering at angles slightly below backscattering. However, with the front-end configuration of Fig. 6.3.4, an optical fiber can be used very effectively over the entire scattering-angle range for dynamic light scattering. If one uses a microlens for L_1, the resulting optical-fiber–microlens configuration can miniaturize the detection optical system and create a new generation of light-scattering spectrometers (see Section 6.9).

6.3.3. RECTANGULAR CELLS

Good spectroscopic rectangular cells are cheaper than cylindrical-type cells of similar quality. The flat entrance and exit windows tend to reduce stray light scattering in the scattering cell. However, the scattering angle θ has to be corrected for refraction of the light as it passes from the scattering medium in the scattering cell to the outside. Thus, the refractive index of the solution must be measured in order to determine θ and K $(=(4\pi/\lambda)\sin(\frac{1}{2}\theta))$. Figure 6.3.7(a) shows a schematic representation of the detection optics for a rectangular cell. The rectangular cell offers several advantages: (1) a long-light path cell may be used in order to increase the effective scattering volume (i.e., L_z); (2) high-optical-quality cells are easily available, so that stray light due to cell imperfections may be minimized; and (3) very low scattering

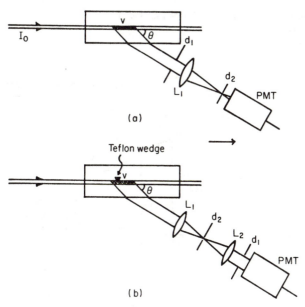

FIG. 6.3.7.   Schematic representations of the detection optics for a rectangular cell. d_2 is an adjustable aperture that determines $\delta\theta$. In (b) the variable aperture d_1, which determines the effective scattering volume, is placed near the face of the photocathode. The Teflon wedge was introduced for the heterodyne configuration by Cummins et al. (1969). The (b) scheme is a compromise. It introduces an additional lens L_2 and takes advantage of a large effective photocathode.

angles are accessible, dependent upon the optical design of the light-scattering spectrometer.

Occasionally, a short-light-path cell (0.1–1 mm thick) has been used for very strongly scattered systems in order to avoid multiple scattering. Furthermore, for systems that scatter light strongly, L_z can be reduced to a very small size, so that the collimating lens L_1 is no longer needed. Figure 6.3.8 shows the scattering geometry for measurements at two fixed scattering angles using a thin flat cell (Lai and Chen, 1972).

In a heterodyne spectrometer, several techniques have been developed to use the incident laser source as an optical local oscillator. The important requirement is to be able to combine the scattered radiation with the local-oscillator signal at the photocathode of the photomultiplier so that the fields are spatially coherent. The local-oscillator signal is usually a portion of the unshifted incident radiation ($\omega_{LO} = \omega_0$), such as a small portion of the transmitted beam (Lastovka and Benedek, 1966a), the scattering from dust or

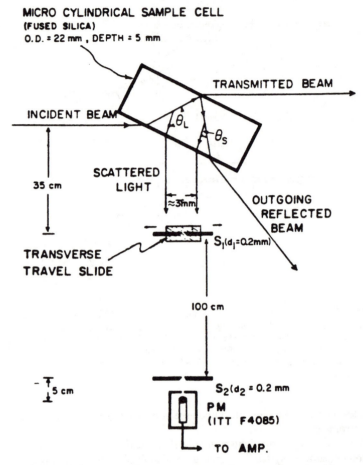

FIG. 6.3.8. Scattering geometry for measurement at two fixed scattering angles. (After Lai and Chen, 1972.)

imperfections on the cell windows (Lastovka and Benedek, 1966b), the scattering from a plate of appropriately cut smoky quartz (Berge, 1967), the scattering from a thin layer of fixed scatterers (Uzgiris, 1972), or the scattering from a Teflon prism in the scattering cell with its edge partly intersecting the incident beam (Cummins *et al.*, 1969), as shown in Fig. 6.3.7(b).

An alternative approach for the local optical oscillator in heterodyne detection is to use a portion of the incident radiation that has been shifted to some other frequency ($\omega_{LO} \neq \omega_0$) (Cummins *et al.*, 1964). The frequency-shifted local oscillator imposes many difficulties, mainly because any optical imperfections will tend to reduce the mixing efficiency, while the scattered field and the unshifted local-oscillator field can be made to traverse identical optical paths and thus have very high mixing efficiencies (Adam *et al.*, 1969). Cummins *et al.* (1969) found that the milky quartz plate did not scatter sufficiently uniformly at all angles to make it generally useful. A small Teflon prism of triangular cross section seemed to work very well. The prism in Fig. 6.3.7(b) should be held rigidly relative to the scatterers. Movement of the prism produces another source of modulation signals that will contribute to the power spectrum, especially at low frequencies. Thus, mechanical disturbances, such as building vibrations, are more crucial in a heterodyne experiment than in a self-beating one. Wada *et al.* (1972) presented a simple modification utilizing a cylindrical collimating lens for the optical heterodyne system in order to reduce the sensitivity of the local oscillator to mechanical vibrations.

6.3.4. CONICAL CELLS

Scattering cells with conical optics have been used successfully by Ford and Benedek (1965) in their observation of the spectrum of light scattered from a pure fluid near its critical point, and by Benedek and Greytak (1965) in Brillouin scattering by water and toluene. Figures 6.3.9 and 6.3.10 show the scattering cells with conical optics. In Fig. 6.3.9, light scattered through an angle θ_c is collected over an azimuth angle of 2π by a conical lens and emerges parallel to the cone axis. The spherical lens L_1 with focal length f then focuses the scattered light at the center of a pinhole aperture d, which controls the aperture $\delta\theta$ of the scattering angle θ_c accepted by the spectrometer. Conical

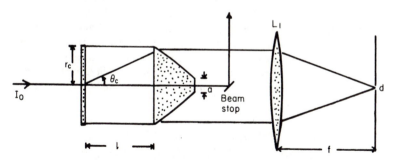

FIG. 6.3.9. Scattering cell with conical optics. The diameter of the flat exit window, a, should be slightly greater than the incident-beam diameter. The maximum light-path thickness of the cell, l, is related to the diameter of the conical lens ($\tan \theta_c = r_c/l$). The pinhole diameter d controls the angular spread $\delta\theta$.

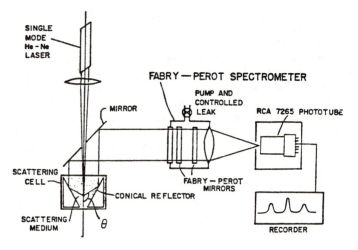

FIG. 6.3.10.   Scattering cell with a conical reflector. (After Benedek and Greytak, 1965.)

optics is a very efficient light collection system, because it collects light over the full length of the illuminated region (l), and it collects over an azimuth angle of 2π. Its disadvantages are as follows: (1) each conical lens permits scattering only at a predesigned fixed angle θ_c, (2) the conical lens, which has to be made specially, is expensive, and (3) the advantage of an azimuth of 2π disappears if the polarized and the depolarized light scattering spectra are appreciably different. Figure 6.3.10 shows a scattering cell with a conical reflector (Benedek and Greytak, 1965). If ϕ is the full apex angle of the conical reflector, the light scattered at an angle $\theta_c = 180° - \phi$ will be collected by the cone and emerge parallel to the axis of the reflector. Benedek and Greytak (1965) have pointed out that if the direction of the incident beam is reversed, one collects light scattered through an angle $\phi = 180° - \theta_c$. Thus, a single conical reflector (or lens) enables a study of scattering at two fixed angles, even though reversing the incident beam direction or the sample cell is usually more involved than it appears.

6.3.5. REFRACTION CORRECTION TO THE SCATTERING ANGLE OF RECTANGULAR CELLS

Refraction at the cell windows may have the following effects: (1) it may change the cross section of the incident beam; (2) it may displace the incident beam; and (3) it changes the effective scattering angle θ as well as its divergence

$\delta\theta$. Figure 6.3.11 shows displacement of the incident beam (a) and the change in dimension of the cross section of the displaced incident beam (b). The basic rule is to use Snell's law ($n_1 \sin \phi_1 = n_2 \sin \phi_2$) If ϕ_1 is the entrance angle from medium 1 with refractive index n_1 and G is the thickness of the cell window glass with refractive index n_2, the parallel incident beam will be displayed by

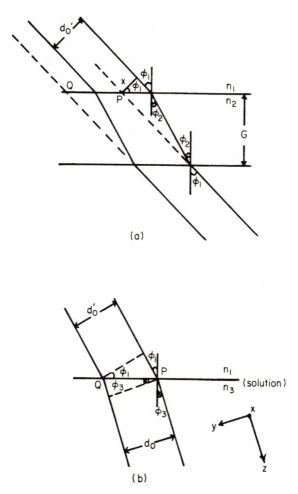

FIG. 6.3.11.   (a) Displacement of a parallel incident beam due to a cell window of thickness G. (b) Change in dimension of a parallel incident beam due to refraction through a flat surface as measured in the y–z plane of the scattering volume.

an amount x given by

$$x = G \cos \phi_1 (\tan \phi_1 - \tan \phi_2). \tag{6.3.2}$$

For simplicity in our discussion, we shall henceforth ignore the beam displacement. Figure 6.3.11(b) shows the change in cross section of the parallel incident beam. We get

$$\frac{\cos \phi_1}{\cos \phi_3} = \frac{d_0'}{d_0}, \tag{6.3.3}$$

where ϕ_3 is the angle between the normal to the cell window and the scattered beam, and d_0' and d_0 are the incident-beam diameters in media 1 and 3, respectively.

If the cell is oriented at Brewster's angle ϕ_1, where

$$\tan \phi_1 = n_{13} = n_3/n_1 \qquad \text{(Brewster angle)}, \tag{6.3.4}$$

then by Snell's law, $(\sin \phi_1)/(\sin \phi_3) = n_3/n_1$, we have

$$\cos \phi_1 = \sin \phi_3 \qquad \text{(Brewster angle)}. \tag{6.3.5}$$

Thus, Eq. (6.3.3) is reduced to

$$\frac{d_0'}{d_0} = \frac{\cos \phi_1}{\cos \phi_3} = \frac{\sin \phi_3}{\cos \phi_3} = \tan \phi_3 = \frac{1}{n_{13}} = \frac{n_1}{n_3} \qquad \text{(Brewster angle)}. \tag{6.3.6}$$

The change in the cross section of the parallel incident beam is proportional to the ratio of the refractive indices inside and outside the scattering medium $(d_0 = (n_3/n_1)d_0'$ at Brewster angle).

Owing to refraction, the scattering angle θ in the scattering medium is changed to θ'. If the scattering cell is oriented perpendicular to the incident beam, as shown in Fig. 6.3.12, a simple application of Snell's law tells us

$$\frac{\sin \theta}{\sin \theta'} = \frac{n_1}{n_3}, \tag{6.3.7}$$

and the changes in the divergence angle correspond to $\delta\theta = (n_1/n_3)\,\delta\theta'$. However, perpendicular orientation leads to a backscattering problem, which can become serious. Orientation of the scattering cell at an angle ϕ_1 changes the apparent scattering angle θ' as follows. From Fig. 6.3.13 we see that

$$\theta = \phi_3^* - \phi_3 \tag{6.3.8}$$

and

$$\theta' = \phi_1^* - \phi_1. \tag{6.3.9}$$

With Snell's law, we have

$$\sin \phi_1 = (n_3/n_1) \sin \phi_3. \tag{6.3.10}$$

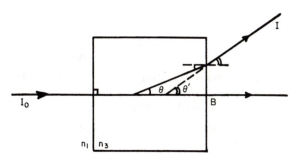

FIG. 6.3.12. Relationship between the true scattering angle θ and the apparent scattering angle θ' for a flat rectangular cell oriented perpendicular to the incident beam. Note: Backscattering at B for the rectangular cell with a glass–air interface can become a serious problem.

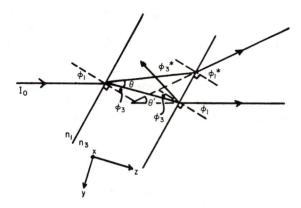

FIG. 6.3.13. Relationship between the true scattering angle θ and the apparent scattering angle θ' for a flat cell oriented at an angle ϕ_1 between the normal of the cell window and the incident beam. Note: In this figure, the reflected beam creates a backscattering problem. The scattering should be viewed on the other side, as shown in Fig. 6.3.14.

By substituting Eqs. (6.3.8) and (6.3.9) into (6.3.10), we get

$$\sin(\phi_1^* - \theta') = (n_3/n_1)\sin(\phi_3^* - \theta). \qquad (6.3.11)$$

Since $\sin(x - y) = \sin x \cos y - \cos x \sin y$, Eq. (6.3.11) can be changed to $\sin\phi_1^* \cos\theta' - \cos\phi_1^* \sin\theta' = (n_3/n_1)(\sin\phi_3^* \cos\theta - \cos\phi_3^* \sin\theta)$.

With Eq. (6.3.10), we get

$$(n_3/n_1)\sin\phi_3^*(\cos\theta - \cos\theta') = (n_3/n_1)\cos\phi_3^* \sin\theta - \cos\phi_1^* \sin\theta',$$

or

$$\frac{n_3}{n_1}(\cos\theta - \cos\theta') = \frac{n_3}{n_1}\cot\phi_3^*\sin\theta - \frac{\cos\phi_1^*}{\sin\phi_3^*}\sin\theta'. \qquad (6.3.12)$$

The angle θ can be determined by knowing $n_3, n_1, \phi_1^*, \phi_3^*$, and θ'. At Brewster's angle, $\sin\phi_1^* = (n_3/n_1)\cos\phi_1^*$, $\sin\phi_3^*/\cos\phi_3^* = n_1/n_3$, and $\cos\phi_1^* = \sin\phi_3^*$. Then, with $n_{13} = n_3/n_1$, Eq. (6.3.12) is reduced to

$$n_{13}(\cos\theta - \cos\theta') = n_{13}^2\sin\theta - \sin\theta',$$

or

$$n\cos\theta - n^2\sin\theta = n\cos\theta' - \sin\theta' \qquad (6.3.13)$$

with the subscript 13 deleted. For small θ, $\cos\theta \simeq \cos\theta' \simeq 1$, $\sin\theta \simeq \theta$, and $\sin\theta' \simeq \theta'$; we then have

$$\theta' = n^2\theta, \qquad (6.3.14)$$

whose divergence angles have the relationship $\delta\theta' = n^2\,\delta\theta$. It should be noted that if the spectrum of scattered light is explicitly dependent upon $\mathbf{K}$, the effect of the finite momentum window on the spectral function should be considered (Yeh, 1969).

Figure 6.3.14 shows the relationship between the scattering angle θ and its apparent scattering angle θ' for the three situations using a rectangular cell oriented at Brewster's angle with respect to the incident beam. Equations (6.3.13) and (6.3.14) correspond to the exit at face a.

From faces b and c we get

$$n^2\cos\theta + n\sin\theta = n\sin\theta' + \cos\theta' \qquad (6.3.15)$$

and

$$n\cos\theta + n^2\sin\theta = n\cos\theta' + \sin\theta' \qquad (6.3.16)$$

respectively.

There are various possible orientations for the scattering cell. We have tried to describe only a few of the many possibilities. The answers can always be obtained through the use of Snell's and Brewster's laws.

For a rectangular cell at normal incidence to the incoming beam, as shown in Fig. 6.3.12, the incident electric and magnetic fields $\mathbf{E}$ and $\mathbf{H}$ are parallel to the interface between air and medium. Then, the distinction between the σ ($\mathbf{E}$ perpendicular to the plane of incidence) and the π ($\mathbf{E}$ in the plane of incidence) cases is lost. With $n_3/n_1 \simeq 1.5$, the reflectivity R is

$$\left|\frac{n_3 - n_1}{n_3 + n_1}\right|^2 = 0.04,$$

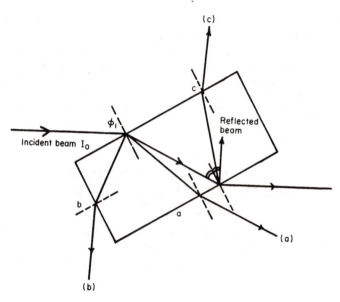

FIG. 6.3.14. Relationship between the true scattering angle θ and its apparent scattering angle θ' using the three exit windows (a, b, c) of a rectangular scattering cell oriented at Brewster's angle with respect to the incident beam. Note: θ and θ' are not shown. The scattered beams have been drawn by viewing the interfaces in order to demonstrate the angular relationships. In practice, the scattering volume should be *away* from the interfaces.

and the transmissivity is $1 - R = 0.96$. However, for an oblique incidence onto a less dense medium, such as for the exit beam, we have $n_3(\text{medium}) > n_1(\text{air})$. In fact, we should take $n_2(\text{glass}) > n_1$. Then we have a critical angle θ_c, which, according to Snell's law, is given by $\sin \theta_c = n_1/n_3$. For $n_3/n_1 = 1.5$, $\theta_c = 41.8°$. At angles greater than θ_c, the light is totally reflected. At angles smaller than θ_c, the reflectivity starts at 0.04 for $\theta = 0$, and approaches 1 as $\theta \to \theta_c$. The important point here is that we have to correct for transmissivity, which depends upon θ in addition to geometrical considerations, if we want to use such optical arrangements for measuring the angular distribution of the time-averaged scattered intensity. Furthermore, the efficiency of the setup for quasielastic light scattering decreases rapidly as θ approaches θ_c. Thus, it is advisable to use the configuration of Fig. 6.3.14 if we want several scattering angles covering a wide angular range. The reader should consult an optics book—e.g., Klein (1970, Chapter 11.3)—for a detailed discussion on reflection and transmission of light at an interface.

The other point to which we should pay attention is that the center of rotation of the detector moves within the rectangular cell because of refraction. It is very convenient to mount the cell on a translational stage (parallel to the incident beam). Such an arrangement permits observation of the scattering in a region with only scattered light for self-beating, or with scattered light and controlled strong interfacial scattering acting as a local oscillator for heterodyning. The use of the latter optical arrangement is often not advisable, however, as dynamical motions can be different near the interface (Lan *et al.*, 1986).

6.3.6. OPTICAL ARRANGEMENTS FOR DEPOLARIZED SPECTRA AND INTRACAVITY CONFIGURATION

Wada *et al.* (1969, 1970) have studied rotary diffusion broadening of Rayleigh lines scattered from optically anisotropic macromolecules in solution. Figure 6.3.15 shows a schematic diagram of the polarization filtering

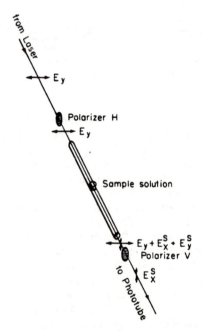

FIG. 6.3.15.   Schematic diagram of the filtering arrangement of depolarized scattered light. **E**, electric vector of light. Superscript s denotes scattered light. (After Wada *et al.*, 1969.)

arrangement for the depolarized scattered light. The polarized incident beam is focused by a lens system into the middle of a long solution cell (150 mm). The incident light must contain polarizations in only one direction. The laser beam first passes through a Glan–Thompson polarizer H before entering the scattering cell as shown in Fig. 6.3.15 in order to remove the remaining other component of polarization and to fix the sense of polarization in space. The light transmitted from the cell contains mainly the very strong incident beam, scattered light polarized in the same direction as the incident beam, and scattered light *depolarized* perpendicular to the incident beam. The last component is retained by passing the transmitted light through a second Glan–Thompson polarizer V, whose optic axis has been carefully adjusted to be perpendicular to the first (polarizer H). In such a setup, a small portion of the incident beam leaks through the depolarization filter even in the absence of the solution cell. The leakage is found to be elliptically polarized, is about 100–1000 times as intense as the depolarized scattered light, and can act as the optical local oscillator. Thus, only a heterodyne spectrum can be observed. More detailed descriptions of measurements for the depolarized component of forward scattering from polymer solutions have been given by Schurr and Schmitz (1973) and by Han and Yu (1974).

It is not essential that depolarized light scattering be limited by the heterodyne technique, even though chances for heterodyning are usually very great, because the depolarized scattered light is much weaker than the polarized scattered light. Figure 6.3.16 shows a schematic diagram for depolarized light scattering using a spherical Fabry–Perot interferometer (Dubin *et al.*, 1971). In Fig. 6.3.16, the effective light path has been increased by passing the incident beam through the scattering cell several times with a pair of high-reflectivity dielectric-coated mirrors. An increase in the collected scattered light by a factor of about 5 was reported. In a multiple-path cell, scattering is restricted to $\theta = 90°$, and depolarization of light, even by dielectric mirrors can change the amplitude, though not the dynamic process. The diaphragm D_1 controls mainly the effective scattering volume, while the pinhole D_2 is used to limit the angular resolution $\delta\theta$. A Glan–Thompson analyzer B permits measurements of the polarized as well as the depolarized scattered light. Removal of the spherical Fabry–Perot interferometer corresponds to changing over the apparatus for intensity-fluctuation spectroscopy. Then, L_2 can be eliminated if we place D_2 at position C and use a photomultiplier with a sufficiently large photocathode area immediately after D_2. There, $\delta\theta \approx D_2/f_1$.

The optical arrangements are such that both the laser and the interferometer system are fixed in space. Occasionally, measurements at different scattering angles are required. Figure 6.3.17 shows one approach to the problem by Fleury and Chiao (1966). When they studied the spectrum of light scattered from thermal sound waves in liquids.

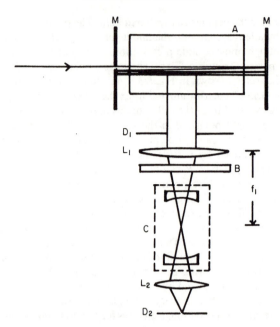

FIG. 6.3.16. A schematic arrangement for depolarized light scattering using a spherical Fabry–Perot interferometer C (Dubin *et al.*, 1971). Lens L_1 has focal length f_1, pinhole D_2 has diameter D_2, A is the scattering cell, and B is the Glan–Thompson prism. M is mirrors.

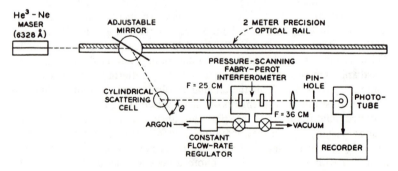

FIG. 6.3.17. Apparatus for observing normal Brillouin scattering as a function of angle. The scattering angle θ is varied by a combination of the adjustable mirror and translation along the precision optical rail. The mirror is mounted on a spectroscopic table with a vernier angular scale for measuring θ. (After Fleury and Chiao, 1966.)

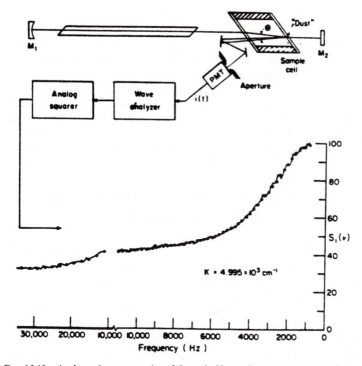

FIG. 6.3.18. A schematic representation of the optical heterodyne spectrometer used to study the spectrum of light scattered from toluene. The Brewster's-angle sample cell is located inside the laser cavity formed by mirrors M_1 and M_2. (After Lastovka and Benedek, 1966a.)

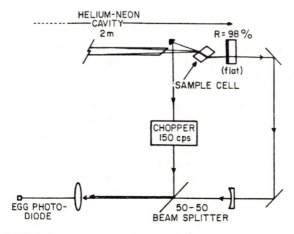

FIG. 6.3.19. Optical system used in the detection of Brillouin-scattered light by opticalsuperheterodyning. (After Lastovka and Benedek, 1966b.)

Lastovka and Benedek (1966a) have devised a rather ingenious method for studying the spectrum of light scattered from toluene using a Brewster's-angle sample cell located inside the laser cavity, as shown in Fig. 6.3.18. An even more demanding optical system was set up for the detection of Brillouin scattered light by optical superheterodyning, as shown in Fig. 6.3.19 (Lastovka and Benedek, 1966b). The detailed design of this optical system, which is based on the optical scheme of a Mach–Zehnder interferometer (Born and Wolf, 1956), can best be appreciated by reading the comprehensive and monumental Ph.D. thesis of Lastovka (1967).

6.4. LIGHT-SCATTERING SPECTROMETERS AND DESIGN CONSIDERATIONS

We shall present a few optical-cell or chamber designs that are preeminent among the many existing optical arrangements for light-scattering spectrometers. In particular, the designs presented here permit access to low scattering angles, high temperatures, or simultaneous measurements in combination with other physical methods such as Raman spectroscopy or differential refractive-index increments. One reason for discussing the merits of those spectrometers is for the reader to be able to appreciate the attempts made by others to alleviate some of the experimental difficulties associated with light scattering experiments. Furthermore, pertinent design considerations are provided so the reader can take advantage of those guidelines when contemplating modifications necessitated by changes of experimental conditions.

6.4.1. PRISM LS SPECTROMETER

The prism LS cell (U.S. Patent 4,565,446, "Prism Light Scattering Cell," issued January 21, 1986) takes advantage of a thick glass block for the entrance window in order to remove the flare produced by the incident laser beam at the air–glass interface of the entrance window. A small pinhole (with diameter larger than the incident-beam cross section and located in the scattering medium immediately behind the entrance window) is used to further reduce the stray light. The channel diameter (a few millimeters) has been optimized so as to reduce internal reflections of the incident beam to tolerable levels. The exit window is a prism permitting viewing of the scattered light at two (or three) surfaces, which cover angular scattering ranges at small (≈ 0) and large ($\approx \pi/4$ and $\approx \pi/2$) scattering angles. The incident beam through the scattering medium strikes the prism at an angle corresponding to the acute angle of the prism. Thus, the incident beam is not back-reflected along its optic axis in the scattering medium. Proper viewing of the scattering volume permits us to avoid the back-reflection problem. The incident beam is refracted if the refractive index of the scattering medium and that of the prism are different. From the refraction together with the refractive index of the prism, we can then determine the refractive index of the scattering medium, equivalent to that determined by a differential refractometer. The prism LS cell, as shown in Fig. 6.4.1(a), utilizes a fairly small scattering volume, is a flow cell, and can

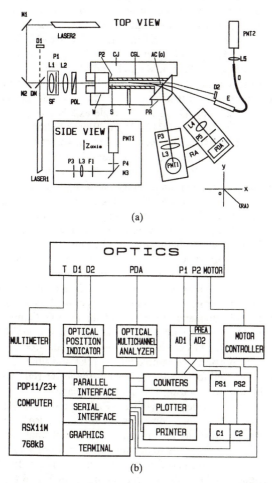

(a)

(b)

FIG. 6.4.1.   Schematic diagrams showing (a) the general design of the optical portion of the spectrometer, (b) the electronics used to interface and control the spectrometer. (After B. Chu and R. Xu *in* OSA Proceedings on Photon Correlation Techniques and Applications, **1**, 1988, pp. 138).

measure the angular distribution of scattered intensity in three angular ranges varying from 2.5° to 20° and in the neighborhood of 45° and 90° scattering angles, as well as its spectral distribution over a range of angles.

The prism light-scattering spectrometer has been described in detail elsewhere (Chu, 1985; Chu et al., 1988b, 1989; Chu and Xu, 1988). We shall outline the description of the spectrometer by Chu and Xu (1988), leaving the details on scattering-angle computations and cell position optimization (Chu et al., 1988b) and the optics of diode array detector, as well as that of the fiber-optics eyepiece (Chu and Xu, 1988; see also Section 6.9) to the original papers.

The parts of the prism light-scattering spectrometer, as shown in Fig. 6.4.1, . are as follows. We have, in Fig. 6.4.1(a): LASER1, He–Ne laser (Spectra-Physics Model 125A), $\lambda_0 = 632.8$ nm; LASER2, argon ion laser (Spectra-Physics Model 165), $\lambda_0 = 488$ nm; M1, M2, and M3, dielectric mirrors (M3 is introduced to shorten the detection arm and for ease of alignment); DM, dichroic mirror permitting transmission of light at $\lambda_0 = 488$ nm and reflection of light at $\lambda_0 = 632.8$ nm; D1 and D2, single-axis lateral position sensors (Silicon Detector Co. Model SD1166-21-11-391 photodiode detectors); SF, spatial filter (Newport Model 900) composed of microlens L1 ($f = 14.8$ mm) and pinhole P1 ($d = 25$ μm); L2, focusing lens for the incident beam ($f = 20$ mm); POL, polarizer; P2, pinhole with hole diameter 1.5 mm; P3, aperture stop to control angular uncertainty $\delta\theta$ ($d = 0.0135$ in.); L3, imaging lens ($f = 10$ cm); F1, interference filter; P4, field stop ($d = 0.008$ in.); CJ, brass cooling jacket; CGL, cobalt glass lining with 9-mm o.d.; AC, apparent scattering center; W, optical glass window, 1-in. dia. $\times$ $\frac{1}{2}$ in. thick; PR, BK-7 glass prism; S, solution; T, thermistor; PMT1 and PMT2, photomultiplier tubes (ITT FW130 and RCA C31014A, respectively); RA, rotating arm; LF, Nikkor camera lens with $f = 50$ mm and a maximum aperture of 35.7 mm; P5, pinhole with hole diameter 0.2 mm; PDA, EG&G PARC model 1422 photodiode-array detector; E, an assembly consisting of a lens ($f = 50$ mm), a pinhole with hole diameter 0.3 mm, and an EG&G Gamma Scientific model 700-10-34A eyepiece with a probe diameter of 450 μm; L5, focusing lens ($f = 30$ mm); O, EG&G Gamma Scientific model 700-3B fiber optic probe. In the cell portion, the sample inlet–outlet and fluid circulating circuit have been omitted for clarity. O–(RA) in the lower right corner represents the position of the rotating arm in the x–y plane.

The linear diode array detector (EG&G PARC model 1422) is connected to a DEC PDP 11/23+ microcomputer via an EG&G PARC model 1461 detector interface, shown as an optical multichannel analyzer (OMA) in Fig. 6.4.1(b). The system is able to acquire, analyze, store, and display data together with computed results or in comparison with theory. In Fig. 6.4.2(b), T, PDA, D1, and D2 have the same meaning as in Fig. 6.4.2(a). P1 and P2 are ITT FW130 and RCA C31034A photomultipliers, respectively; the motor, Superior Electric Co. model MO63-FD09; the motor controller, Modulynx Superior Electric Co.; the multimeter, Keithly model 195A; the optical position indicator, United Detector Technology model OP-EYE-4; the optical multichannel analyzer, EG&G PARC model 1461 detector interface; PREA, an Ortec model 9301 preamplifier; AD1 and AD2, SSR Instrument model 1120 and Ortec model 9302 amplifier–discriminators, respectively; the counters, HP model 5345A 500-MHz electronic counter and HP model 5328A universal counter; PS1 and PS2, ECL laboratory-built pulse stretchers; C1 and C2, Brookhaven Instruments BI2030AT 136-channel $4 \times n$ correlator and Malvern model 7027 64-channel single-clipped correlator; the plotter, IBM 7372; the printer, DEC LA100; the graphics terminal, DEC VT240. The model 1461 is not an OMA. However, it serves as an OMA when utilized in combination with the microcomputer. Time-resolved light scattering intensity measurements with an angular range reaching a maximum external scattering angle of 17.5° and a time sequence down to ≈ 20 msec per scattering pattern can be achieved for kinetic light scattering studies.

From the combination of detection schemes as shown in Fig. 6.4.1, we can determine (1) the solution concentration from the refractive index of the solution, or, if the solution concentration is known, the refractive-index increment, (2) the weight-average molecular weight (M_w) from the absolute scattered intensity at small scattering angles, (3) the radius of gyration (R_g) from the angular distribution of absolute scattered intensity, (4) the second virial coefficient (A_2) if several concentrations of the same solution can be measured, (5) the translational diffusion coefficient and its corresponding hydrodynamics radius (if the solvent viscosity is known), (6) the polydispersity index of the polymer molecules in solution or colloidal particles in suspension from moment analysis of the spectral distribution of scattered intensity, and possibly (7) the (hydrodynamic) size distribution function from Laplace inversion (see Chapter 7) of the intensity–intensity time correlation function or power spectrum, as well as internal motions at large scattering angles provided that we can reach $KR_g \gg 1$. If the volume capacity of the prism light-scattering cell can be made comparable to that in size-exclusion chromatography (SEC), then its use as a detector, in addition to the on-line viscosity detector (Yau, 1990) and other more standard detectors such as UV absorption detectors, could make SEC an even more powerful analytical tool.

6.4.2. PHOTOMETER WITH FIXED SCATTERING ANGLES (HALLER ET AL., 1983)

A photometer employing 18 fixed scattering angles ranging from ≈ 3° to 163° has been reported. The apparatus has a complex thermostat which can be controlled to ±0.1 mK near 30°C and uses optical fibers to transmit the

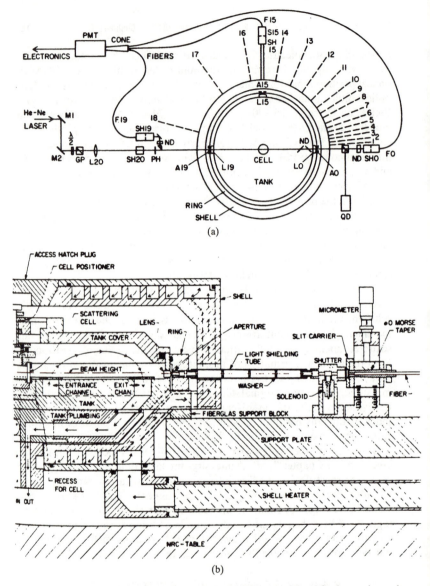

(a)

(b)

FIG. 6.4.2. (a) Schematic diagram showing the optical layout of the 18 fixed scattering angles (with K^2 spacing) as well as provisions for transmitted and reference intensities. Variable attenuator consists of $\lambda/2$ plate and Glan–Thompson prism (GP); spatial filter consists of lens (L20) and electron-microscope aperture (PH). Aperture (A19) permits the central spot and first ring of the spatially filtered incident beam to pass through. The incident beam is then focused to the center of the scattering cell by means of L19. L0 collects the transmitted beam, while lens (L1–L18) images the central region of the scattering cell onto the corresponding slit (S1–S18), which acts as a field stop for the scattered light. F, SH, A, ND, and QD denote optical fiber, shutter, aperture, neutral-density filter, and quadrant detector, respectively. (b) Cross-sectional view of one collection channel. (After Haller et al., 1983.) He–Ne laser beam propagates in the x-direction.

scattered light from the 18 scattering angles to a single photomultiplier tube. Measurements of the scattered intensity at each fixed scattering angle are controlled by the 18 shutters (SH) as shown schematically in Fig. 6.4.2(a). Details of the optical design, electronics, temperature control, testing, and calibration, as well as the results on the capabilities of this apparatus, are best left for the reader to consult the original paper published in the *Review of Scientific Instruments* (Haller *et al.*, 1983). While any fixed-scattering-angle instrument tends to have high measurement precision, its rigidity becomes a hindrance to instrument flexibility (see Section 6.4.3). It should also be noted that the complex thermostat that forms an integral part of the light-scattering spectrometer has a limited temperature range, and the ± 0.1-mK control is overkill for most experiments not dealing with temperature-dependent phase-transition studies.

6.4.3. HIGH-TEMPERATURE LIGHT-SCATTERING SPECTROMETER (CHU AND WU, 1987; CHU ET AL., 1988A)

The fixed-angle spectrometer designed and constructed by Cannell and his coworkers can provide very broad angular ranges (2.6–$163°$) with a K^2 spacing and very good temperature control. However, there can be other needs, which favor variable-angle instruments. For example, if we are interested in determining the radius of gyration of a large particle, the number of scattering angles in the angular range covered by the fixed-angle arrangement may not be sufficient. If we want to examine the relationship between translational diffusion and internal motions, the fixed K^2 spacing is not optimized. The specifications for the high temperature LS spectrometer are as follows:

(1) The high-temperature chamber permits temperature control above the melting point of Teflon* at $330°C$. It has a small temperature gradient ($\lesssim 0.1°$/cm) in the solution cell and can be maintained over a broad temperature range from $\approx 15°C$ to $\approx 400°C$. Short-term (20 min) control of $\pm 0.005°C$ and intermediate-term (60 min) control of $\pm 0.01°C$ can be achieved at $\sim 300°C$.

(2) In view of the high-temperature operation, no refractive-index-matching fluid can be used. Thus, the high-temperature chamber is designed to accommodate a cylindrical light-scattering cell with diameters up to 27 mm o.d., which can provide serviceable angular ranges from $\approx 20°$ to $\approx 160°$ without using an excessive amount of the solution (or suspension). At 27 mm o.d., we need about 4 ml of liquid for each run.

* DuPont trademark.

(3) The dual-detector instrument permits simultaneous measurements of scattered intensity at two scattering angles, tentatively set 90° apart. Thus, it can be used for fast determinations of the radius of gyration in kinetic processes.

(4) The instrument has analyzers (item 6 in Fig. 6.4.3) and is capable of depolarized light scattering measurements.

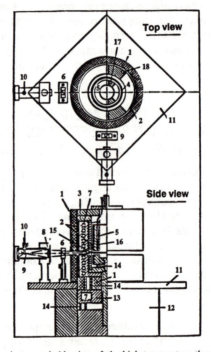

Fig. 6.4.3. Schematic top and side view of the high-temperature thermostat and detection system of the light-scattering spectrometer: (1) silicone-rubber insulation; (2) heating wires for the brass thermostats; (3) outer brass thermostat with fluid circulation facilities; (4) stainless-steel temperature shield with precision polished glass windows (15) of 2.25-in. o.d., machine-centered to coincide with the center of rotation of the turntable (12) to 0.001 in.; (5) inner brass thermostat, which has a separate temperature controller and thermometer and can accommodate light scattering cells up to 27-mm O.D.; (6) Glan–Thompson polarizers; (7) fluid circulation paths; (8) lens; (9) field aperture; (10) optical-fiber bundle; (11) rotating plate for multiple detectors; (12) Klinger RT-200 rotary table with 0.01° step size; (13) cooling plate to isolate the outer thermostat from the rotary table; (14) stainless-steel standoffs for thermal isolation; (16) location of platinum resistance thermometer; (17) location of temperature sensor for temperature control of inner thermostat (5); (18) location of temperature control of the outer brass thermostat (3). (Reprinted with permission from Chu and Wu, ©1987, American Chemical Society Chu et al., 1988a.)

(5) The dual detectors are mounted on a rotary table, permitting light scattering measurements at variable scattering angles, which is more flexible than the previous spectrometer (Section 6.4.2), having even 18 fixed scattering angles with K^2 spacing. For example, if the static structure factor exhibits maxima and minima, or if we have long semiflexible polymer chains and are interested in examining the onset of internal motions, then the variable-scattering-angle arrangement permits more detailed examinations in the angular range of interest.

(6) The spectrometer is compact, being mounted on a 1-ft^2 platform. With closely packed mechanical components, it is very stable when compared with larger-size spectrometers. Many commercial spectrometers employ a refractive index-matching fluid, have a limited temperature range, and use one detector on a rotary table. The detection optics of the Malvern System 4700 (Malvern Instruments Ltd., Spring Lane South, Malvern, Worcestershire, WR14 1AQ, England) and that of the Brookhaven BI-200SM goniometer (Brookhaven Instruments Corporation, Brookhaven Corporate Park, 750 Blue Point Road, Holtsville, New York, 11742, U.S.A.) are equivalent to Fig. 6.3.3(b), while the present instrument, with an optical-fiber eyepiece, has the detection optics schematically represented by Fig. 6.3.3(a).

6.4.4. OTHER COMMERCIAL LS PHOTOMETERS

Four additional commercial LS photometers, shown schematically in Figs. 6.4.4, 6.4.5, 6.4.6, and 6.4.7, are presented for completeness.

Two main features of the Chromatix low-angle LS photometer are (1) its cell design and (2) its detection optics configuration. The cell design takes advantage of two thick high-quality optical-glass blocks as the entrance and exit windows for the incident and scattered beams as shown in Fig. 6.4.4. By virtue of the very thick optical-glass windows, stray light from the glass–air

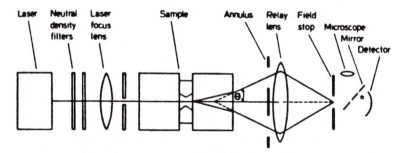

FIG. 6.4.4.   Schematic of the Chromatix KMX-6 low-angle laser LS photometer. (Reprinted with permission from Springer-Verlag, After Huglin, 1978.)

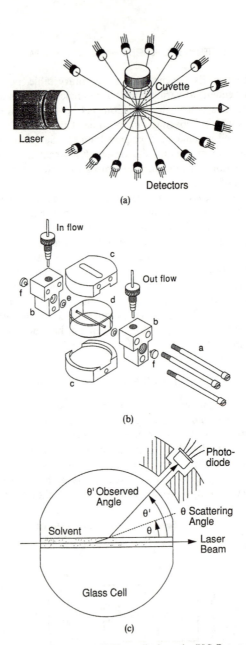

(a)

(b)

(c)

FIG. 6.4.5. (a) Dawn detector array. (b) Flow cell schematics (U.S. Patent 4,616,927, October 14, 1986). The flow cell includes a high-refractive-index glass cell, d with its highly polished 2-mm-diameter throughbore. The inflow and outflow manifolds b are stainless steel, and house optical windows, f. The ultrafine illuminating laser beam passes through the entire cell a, retaining screws; c, cell block; e, o-ring seal; (c) Flow-cell refractions θ, actual scattering angle; θ', apparent (and instrument) scattering angle. (Wyatt Technology Corp., with permission.)

Optical System

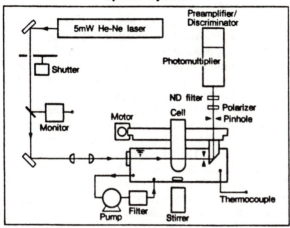

FIG. 6.4.6. Optical system of DLS-700 LS spectrophotometer manufactured by Union Giken Co. (With permission.)

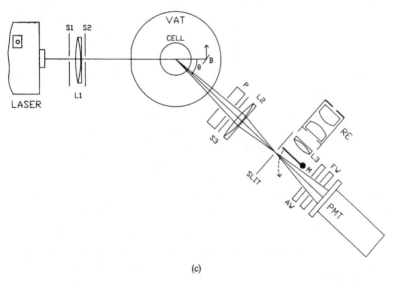

(c)

FIG. 6.4.7. Optical system of BI-200SM laser light-scattering system. (Brookhaven Instruments, with permission.) L1, L2, and L3 are lenses; S1, S2, and S3 are apertures; B is a beam stop; P is an optional Glan–Thompson polarization analyzer; M is a mirror; Re is a Ramsden eyepiece; AW is an aperture wheel with 100-, 200-, and 400-μm and 1-, 2-, and 3-mm positions; FW is a filter wheel with 632.8-, 514.5-, and 488.0-nm filters and one open and two blank positions.

interface for both the incident beam and the scattered beam is greatly reduced. Thus, for a well-defined incident beam, small-angle light scattering (down to a scattering angle of a few degrees) becomes feasible. The detection optics, like that of Fig. 6.3.9, permits measurements of the scattered light from θ to $\theta + \delta\theta$, but integrated over the entire annular ring. It has a very efficient light-collecting geometry. Over a small angular range, as limited by the relay lens aperture, only a low-power (≈ 5 mW) He–Ne laser is needed for most light scattering intensity measurements. The cell design also requires only a small solution volume and can be used as a molecular-weight detector in size-exclusion chromatography. The chromatic photometer (1) can perform small-angle LS measurements, (2) uses a small solution volume, which permits LS measurements from the effluent of a chromatographic column, and (3) has a highly efficient light-collection geometry. However, the collection geometry ignores polarization properties of the scattered light, as discussed in Section 6.3.4, and the field stop controls only the beam cross section, defining the cross section of the scattering volume as viewed by the detector. As the length dimension along the incident beam axis is limited by the space between the two glass blocks, the coherence properties related to the scattering volume, as discussed in Chapter 3, have not been optimized; but they can be improved according to the discussions presented in this book. In measurements of angular distribution of absolute scattered intensity at small scattering angles, limited variation of scattering angles can be accomplished by moving the annulus along the optic axis of the incident beam (Chu, 1977). It should be noted that the dual-block approach has a problem due to the back reflection ($\pi - \theta$) of the incident beam at the glass–air interface of the exit block as the incident beam leaves the light-scattering cell. The back-reflection problem can, however, be alleviated by using antireflection coating at the glass air interface of the exit window.

Figure 6.4.5 shows the essential features of a Dawn laser light-scattering photometer (Wyatt Technology Corp., Santa Barbara, California, U.S.A.). The Dawn laser LS photometer uses a multiple high-gain hybrid photodiode detector array (18 angles equidistant in $\cot\theta$ between -1.6 and 2.0 for model F, and 15 angles equidistant in $\sin(\theta/2)$ from 0.2 to 0.9), as shown schematically in Fig. 6.4.5, and is primarily a light-scattering photometer designed for measurements of the angular distribution of time-averaged scattered intensity. The virtue of this instrument is its simplicity. The flow cell, as shown in Fig. 6.4.5(b) (U.S. Patent 4,616,927, October 14, 1986) has a unique glass cell (d), which permits connection as a LS detector similar to that of the Chromatix LS photometer. The Dawn cell has a scattering angular range that depends partially upon the refractive index of the scattering medium. It covers a much broader angular range than the Chromatix LS cell, but does not have low-angle accessibility comparable to those of Figs. 6.4.1, 6.4.2, 6.4.3, and 6.4.4. In the Dawn LS photometer, the detection optics is controlled by apertures only, and the scattered light is refracted at the liquid glass interface, as shown in Fig. 6.4.5(c). Thus, LS measurements require knowledge of the refractive index of the scattering medium. The scattering-angle correction is similar to that of Section 6.3.5. The total volume of the flow cell is about 75 μl with a measured or detected scattering volume of about 0.75 μl. Geometric volume differences seen at each detector location, as well as gain variations of the detectors, have to be calibrated. Such an arrangement often makes absolute calibration more difficult, especially when the entrance window has not been located reproducibly.

Figure 6.4.6 shows the optical system of a DLS-700 dynamic light scattering spectrophotometer manufactured by Union Giken Co., Ltd. (3-26-2, Shodai-Tajika Hirakata, Osaka, Japan). The detection optics is similar to a modified Sofica LS photometer (Huglin, 1978). Although it has a reported angular range of 5–150°, it is not clear how well the stray light is excluded in the 5–10° scattering angular range when compared with data at higher scattering angles, especially for a 12-mm-o.d. cylindrical light-scattering cell. It should also be noted that the detection optical arrangement for the LS spectrophotometer utilizes a double-pinhole approach. Without a lens, the angular and field apertures are not as conveniently controlled. Furthermore, the use of a 2-in. photomultiplier tube appears to be unnecessary, as it would increase the dark count rate with such a large photocathode area.

The optical system of the Brookhaven Instruments BI 200SM light-scattering system, as shown in Fig. 6.4.7, is typical of what a laser light-scattering spectrometer should be. The detection optics is identical to Fig. 6.3.3(b). Lens L1 focuses vertically polarized laser light into the sample cell. Apertures S1 and S2 reduce stray light. The cell is temperature-controlled. An index-matching liquid, contained in a vat, surrounds the cell. The entrance of the vat is a polished and optically flat window to reduce lensing effects and stray light. A beam stop, B, consists of a mirror at 45° and dark glass. Scattered light is focused by lens L2 onto a vertical slit. The (aperture) stop S3 reduces stray light. A Glan–Thompson polarizer, P, is optional. With the mirror M rotated down, the objective lens L3 focuses light into the Ramsden eyepiece, RE. Since the scattering volume is imaged onto the slit, the image in the eyepiece is the same as the light detected by the PMT. An aperture wheel, AW, consists of three pinholes (100, 200, 400 μm) for DLS measurements and three apertures (1, 2, 3 mm) for SLS measurements. A filter wheel, FW, contains narrow-band interference filters for 632.8-, 514.5-, and 488-nm wavelengths as well as one open and two blank positions. The blank positions are useful as shutters.

6.4.5. DESIGN CONSIDERATIONS

From the discussions of previous sections on the optics of some of the LS spectrometers, we see that it is not so easy—indeed, nearly impossible—to construct a universal instrument that has all the parameters optimized. Accessibility of very low scattering angles and/or high or low temperatures, variable vs fixed scattering angles, ease of sample-cell exchange vs small sample-cell volume and flow-cell operations, and simultaneous multiple detector vs multiple-angle single-detector geometry are some of the choices. Some capabilities are exclusive of each other, and others have conflicting requirements. Nevertheless, in examining the optical design of LS spectrometers, we have some general considerations that it may be useful to summarize here, whether for modification of existing LS spectrometers, for planning of future designs, or for evaluation of existing ones. The optics of the spectrometer may be subdivided into three sections: (A) collimation of the incident beam, (B) sample-chamber design, and (C) detection optics.

6.4.5.A. COLLIMATION OF THE INCIDENT BEAM

(1) The incident laser beam should be adjusted to pass through the center of rotation of the goniometer (e.g. see Fig. 6.4.3) or some other reference point (e.g. Fig. 6.4.1). However, a laser, especially an argon ion laser, is quite bulky. Thus, it is usually more convenient to mount the laser at the proper height and use M1 (Fig. 6.4.2), which is mounted on a gimbal mount (θ, ϕ adjustments) with translational adjustment along the optical (x) axis of the laser. Adjustments for M1 should be precise and stable, because it is located furthest away from the center of rotation.

(2) The variable attenuator, consisting of a half-wave plate and the Glan–Thompson prism (on a gimbal mount with θ, ϕ adjustments), is mounted on an optical rail, not only for convenience, but also for providing a well-defined polarization for the incident laser beam.

(3) Pinhole P2 in Fig. 6.4.1(a) is an extra stray-light-limiting piece. It is not an essential component and should permit most of the laser light to pass through. According to Eq. (6.2.3), if $2r_0 = 1$ mm, a 1.5-mm pinhole permits 99% of the laser power to pass through.

(4) The spatial filter, consisting of lens L1 (with x, y, z adjustments) and pinhole P1 of Fig. 6.4.1(a) (with fine y, z adjustments), is essential for small-angle ($\theta \lesssim 10°$) laser light scattering, but can be omitted for higher-scattering-angle ($\theta \gtrsim 12°$) measurements. Equation (6.2.4) can be used to compute the spatial-filter pinhole size. Usually a short-focal-length lens is used. For example, with a Spectra Physics model 125A He–Ne laser, we used a lens (L1) with $f = 14.8$ mm and a pinhole (P1) size of 25 μm. If we take D = 1.5 mm,

then d_s due to the diffraction angular radius should be $2 \times 1.22 \times 0.6328(14.8/1.5) \cong 15$ μm according to Eq. (6.2.4) while according to Eq. (6.2.5), d_s due to aberration is about 2 μm if a planopositive lens with $n = 1.52$ is used. Thus, we note that, in practice, P1 should be somewhat larger than the computed value based on Eq. (6.2.4). We could try to take into account the aberration angular radius as expressed by Eq. (6.2.5) for both effects if a simple lens is used. However, θ_a is unimportant whenever f is large. With the above considerations, we can produce a clean incident laser beam and perform light scattering experiments down to $\approx 2.5°$ scattering angle.

(5) The spatially filtered incident laser beam is refocused using lens L2, as shown in Fig. 6.4.1(a). The beam depth of focus can be estimated using coarse x, y, z adjustments from the rail and θ, ϕ adjustments from the gimbal mount and Eq. (6.2.8). With a spatial filter, Eq. (6.2.9) is no longer applicable. For small-angle LS measurements, we want to use a longer-focal-length lens, while for larger-scattering-angle measurements, a shorter-focal-length lens can be used. For example, without the use of spatial filtering, a lens with $f = 60$ cm and $d_s = 1$ mm has $\delta\theta_{INC} = 1.67$ mrad (or $\approx 0.1°$), which is quite acceptable for most light scattering linewidth measurements, except those at the smallest scattering angles ($\theta \lesssim 2°$). It should be noted that the diameter d_s at the scattering center is proportional to $\lambda f/D$ according to Eq. (6.2.4) if no spatial filter is used. As the increase in power density and the decrease in beam divergence of the incident beam are conflicting requirements, we shall use the largest $\delta\theta_{INC}$ that can be tolerated in order to achieve the smallest beam cross section for improved signal-to-noise ratio and the coherence requirements of Section 3.1. With a spatial filter, we project the pinhole (P1) to the scattering center by means of lens L2 of Fig. 6.4.1(a) using a simple lens formula: $1/f = 1/s_o + 1/s$ where s_o and s are conjugate distances and the magnification ratio is s/s_o with subscript o denoting the object. The conflicting requirements have always to be compromised. What we have presented here is a procedure for achieving a reasonable compromise.

(6) An additional pinhole (e.g., P2 in Fig. 6.4.1(a) and A19 in Fig. 6.4.2(a)) is used to keep out the stray light. The size of P2 is determined by simple geometry using Eq. (6.2.8) as an estimate of the lower limit. Haller et al. (1983) permitted the diffraction pattern of the incident laser beam produced by the pinhole of the spatial filter to be apertured by A19 so as to pass the central spot and the first ring.

In the Dawn LS photometer, a He–Ne laser selected for its natural small beam diameter is aligned directly without additional optics. Therefore, it is quite acceptable for scattering intensity measurements but can suffer an appreciable loss of the signal-to-noise ratio in linewidth measurements because the power density has not been optimized.

6.4.5.B. SAMPLE-CHAMBER DESIGN

A variety of cell designs have been presented. One should consider the alternatives: (1) cylindrical cell, (2) flow cell, or (3) cells of other geometry. Cylindrical cells with typical diameters ranging from 10–40 mm o.d. and a cell wall thickness of 1 mm is very popular. By using a refractive-index matching fluid, thick-walled LS cells with 10-mm o.d. and a wall thickness of up to 3 mm have been used successfully over a fairly broad scattering-angle range. The scattering cell should be prealigned with the center of rotation of the goniometer to $\sim \pm 0.01$ mm.

Several flow-cell configurations exist. Each has different capabilities. The prism light-scattering cell (Fig. 6.4.1) is versatile but also elaborate. The Dawn flow cell is simpler to use, but has other limitations, e.g., larger scattering angles, suitability mainly for time-averaged scattered intensity measurements, and dependence of the scattering angle on the refractive index of the scattering medium.

Cells of other geometry may be useful to alleviate multiple-scattering problems. Often, they are not worth the effort except under special circumstances.

6.4.5.C. DETECTION OPTICS

The detection optics have been discussed in detail in Section 6.3. It suffices to mention here that the spectrometer should have

1. a monitor to measure the incident laser-beam intensity,
2. a monitor to measure the transmitted laser-beam intensity, and
3. a neutral-density filter (or light trap) after the light-scattering cell in order to eliminate the back-reflected incident laser beam.

It may also be useful to have

4. provisions to monitor the incident beam direction using a quadrant or lateral sensor, as shown in Fig. 6.4.1(a).

The light-scattering spectrometer should have a minimum number of components, and the mechanical elements should be packed as close as possible in order to increase overall stability for the instrument. Finally, in the alignment for the detection optics, it is often advisable to view the scattered intensity a few degrees above or below the usual scattering plane in order to further reduce stray-light detection.

6.5. PHOTON-COUNTING TECHNIQUE

6.5.1. INTRODUCTION

In laser light scattering the techniques of optical mixing spectroscopy and, to a lesser degree, of optical interferometry require us to pay particular attention in selecting photomultipliers (PM or PMT) and their housing designs. Furthermore, the same system can be used, in a similar manner, for measuring the time-averaged intensity of scattered light by means of photon counting. Using photomultipliers of an advanced type as well as fast, low-noise electronic devices capable of wide-band amplification and high-speed digital processing, the photon-counting technique covers a dynamic range for the integrated intensity extending from very low (statistically or quantum-limited) light levels to high light levels where the PM linearity becomes marginal. In our studies, we are mainly interested in photon counting and photon correlation, though at low scattering angles and high light levels, photocurrent and FFT power spectral analysis can be useful. Thus, the selection of an appropriate detector and its housing is the first crucial step in achieving good instrumentation in laser light scattering. Beginning chemists and biologists are usually not familiar with the use of photomultipliers. In this section, we shall outline some of the pertinent features of good housing design and the PM-selection criteria for application to photon counting and correlation.

When light falls on the photocathode of the PM, single photoelectrons are ejected. These photoelectrons are then multiplied by a cascaded secondary emission process to produce pulses of charge at the anode. At high light levels, the pulses overlap and the light intensity is measured by the anode current. At low light levels, the pulses no longer overlap and the light intensity is proportional to the number of those pulses. Aside from photon correlation, the principal modes of operation in detecting the time-averaged intensity are dc current or voltage measurements, charge integration, synchronous detection, pulse counting, and the shot-noise method of Pao $et\ al.$ (1966) and Pao and Griffiths (1967). Young (1969) noted that dc and charge-integration detection are equivalent. Chopping, followed by phase-sensitive (i.e., synchronous) detection, is similar to dc measurements, except that (1) chopping often throws away half of the incident signal power, and (2) the final bandpass is shifted from dc (zero frequency) to the chopping frequency. Thus, the signal-to-noise ratio for phase-sensitive detection is $\sqrt{\frac{1}{2}}$ that of the dc method if the noise is independent of frequency. There are then three basic schemes: dc, pulse counting, and the shot-noise-power method (Pao and Griffith, 1967).

The three basic methods of detection correspond to utilizing three different weighting functions of the signal pulse height. In pulse counting, all pulses

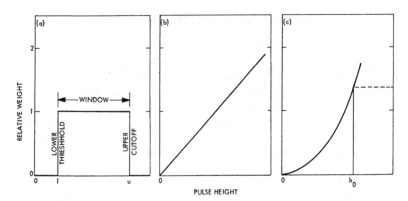

FIG. 6.5.1.   Weighting functions for (a) pulse counting, (b) dc detection, and (c) shot-noise-power detection. The dashed line in (c) indicates the effect of clipping at height h_0. (After Young, 1969.)

between two pulse heights are counted equally (weight 1) and all others are ignored (weight 0). The discrete single photoelectron pulses with random amplitudes on a wandering baseline are first amplified and put on a stable baseline. The single-channel analyzer (or discriminator) selects the pulses with heights E to $E + \Delta E$, and all the pulses between those two heights are counted with equal weight. In dc or charge integration, each pulse is weighted by its height (charge), while in shot-noise-power detection, each pulse is weighted by the square of its height, as shown schematically by Young (1969) in Fig. 6.5.1.

6.5.2. PHOTOMULTIPLIER SELECTION (NOISE AND TIME-DEPENDENT STATISTICAL PROPERTIES)

The noise in photomultipliers influences the sensitivity with which the photoelectrons can be detected. Various sources of noise are present (Eberhardt, 1967), the main ones being (Akins et al., 1968): (1) noise from thermionic emission at the cathode, (2) temperature-independent dark noise, (3) gain noise of the multiplier chain, and (4) statistical noise in the signal (shot noise).

The primary source of noise in an uncooled photomultiplier is thermionic emission from the cathode, described by the quasiexponential dependence of anode dark current on temperature. Cooling the photomultiplier will greatly reduce the thermal electrons, but may not enhance the signal-to-noise ratio (S/N), because the overall quantum counting efficiency of the tube (counts out per photon in) depends, in general, upon temperature as well as wavelength. Harker *et al.* (1969) have shown that for a (nonselected) ITT FW 130 (S-20) photomultiplier, a larger S/N is obtained with the tube cooled to 205 K if the counting rate is less than about 8000 photons/sec, while for counting rates greater than 8000 photons/sec room-temperature operation gives the larger S/N. Thus, the optimum temperature in photon-counting systems depends upon the counting rate and the tube characteristics. In other words, optimum performance is achieved at a specific tube temperature, and it is not true that cooling the PM will always improve its performance. In fact, performance in the far red for most photocathodes will be impaired by excessive cooling. In some applications, a PM with a small photocathode area should be selected. The requirement for coherence-area consideration in signal correlation makes fast photon-counting PMs with small effective photocathode areas particularly suitable for laser light scattering. The FW 130 PM can be ordered with an effective photocathode area of about 0.1 mm^2, which reduces the dark count rate to a few counts per second even with the PM operated at room temperatures.

When the dark pulses are reduced to a few counts per second, the bulk of those nonthermal dark pulses have other origins. The residual pulses originating in the PM (not the amplifier noise or Johnson noise in the load resistor) have been ascribed to photoemission in the PM window from light generated by cosmic rays (Young, 1966; Chodil *et al.*, 1965; Jerde *et al.*, 1967) and to radioactivity in the tube (Krall, 1967; Dressler and Spitzer, 1967). Electroluminescence of the glass and dynode glow are, at most, only weakly temperature-dependent and are usually made worse by cooling the tube. After applying voltage to a PM, the dark count rate will gradually decrease to an equilibrium value over a period of several hours or days. Phosphorescence is partially responsible for the temporary increase in dark count rates after the PM has been exposed to light. The glass envelopes of PMs contain enough potassium 40 to produce one or more dark counts per second. PMs with pure silica windows are less radioactive but are more sensitive to cosmic rays. For some tubes, there is an additional component of the dark current that is due to the ionization of gases by energetic electrons (Young, 1969).

The temperature dependence of the dark count rate has been extensively studied (Morton, 1968; Oliver and Pike, 1968; Gadsden, 1965; Rodman and Smith, 1963). The dark count rate of most photocathodes, with the exception of type S-1, asymptotically approaches a minimum at about -30 to $-40°C$ with the higher-temperature region obeying the Richardson–Dushman equation for thermionic emission, $\ln(\text{count rate}/T^2) \propto 1/T$ (Kittel, 1968). The type S-1 photocathode requires cooling to below $-100°C$ in order to reduce the temperature-dependent component of the dark count rate.

In photon correlation, it is particularly important to select a PM tube whose dark counts exhibit Poisson photon-counting statistics, i.e., whose dark pulses are randomly distributed in time. Oliver and Pike (1968) have shown that the dark counts of an ITT FW 130 PM have Poisson statistics both at room temperature and cooled, while a Mullard 56 TVP PM has about two standard deviations from the correct value at room temperature and major correlations when the tube is cooled.

The variation of photon-counting correlations with discriminator dead time for the ITT FW 130 PM, when the cathode is illuminated with coherent light, is shown in Fig. 6.5.2 (Foord *et al.*, 1969). The measurements show that the ITT FW 130 tube can be used for photon correlation studies with a discriminator dead time of 75 nsec. The shorter the dead time, the higher the permissible repetition rate. So PMs with very short dead times permit measurements of very short correlation times. The RCA C70045C gives good photon-counting performance with a dead time of only 10 nsec, while the EMI

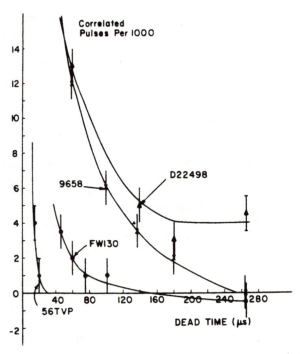

FIG. 6.5.2. Variation of the photon-counting correlations with discriminator dead time for the ITT FW 130 PM tube. (After Foord *et al.*, 1969.)

Table 6.5.1

Typical Photomultipliers Suitable for Photon-Counting
and Photon-Correlation Measurements[a]

Manufacturer	Hamamatsu	EMI		ITT	RCA
Type no.	R464	9863/100	9893/100	FW130[b]	C31034[c]
Spectral response	Bialkali	S20	Bialkali	S20	300–800 nm (GaAs)
Effective cathode (mm)	5×8	2.5 dia.	2.5 dia.	1.0 dia.	4×10
Cathode sensitivity, typ. (μA/lm)	50	180	60	160	975
Anode sensitivity gain, typ.	6×10^6	2.5×10^7	8×10^7	3×10^5	3×10^5 (4×10^6)
V_{k-a} (V)	1000	2000	2250	1800	1500(2000)
Dark count (sec^{-1})	~150	~40	~20	~10	<5
Rise time (nsec)	16	2.5	2.5	15	<7

(a) Listed from left to right roughly in order of increasing cost. Cooling (e.g. -20 to $-30°$C) is recommended for S20 and RCA photomultipliers.

(b) This tube may be purchased with effective photocathode areas down to 10^{-4} cm^2 with correspondingly lower dark count rates of ~1 sec^{-1} at room temperatures.

(c) Er. Gulari and B. Chu, *Rev. Sci. Instrum.* **48**, 1560 (1977); the rise time is due mainly to pulse pair resolution of the Ortec 9302 100-MHz amplifier–discriminator used. It should be noted that C31034 photomultipliers are fairly short (≈ 90 mm) and have only 11 dynodes, which reduces the transit time and the spread of the pulses. Thus, the actual rise time is probably less than 7 nsec.

(d) V_{k-a}: Voltage between cathode and anode.

9558 tube has a dead time of 200 nsec. Table 6.5.1 shows a summary of photomultiplier photon-counting and dark-count properties of current usage. For a more detailed discussion on the use of photomultipliers for photon counting, the reader is advised to read the excellent article by Foord *et al.* (1969). We shall restate the pertinent properties of the PM suitable for laser light scattering as summarized by Foord *et al.* (1969):

1. There must be few correlated pulses for coherent illumination. This requirement is satisfied by the EMI 9863, 9893, the Hamamatsu R464, the ITT FW 130, and the RCA C31034, C70045C.

2. The anode pulses must be as narrow as possible for the PM to take high pulse counting rates without pulse pileup, and there should be no correlated afterpulsing. The most suitable PMs, in order of merit, are the RCA and EMI. The FW 130 is slower, but it has a small photocathode, which is uniquely suitable for photon correlation measurements.

3. For low-light-level detection, the dark count rate per unit quantum efficiency must be as low as possible. In this respect, FW 130 and RCA C31034 tubes have the edge.

Coherence-area limitations often involve the use of apertures of less than 0.05 cm. Thus, the FW 130 tube has a unique advantage, since effective cathode diameters of 0.02 cm can be achieved. The room-temperature dark count rates for such a tube are about 1 count/sec. On the other hand, the FW 130 tube has a relatively low quantum counting efficiency and collection efficiency. The search for better PM tubes goes on, but most such studies are not concerned with correlated afterpulses (Lakes and Poultney, 1971; Birenbaum and Scarl, 1973; Poultney, 1972; Reisse et al. 1973). It should be noted that evaluation of PM tubes is more complex than it appears (Young and Schild, 1971).

In conclusion, several PMs are suitable for photon counting and photon correlation. The RCA C31034, the EMI 9863, and the ITT FW 130 PMs are appropriate for laser light scattering using visible wavelengths $\lambda_0 > 600$ nm, while the same tubes, as well as the EMI 9893 and the Hamamatsu R464 PMs, are suitable for blue–green wavelengths. The general rule is to use a bialkali photocathode tube for the blue–green light, because it is often cheaper and has a lower dark count rate without any need to cool the PM. At low light levels, it is advisable to buy a PM selected for low dark count rate, as well as for low afterpulsing.

6.5.3. PHOTOMULTIPLIER HOUSING DESIGN

Zatzick (1971) has written a very good discussion of the requirements for an effective PM housing to be used with a high-gain wide-band photon counting system. PM housings and circuits designed for analog measurements are usually not suitable for use in photon counting. On the other hand, housings and dynode voltage-divider circuits designed for photon counting can be used for current measurements, the only precaution being that much lower anode currents are permissible for linear performance. The essential features for good PM housing are summarized as follows.

6.5.3.A. VOLTAGE-DIVIDER CIRCUIT

A typical schematic and wiring diagram of an ITT FW 130 PM-tube voltage-divider circuit is shown in Fig. 6.5.3. The reader may worry that the circuits for the voltage divider are very old and wonder what has happened to more recent versions. The fact is, nothing much has been changed. The old circuits are retained so that when the "new" divider circuit is obtained from the manufacturer, one can make a comparison to see that they are about the same. Furthermore, photomultiplier tube housings with the voltage dividers are directly available from PM manufacturers such as EMI. The input high-voltage line is filtered with a 1-MΩ resistor and a 0.001–0.1-μF capacitor, giving an attenuation of 10^{-4}–10^{-6} at 1 MHz. The zener diode (IN 992B) is

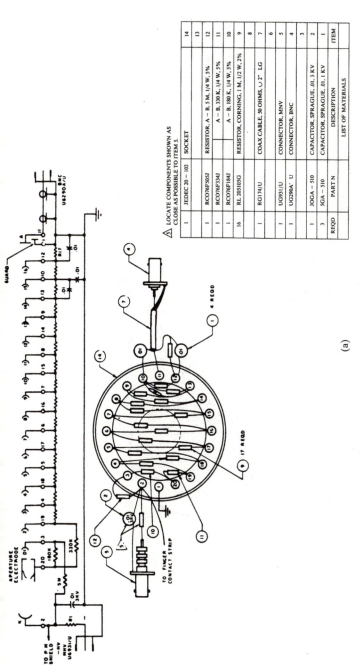

		SOCKET	
			14
			13
1	RCO76F505J	RESISTOR, A — B, 5 M, 1/4 W, 5%	12
1	RCO76F334J	A — B, 330 K, 1/4 W, 5%	11
1	RCO76F184J	A — B, 180 K, 1/4 W, 5%	10
16	RL 20510SG	RESISTOR, CORNING, 1 M, 1/2 W, 2%	9
			8
1	RG174/U	COAX CABLE, 50 OHMS, ∪ 2" LG	7
1	UG931/U	CONNECTOR, MNV	6
1	UG290A' U	CONNECTOR, BNC	5
			4
1	JOGA – 510	CAPACITOR, SPRAGUE, .01, 3 KV	3
3	SGA – 510	CAPACITOR, SPRAGUE, .01, 1 KV	2
			1
REQD	PART N	DESCRIPTION	ITEM
		LIST OF MATERIALS	

⚠ LOCATE COMPONENTS SHOWN AS
CLOSE AS POSSIBLE TO ITEM 5.

JEDEC 20 – 102

(a)

Fig. 6.5.3.   Suggested schematic and wiring diagram for ITT FW 130 photomultiplier when used with the SSR 1100 series photon-counting system (M. R. Zatzick, SSR Instrument Co., Application Note 71021). Dropping resistors are 1 MΩ, $\frac{1}{2}$ W, 2%; tin oxide; capacitances are in microfarads; K means kΩ; lead lengths to be as short as possible. (a) Typical grounded-anode configuration. (b) Typical grounded-cathode configuration: can be used only in the photon-counting mode and eliminates the sensitivity of the PMT focusing to potential differences between the photocathode and the grounded assembly at the PMT faceplate (M. R. Zatzick, private communication).

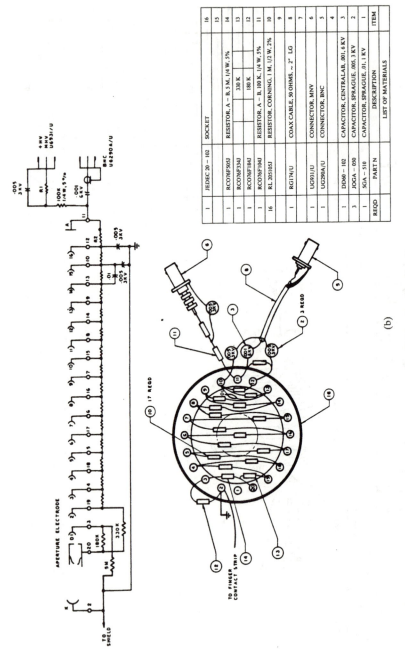

16		
15		
14	RCO76F505J	RESISTOR, A – B, 5 M, 1/4 W, 5%
13	RCO76F334J	330 K
12	RCO76F184J	180 K
11	RCO76F104J	RESISTOR, A – B, 100 K, 1/4 W, 5%
10	RL 2051051	RESISTOR, CORNING, 1 M, 1/2 W, 2%
9		
8	RG174/U	COAX CABLE, 50 OHMS, ~ 2" LG
7	UG931/U	CONNECTOR, MNV
6	UG290A/U	CONNECTOR, BNC
5		
4		
3	DD60 – 102	CAPACITOR, CENTRALAB, .001, 6 KV
2	JOGA – 050	CAPACITOR, SPRAGUE, .005, 3 KV
1	SGA – 510	CAPACITOR, SPRAGUE, .01, 1 KV
1	JEDEC 20 – 102	SOCKET
REQD	PART N	DESCRIPTION
		LIST OF MATERIALS
		ITEM

Fig. 6.5.3. (continued)

used to establish a constant potential between the cathode and the first dynode. Older-type zeners are high-temperature blackbody emitters and should therefore be coated with opaque material. The bypass capacitor is highly recommended, since zeners operating in the avalanche mode are noise sources. The use of 1-MΩ resistors in each dynode stage assures a divider current of about $10^3/(10 \times 10^6) = 100$ μA for tubes with ten (or more) stages. Thus, a typical maximum anode current with stable linear performance and without tube fatigue should be less than 5 μA.

One can maintain linear response with increasing maximum anode current by reducing the dynode resistance per chain. However, dissipation of heat in the base region will then increase and may result in higher dark count rates. Corning glass–tin oxide resistors are often used because of their good temperature stability and low noise. Disc ceramic or equivalent high-frequency capacitors of approximately 0.01 μF with an excess voltage rating of 1 kV are used to bypass the last three dynodes to ground in order to minimize radio-frequency interference pickup via the dynode-to-anode capacitance and to sustain a constant potential at the latter dynodes during high counting rates. The capacitors are grounded at the shield of the anode coaxial lead. The lead length for all components should be kept as short as possible. Double-shielded RG55A 50-Ω coaxial cable, matched at both ends to reduce reflection, is used to connect the anode with the (pre)amplifier. The base design should give the optimum pulse shape without a long-time-constant back edge to the pulse. Variation in the position of the same components has been known to change the pulse-counting properties, or even the pulse-height distribution. Thus, good base construction is very important.

6.5.3.B. ELECTROMAGNETIC SHIELDING

Each penetration of the housing for signal and power decreases its isolation from electrical interference. A fairly heavy-wall electrostatic shield with an inner magnetic shield can provide good radio-frequency shielding. However, sharp edges in the electroshield should be avoided, and good insulation between the Mu-metal shield and the outer ground shell is essential, as corona may become a problem.

6.5.3.C. CHOICE OF HOUSING MATERIAL

Aluminum for the outer shell and Plexiglas or Teflon for the insulators are recommended. The polymers should be tested for electroluminescence properties, and use of lint-producing materials should always be avoided. The housing subassembly should be properly grounded without electrical contacts through paint or anodized material, since oxides are poor conductors.

6.5.3.D. CLEANLINESS AND OTHER CONSIDERATIONS

Leakage is generally created by a film of moisture, often caused by finger marks, on the tube base or socket. The leakage current of any good tube can be made negligibly small by keeping the tube and socket assembly dry and clean. The socket assembly can be potted to avoid contamination by moisture and other material. It is advantageous to avoid using lenses and windows in order to reduce scattering, reflection, and transmission losses.

The above considerations clearly demonstrate that the construction of a good tube housing is a tricky undertaking. Fortunately, reasonable commercial housings, such as those made by Thorn EMI and Products for Research, are available (not an endorsement). Home construction of coolable tube housings that require additional penetrations from electrical isolation is more difficult. Therefore, my advice is simply to buy from a reputable manufacturer.

6.5.4. AMPLIFIERS, DISCRIMINATORS, AND SCALERS

The configuration of a detection system for measurements of time-averaged intensity is shown in Fig. 6.5.4. Generally, a single-channel analyzer leads to lower count-rate capability than a discriminator. The amplifier should have a rise time comparable with that of the tube, a gain sufficient to make the

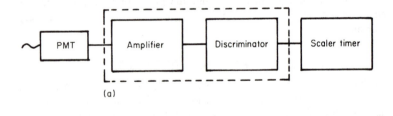

(a)

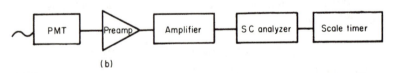

(b)

Fig. 6.5.4. Block diagram for photon counting of time-averaged intensity. Typical manufacturers for fast (~100 MHz) systems: LeCroy Research Systems, EG&G/Ortec, EG&G/PAR. (a) Amplifier–discriminator in combined unit. (b) A preamplifier is used. The cable between PMT and preamplifier should be very short and doubly shielded. With the preamplifier, the amplifier–discriminator can be located at large distances from the PMT ($\equiv$ PM).

single-channel analyzer or discriminator operational, and low correlations or distortions in its output. If the amplifier gain is insufficient, two amplifiers can be cascaded. In cascading amplifiers, the amplifier noise can become a problem. Foord et al. (1969) pointed out that the PM, which provides the best high-gain, wide-band amplification, should be operated at high supply voltage in order to minimize external amplification. Photomultiplier gains in the range $10^6 - 10^8$ with amplifier gains of 10–100 yield appropriate pulse heights for most single-channel analyzer or discriminator use.

The discriminator should be set near maximum sensitivity so that the overall gain of the combined photomultiplier–amplifier system is kept at a minimum. Young (1969) mentioned that any nonlinear weighting function results in limited dynamic range. In photon counting, the nonlinearity after corrections for background arises from pulse overlap. Present-day technology permits 100-MHz system without excessive expense, and such a system is quite adequate for most purposes.

Jonas and Alton (1971) observed that, in the photon counting mode at a constant discriminating bias, the signal-to-noise ratio (S/N) improved with increasing operating voltage in a box-and-grid-structure PM such as the ITT FW 130, but changed little in a venetian-blind-structure tube such as the EMI 6256S. On the other hand, S/N remained relatively constant over a supply-voltage range of 1800–2000 V with the dc method of detection for the FW 130 PM.

In depolarization studies, it is important to realize that the photocathode sensitivity, which usually varies across the photocathode surface, can depend upon the polarization of the light. The surface polarization effects are wavelength-, temperature-, and voltage-dependent. Furthermore, the polarization also depends on the angle at which the light strikes the photocathode surface.

6.5.5. SEARCH FOR OPTIMUM OPERATING CONDITIONS IN A PHOTON COUNTING SYSTEM

The photomultiplier is a low-noise, high-gain, broad-band amplifier. In a photon counting system, we consider the photomultiplier–amplifier–discriminator as an integral unit, because our final aim is to change photons to photoelectron pulses of prespecified width and pulse height so that they can be counted by a scaler over some fixed time intervals. Efficient operation of the photomultiplier requires us to select (1) an appropriate operating voltage V for the photomultiplier, together with (2) an amplifier gain G so that the signal due to photoelectron pulses can be distinguished from other noises, using (3) a discriminator at an appropriate level D_V, and counted by a scaler. The three adjustments (V, G, D_V) are coupled, and a variety of procedures exist for searching for an optimal setting for the photon counting system. The following procedure represents one approach, which was devised by H. Dhadwal and used satisfactorily in our laboratory.

We want to be able to achieve the following two general conditions: The photomultiplier should be set at fairly high operating voltages because it is a very good low-noise amplifier. The discriminator should be set near maximum sensitivity so that the overall gain of the combined photomultiplier–amplifier system is kept at a minimum in order to reduce unwanted electronic noise. However, when we start the adjustments, we set the discriminator level fairly low (≈ 200 mV) so that all the pulses are counted by the scaler. In a plot of $\langle n(t)\rangle$ versus V, as shown schematically in Fig. 6.5.5, there is usually a plateau region, in which $\langle n(t)\rangle$ is relatively insensitive to the supply-voltage fluctuations. In the plateau region, there is a range of supply voltage that gives the lowest dark count, as shown schematically by V_0. This will then be the first approximation to the operating voltage of the photomultiplier for the present amplifier gain.

The discriminator setting can best be examined using a multichannel analyzer. It is easier, however, to examine a plot of $\langle n(t)\rangle$ versus discriminator level and note that there are three distinct regions, as shown schematically in Fig. 6.5.6. In region I, the counts are excessive because shot-noise pulses are being counted; II is a plateau region, and in III the signal pulses are being clipped. One can usually observe regions I and II. The ideal setting is to set the discriminator level close to I but in region II. The two steps may be repeated by using a different amplifier setting. The optimal setting is for V and D_V

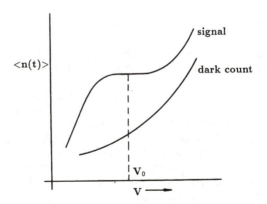

FIG. 6.5.5.   Intensity vs supply voltage for a PMT exposed to white light (signal) and in the dark. $\langle n(t)\rangle$, average number of counts per sample time.

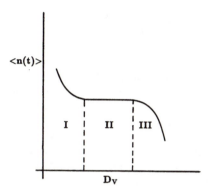

FIG. 6.5.6.　Intensity vs discriminator level setting. $\langle n(t) \rangle$, average number of counts per sample time.

to be insensitive to supply-voltage or discriminator-level fluctuations, using an amplifier gain that is just enough to have the scaler count all the signal pulses.

For photon correlation, a white-noise correlation from a dc lamp is accumulated using various delay-time increments in order to assure a flat response in the intensity–intensity time correlation function.

6.5.6. AVALANCHE PHOTODIODES

Avalanche photodiodes (APDs) have been used successfully in the photon-counting mode for photon correlation measurements. They are usually manufactured from silicon or germanium, depending on the required spectral sensitivity. Si diodes are selected because their spectral range—from mid-visible through ~ 1-μm wavelength—is especially suitable for laser light scattering experiments. Laser diodes and APDs are closely matched in emission wavelength and spectral response, whereas most photomultipliers are not efficient in the 750–800-nm range of convenient operating wavelengths of present-day laser diodes. More importantly, laser diodes, fiber-optic probes and Si APDs together form the essential small, (eventually) cheap, and rugged components for miniaturization of future light-scattering spectrometers. Thus, its development cannot be ignored, even though the use of APDs for photon correlation as a stock item is only in its initial stage.

APDs have been available commercially for some years. They have already found applications in the telecommunications industry and may be purchased from RCA, NEC, Hamamatsu, and other sources. Characterizations of (1) passively quenched (Brown *et al.*, 1986) and (2) actively quenched (Brown *et al.*, 1987) silicon avalanche photodiodes for photon correlation measurements have been reported. Brown and his coworkers used a selection of RCA C30921S silicon APDs and examined their performance in terms of magnitude of dark counts, correlations, afterpulsing effects, quantum efficiency, and stability of counting rates as a function of temperature and operating voltage. The interested reader is advised to study the two papers by Brown *et al* (1986, 1987) and references therein for details.

The outcome of their investigations is summarized in Table 6.5.2. The conclusion is that an actively quenched APD, when cooled and operated at ~ 4 V beyond breakdown voltage, has a photodetection performance similar to photomultipliers. Brown *et al.* (1987) could achieve afterpulsing performance of better than 0.047% and a dead time of ≈ 40 nsec using a quench delay time of 6 nsec. Further improvements can be expected, though the present setup is already adequate and competitive with photomultipliers for laser light scattering experiments under most operating conditions (with $\Delta\tau > 50$ nsec). Higher quantum efficiency can be achieved by operating the diode at an increased voltage, in excess of breakdown. Thus, efforts will be made for adequate thermally stabilized cooling in order to reduce the dark count rate, preferably by another factor of ten or more.

Table 6.5.2

APD Characteristics

	Passively quenched[a]	Actively quenched[b]
Dark count rate, $-20°C$ (counts per second)	A few hundred	A few hundred
Quantum efficiency, 633 nm (%)	≈ 7.5[c]	≈ 9[d]
Dead time (μsec)	≈ 1	≈ 0.04
Maximum photodetection rate (MHz)	≈ 0.25	≈ 1

(a) Passive quenching: APD is placed in series with a ballast resistor, and, on photodetection, the voltage drop across the resistor quenches the breakdown. In the passive quenching operation, the APD is recharged through the ballast resistor with time constant RC, where C is the APD capacitance.

(b) Active quenching: To reduce the dead time, the APD capacitance should be recharged as quickly as possible after the avalanche has been quenched. Active quenching circuits can be used to increase the recharging rate.

(c) APD operating voltage was 3 V in excess of its breakdown voltage.

(d) APD operating voltage was 4 V in excess of its breakdown voltage.

6.6. CURRENT DETECTION

6.6.1. CURRENT DETECTION VS PHOTON COUNTING

The photon-counting technique is the most efficient detection method (Young, 1969; Jones *et al.*, 1971), though some, such as Robben (1971) and Alfano and Ockman (1968), have indicated otherwise. The work by Jones *et al.* (1971) was stimulated by the results of Rolfe and Moore (1970), who draw experimental conclusions with respect to the merits of various techniques not acceptable to Jones *et al.* When the PMs are operated under noise-in-signal-limited conditions with negligible dark count rates, Jones *et al.* (1971) showed that the photon-counting technique was superior to current measurements by a factor of 2.6 in the duration time required to achieve the same variances. They operated the ITT FW 130 with a cathode-to-dynode-1 voltage of 300 V (Barr and Eberhardt, 1965) and a dynode-1-to-anode voltage of 1800 V in order to achieve an optimum pulse-height distribution. Regardless of current arguments and discussion on the advantages and disadvantages of various techniques, it is advisable to set up the detection system for pulse counting and photon correlation rather than current measurements. However, there is an exception, i.e., at low scattering angles, when the scattered intensity from large-particle scattering is strong, the photoelectron pulses tend to overlap, and characteristic times are long. Then, current detection together with FFT power spectral analysis can be more appropriate.

A photomultiplier is essentially an ideal current source, the current being proportional to the incident photon flux. It has a gain of about 10^6 with a bandwidth of $\gtrsim 100$ MHz. It must be emphasized that this high gain is essentially noise-free; nothing flows through the output load resistor unless that first electron starts down the multiplier chain. For a detailed discussion on noise processes, the reader is referred to van der Ziel (1986).

For dc and low-frequency applications, a load resistor can be used to convert the photocurrent into a voltage as shown in Fig. 6.6.1. There are

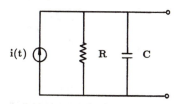

FIG. 6.6.1. Equivalent circuit for a photomultiplier. R is effectively the load resistance, since the output impedence of the photomultiplier is $\sim 10^{12}$ Ω. C is the stray shunt capacitance (5–20 pF, depending on tube type and circuit layout).

practical limitations on the size of the load. Two undesirable effects can arise due to large values of the load resistor. Firstly, if the voltage drop across the load resistor becomes too large, the potential between the last dynode and the anode can fall and thus affect the linear relationship between anode current and incident photon flux. Secondly, the load resistor affects the frequency response of the photomultiplier. This is due to the unavoidable stray capacitance that exists across the load, the bandwidth being

$$\Delta f = \frac{1}{2\pi RC}.$$

In the photon-counting regime, an amplifier designed for an analog-frequency bandwidth encompassing the highest frequency component in the signal will lead to distortions of the photoelectron pulses.

For a better understanding of the effects loading the photomultiplier, the pulse response of the circuit of Fig. 6.6.2 should be examined. In order to determine the pulse response it is necessary to assume that the incident light decays with a single time constant, τ_s. The anode photocurrent is given by

$$i(t) = \frac{NGe}{\tau_s} \exp\left(-\frac{t}{\tau_s}\right). \tag{6.6.1}$$

where G is the gain of the photomultiplier; N, the number of photoelectrons per second; and e, the electronic charge. The impedance of the loading network is

$$Z(S) = \frac{R}{1 + SRC}, \tag{6.6.2}$$

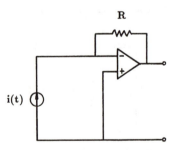

FIG. 6.6.2.   Equivalent circuit for dc or low-frequency current operation.

where S is the Laplace operator ($\equiv i\omega$). If $I(S)$ and $V(S)$ are the Laplace transforms of $i(t)$ and $u(t)$, respectively, then

$$V(S) = I(S)Z(S) = \frac{R}{1 + SRC} \frac{NGe}{\tau_s} \frac{1}{S + (1/\tau_s)} \tag{6.6.3}$$

If $\tau\,(=RC)$ is the output time constant, Equation (6.6.3) is reduced to

$$V(S) = \frac{NGeR}{\tau_s\tau} \frac{1}{S + (1/\tau_s)} \frac{1}{S + (1/\tau)}. \tag{6.6.4}$$

Inverse Laplace transformation of Eq. (6.6.4) yields

$$V(t) = \frac{NGeR}{\tau - \tau_s}(e^{-t/\tau_s} - e^{-t/\tau}), \tag{6.6.5}$$

where τ includes the anode load resistance and any external combination of R and C in parallel and coupled to the photomultiplier output. Equation (6.6.5) has several regions of interest:

(1) For $\tau \ll \tau_s$, $V(t) = (NGeR/\tau_s)e^{-t/\tau}$, i.e., the output-voltage time profile follows the input-current time profile because $e^{-t/\tau}$ has a much faster rise time than e^{-t/τ_s}. The anode current pulse is faithfully reproduced as a voltage pulse across the load resistor.

(2) For $\tau \gg \tau_s$, the pulse charge is integrated. In the extreme case of R very large, $V(t) = (NGe/C)(e^{-t/\tau_s} - 1)$; $i(t)$ is simply charging C, and the voltage rise time is equal to the input decay time. In the limit $R \to \infty$, the maximum signal amplitude is obtained.

(3) A long time constant τ is suitable for low count rates; but if the rate $\sim 1/\tau$, pulse pileup occurs. Thus, for dc or low-frequency current operations, R should be chosen such that the time constant provides the correct integration time.

A voltage-to-frequency converter with a 1-μsec sampling time, as shown in Fig. 6.6.3, has been used to process analog signals for correlation-function measurements by means of a digital photon correlator. The reader is referred to the original paper (Terui et al., 1986) for details. It is sufficient to note that Fig. 6.3.3(b), working as a V/F converter at 25 MHz with a 1-μsec sampling speed and a higher cutoff frequency, works much better. A limiting factor in using a V/F converter is the slow rate of the device.

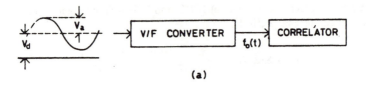

(a)

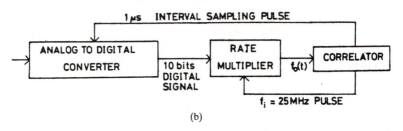

(b)

FIG. 6.6.3. Methods of using a digital photon correlator as an analog correlator. To convert the fluctuations of input analog signals to the fluctuations of the time density of TTL pulse signals, two methods were used. A commercial V/F converter was used in the case (a). In case (b), a fast A/D converter and rate multiplier were combined to work as a V/F converter. (Reprinted with permission from the Japanese Journal of Applied Physics, after Terui *et al.*, 1986.)

6.7. METHOD TO COMPENSATE LASER FLUCTUATIONS IN PHOTON CORRELATION SPECTROSCOPY (OHBAYASHI *ET AL.*, 1986)

Two types of unwanted fluctuations, i.e., the incident laser power fluctuation and the scattered light intensity fluctuation due to the presence of dust particles passing through the scattering volume, may deform the measured time correlation function, especially at slow delay-time increments. This *instability of the laser power* can be compensated at relatively high speeds using a rate multiplier that divides the scattered intensity I_s by the incident laser intensity over the sampling time interval of interest.

Figure 6.7.1 shows a schematic block diagram of photon correlation spectroscopy (PCS) with a circuit to compensate the laser power fluctuation. In the present experimental condition, we emphasize that the scattered intensity $I_s(t)$ is proportional to the incident laser intensity $I_i(t)$ ($\equiv I_{INC}(t)$), which is no longer constant, i.e.,

$$I_s(t) = \alpha(t)I_i(t), \tag{6.7.1}$$

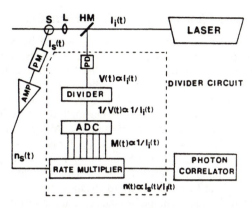

FIG. 6.7.1.   A schematic block diagram of photon correlation spectroscopy with a circuit to compensate the laser power fluctuation. HM, glass plate used as a half mirror; L, lens; S, sample; PM, photomultiplier tube; AMP, amplifier and discriminator; PD, photodiode. For details of the divider circuit enclosed in the broken lines, see original reference. (After Ohbayashi *et al.*, 1986.)

where $\alpha(t)$ is the relative scattered intensity, involving information concerning the desired local fluctuations. Without laser power fluctuations, PCS measures $\langle I_s(t)I_s(0)\rangle = \langle \alpha(t)\alpha(0)\rangle$. With laser fluctuations, PCS measures $\langle I_s(t)I_s(0)\rangle$; but what we want is $\langle \alpha(t)\alpha(0)\rangle = \langle [I_s(t)/I_i(t)][I_s(0)/I_i(0)]\rangle$. The compensation divider circuit in Fig. 6.7.1 tries to perform this operation and has a bandwidth of ≈ 180 kHz. It should be noted that the scattered intensity fluctuation is compensated in the circuit before correlation-function computations. The reader should consult the original reference for the details of such a circuit, which may be useful for different types of correlators with appropriate adjustments.

There are routines for dust rejection, especially when the correlator is controlled by a computer. The idea is that when a dust particle enters the scattering volume, the measured scattered intensity, which now includes the additional scattering from the dust particle, is increased substantially. When the counter sees this increase, the computer instructs the correlator to stop the measurement. A delay time, which is determined by the dust-particle size and the scattering-volume dimensions, is set for the correlator to start again. Thus, dust discrimination can work, but requires knowledge of the dust particles, the dimensions of the scattering volume, and the nature of the time correlation function to be measured. It does not work for really dirty solutions and involves subjective discrimination. One should always try to clarify the solution first whenever possible.

6.8. FABRY–PEROT INTERFEROMETER

This section is included for completeness. It should be passed over by readers who are primarily interested in photon correlation spectroscopy. The discussion centers on work dating back to the 1970s. There has been little new development in instrumentation principles on Fabry–Perot (FP) interferometry. However, the reader may wish to consult the latest products from Burleigh (New York) and Queensgate Instruments Ltd. (Berkshire, England). FP interferometers can be purchased ready made. Many commercial firms make FP interferometers that are quite adequate for laser light scattering.

In interferometry, as in optical homodyne or heterodyne spectroscopy, rigidity of the entire optical setup, isolation from mechanical vibrations, and high-quality optics are helpful in order to ensure the success of the experiment. The FP interferometer should be made of materials with a low thermal coefficient of expansion, such as Invar or Superinvar. In this respect, stainless steel is better than aluminum. Sometimes, the FP interferometer is thermally compensated, or situated in a constant-temperature environment. The FP etalon spacer *must* be made of materials with a low thermal coefficient of expansion.

Wavelength variations of a scanning FP interferometer can be achieved by changing (1) the index of refraction between the interferometer plates and (2) the separation distance between them. In fact, many systems have been used or proposed for the scanning of etalons (see *J. Phys. Radium* 19, 185–436 (1958), and Jacquinot, 1960), including variation of the plate separation distance by thermal expansion of the spacer (Burger and van Cittert, 1935), electromagnetic attraction (Bruce and Hill, 1961), magnetostriction (Bennett and Kindlmann, 1962), piezoelectricity, bellows action under pressure, and precision screws. The other common approach varies the ambient gas pressure and thus the refractive index between the etalon plates instead of the etalon geometry. These two methods are by far the simplest and most widely used.

The principle of scanning the wavelength by changing the index of refraction between the interferometer plates was first used by Jacquinot and Dufour (1948), and later by Connes (1956, 1958) and Hindle and Reay (1967). The details may be found in a paper by Chabbal and Jacquinot (1961). Earlier, Biondi (1956) designed a high-speed direct-recording FP interferometer. In the pressure scanning technique, the spacers for the FP etalon can be very crucial. Phelps (1965) described procedures for making spacers, but it is advisable to purchase such spacers, which are usually made of quartz or Invar, from reliable commercial sources.

Piezoelectric scanning of FP interferometers has become the standard method. Piezoelectric elements made of barium titanate (Cooper and Greig, 1963; Peacock et al., 1964) and of lead zirconate (Fray et al., 1969) have been used successfully. Fork *et al.* (1964) have constructed a scanning spherical mirror interferometer for spectral analysis of laser radiation. Their interferometer used a moving-coil driving system of the Tolansky–Bradley (1960) type, which permitted mirror motions to over 70 μm. Stacking of piezoelectric elements also permits sufficient scanning range, but care should be exercised in testing that the movable mirror can be translated in the axial direction without tilting. In addition, piezoelectric scanning—unlike pressure scanning, which needs an etalon spacer for each fixed cavity length—utilizes the same scanning elements over large ranges of cavity length.

Jackson and Pike (1968) used a stack of six 50-mm-diameter lead zirconate titanate disks (Mullard MB 1019 PXE5) with a 2-cm aperture as their scanning elements. The piezoelectric rings, separated by pyrophylite or glass annular spacers, are glued together with Araldite 103 and are wired in parallel as shown in Fig. 6.8.1. The FP mirror is mounted on three small feet glued to the free end of the stack. The system can be scanned at low sweep rates (< 50 Hz) in order to avoid mechanical resonances. The spacings of Brillouin doublets scattered from liquids are references for calibration. The dc bias controls the fine adjustments in plate separation distance, while the MCA

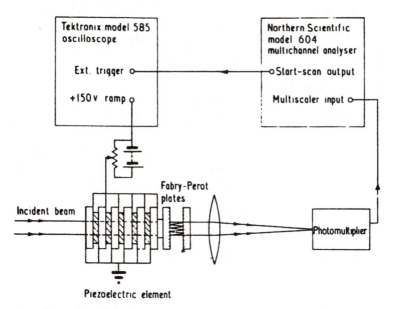

FIG. 6.8.1. Block diagram of a multichannel scanning system. (After Jackson and Pike, 1968.)

can be internally programmed so that it scans through a preset number of channels and returns to channel 1 in a time interval just shorter than the ramp-plus-flyback time of the oscilloscope. Repetition is achieved by subsequent triggering of the system.

Further improvements, including servo control of drifts of the laser frequency and of the FP cavity length, have been achieved (Fray *et al.*, 1969). Figure 6.8.2 shows the schematic block diagram of their servo-controlled digital FP interferometer. A dc level is applied from a servo loop to stabilize the spectrum. This servo loop is controlled by obtaining a reference spectrum of the laser line on alternate cycles of the sawtooth that drives the piezoelectric system. The laser beam, after passing through the sample, is returned through the FP interferometer via a mechanical chopper every second cycle.

The servo loop has been very cleverly designed. *A* and *B* can be logic units (flip-flops) that turn on either the MCS scan or the laser beam. When the laser

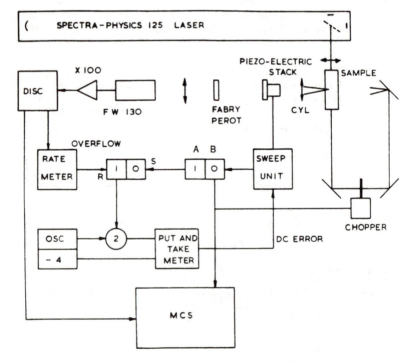

FIG. 6.8.2. Block schematic of a scanning servo-controlled digital Fabry–Perot interferometer. (After Fray *et al.*, 1969.)

beam is on, the oscillator signal and the oscillator signal divided by 4 are started by the flip-flop trigger as shown in Fig. 6.8.3. Then the oscillator signal is stopped by the rate-meter overflow, which is set at maximum of the laser spectrum. We visualize the function of the put-and-take meter as follows. The oscillator signal counts accumulated over the time interval $L - S$ are compared with the (oscillator signal)/4 counts accumulated over the time interval $T - S$. The difference signal can be latched, converted to a voltage by a digital-to-analog converter (DAC), amplified, and fed back to the piezo-electric stack on top of the constant-amplitude sawtooth with the correct polarity (negative feedback) to complete the servo loop. The dc output of the put-and-take meter corresponds to $V/4$ of the oscillator signal when the laser peak is located at $\frac{1}{2}(T - S)$. This output is integrated over a number of cycles for better S/N and will move in one direction or the other depending upon whether the laser line appears later or earlier than halfway across the reference sweep.

Such a setup removes drifts of the laser frequency and the FP cavity length at rates lower than the reciprocal of the integration time of the servo loop. It should be noted that variation of the incident laser intensity during each sweep will be viewed by the servo loop as drifts of the laser frequency and the FP cavity length because of the overflow setting at the rate meter. In fact, a significant decrease in laser intensity should make the servo loop inoperative by causing the overflow not to be triggered at all. Thus, the laser must be stabilized for constant intensity.

An improvement of the above scheme has been accomplished by Pike and his coworkers at Malvern, England. Basically they increase the scan rate during the reference sweep by a factor of 5–10. Thus, an 80–90% efficiency can be achieved, and the Rayleigh line also acts as a trigger for the signal sweep. The variable scan rate permits magnification of the spectrum over regions of interest.

Details of a simple zero-crossing signal-averaging system for FP inter-ferometry are available (Lao *et al.*, 1976). In addition, Sandercock (1970) has designed a servo system in which the cavity length as well as the tilt of the FP mirrors can be adjusted automatically. The servo for the tilt uses a four-stroke approach, where the criterion of the maximum Rayleigh peak intensity controls the parallelism of the FP mirrors. Hicks *et al.* (1974) have developed a servo-controlled FP interferometer using capacitance micrometers to sense departures from parallelism and variations in the mean cavity length to an accuracy better than the surface quality of the plates ($\approx \lambda/150$ at 500 nm). McLaren and Stegeman (1973) have reported a digital stabilizer specifically designed to eliminate the effects of long-term frequency shift.

The application of piezoelectric transducers to spectral scanning using FP interferometers has also been reported in the reviews of Vaughan (1967) and

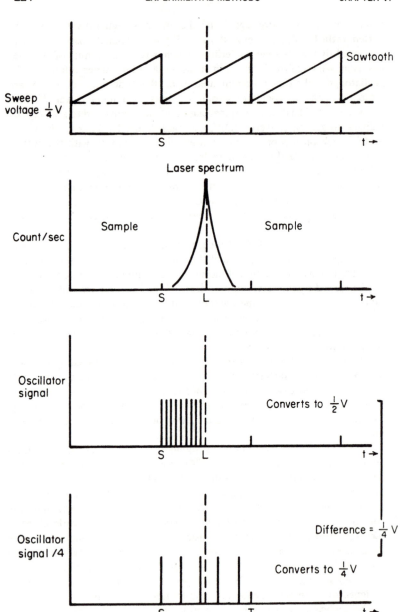

FIG. 6.8.3. A schematic indication of the servo loop that removes drifts of the laser frequency and of the Fabry–Perot cavity length.

Greig and Cooper (1968). Astronomers have made valuable contributions in improving the parallelism adjustment as well as the long-term stability of FP spectrometers. In the servo loop designed by Fray et al. (1969), the two etalon plates are mounted independently and parallelism adjustment is achieved by alignment of the stationary etalon plate. While finely controlled mirror mounts with high accuracy in alignment and good stability exist (Simic-Glavaski and Jackson, 1970), automatic control of both the parallelism and the spacing of FP interferometers has been devised (Ramsay 1962, 1966; Ramsay and Mugridge 1962a, b; Hicks et al., 1974). The reader should consult the references for practical details of the various components in constructing such electronically controlled FP spectrometers. Smeethe and James (1971) have successfully adopted the active servo system of Ramsay with an optical–electrical loop where the optical path of the servo loop requires a truncated pyramid at the rear edge of one of the two etalon plates. Bates et al. (1971) have described a stable, high-finesse scanning FP interferometer with different piezoelectric ceramics for plate parallelism adjustment and spectral scanning. Control of the mirror position to within 5×10^{-5} wavelength has been achieved by the use of an auxiliary FP interferometer (Tuma and Van der Hoeven, 1973). Hernandez and Mills (1973) have been able to stabilize their finished instrument with respect to the reference wavelength and parallelism to approximately $\lambda/1000$ at 546 nm for a 6-hr period. Problems exist concerning the linearity of the sweep, the constancy of finesse over the sweep, and long-term stability, all of which are essential for high-resolution studies of weak lines. Cooper et al. (1972) have constructed a digital pressure-scanned FP interferometer where repetitive scans are reproduced within $\frac{1}{640}$ of the spectral free range. Their spectrometer is simpler in electronics and design and has other virtues that should not be overlooked in comparison with the more sophisticated piezoelectric scanned systems.

Double- or multiple-pass FP interferometers with very great contrasts, instead of FP interferometers operated in tandem (Cannell and Benedek, 1970), have been developed to detect weak lines in a spectrum in the presence of nearby intense lines. Sandercock (1971) used retroreflective cubes to pass the light back and forth in different portions of the plates of a flat FP interferometer. However, the technique requires extremely precise alignment of the interferometer. An alternative approach is to place the interferometer in an optical isolator (Hariharan and Sen, 1961; Cannell et al., 1973). Cannell et al. have constructed a double-pass spherical FP interferometer that has an instrumental profile with a full width of 4.8 MHz.

In recent years, piezoelectric scanned FP interferometers, including multiple-pass options, have become readily available and can be purchased commercially; e.g., Burleigh Instruments, Inc., Burleigh Park, Fishers, New York, produces several high-quality models. Furthermore, the FP interferometer has been extended to the infrared range (Lecullier and Chanin, 1976; Belland and Lecullier, 1980) and used in a variety of applications (e.g., see bibliography compiled by Burleigh, 1986).

6.9. FIBER OPTICS

Optical fibers, whether single-mode, multimode, or in bundles, are important tools for the practice of laser light scattering (LLS) and will play an increasingly important role in LLS instrumentation. Although the use of fiber-optic probes has been discussed in Section 6.3.2, the early developments (Tanaka and Benedek, 1975; Dyott, 1978; Ross et al., 1978; Dhadwal and Ross, 1980; Auweter and Horn, 1985) involve mainly LLS in the backscattering ($\theta \approx 180°$) mode. The basic principles for using (multimode or bundle) optical fibers at scattering angles other than backscattering have been presented briefly in Section 6.3.2. Macfadyen and Jennings (1990) have reviewed fiber-optic systems for dynamic light scattering. However, in view of the importance of optical fibers in LLS instrumentation, a separate section is presented here. In this section, the use of monomode and multimode optical fibers, in combination with microlenses, is disucssed. Brown and his coworkers have also attempted to miniaturize LLS instrumentation as a whole from the light source to the detector, including fiber-optic detection schemes. Instead of standard cw argon ion (or He–Ne) lasers and photomultipliers, the use of laser diodes (see Section 6.2.4 and also Brown and Grant, 1987) and silicon avalanche photodiodes (see Section 6.5.6 and also Brown et al., 1986, 1987), respectively, as compact and reliable light sources and detectors has been examined. While laser diodes are not likely to exceed the specifications of argon ion lasers, their small cost and size may often recommend them for measurements where the power of an argon ion laser is not needed. Rather than He–Ne lasers, future visible-light laser diodes may very well be the dominating laser light source used in LLS instrumentation. Silicon avalanche photodiodes (APDs) look very promising and should be available on the market for photon-counting purposes very soon, if not by the time this book is published. At present, APDs still have higher dark count rates than good photomultiplier tubes; but they have very high quantum efficiencies ($\approx 40\%$). Thus, both visible-light laser diodes and APDs are likely to occupy useful roles in LLS experiments.

For the detection optics, the use of optical fibers was first introduced in laser anemometry (Knuhtsen et al., 1982; Jackson and Jones, 1986). In LLS, there are two complementary approaches to fiber-optic detection schemes. Brown (1987) and his coworkers (Brown and Jackson, 1987) prefer to use monomode optical fibers, while Dhadwal and Chu (1989) prefer to define the optics immediately before the light enters the fiber and consider the fiber mainly as an extremely low-loss transmission device ($\approx 8\%$ reduction in intensity over 1 km). The second approach does not preclude the use of monomode fibers, but uses them only if the optical geometry (e.g., the fiber diameter) happens to be suitable.

In the Brown (1987) approach, as shown in Fig. 6.9.1, the clean cleaved end face of a monomode fiber is placed at the focal point of the lens. The point-spread function (PSD) Airy disk of the lens ($d_s = 2.44\lambda f/D$ of Eq. (6.2.4)) is matched to the numerical aperture (NA) of the fiber. D is defined by the (field) stop in front of the lens, also shown as d_1 in Fig. 6.3.7(a). If the diameter of the microlens is sufficiently small, then the lens diameter acts as an effective D. When the NA of the fiber is matched to the lens, the fiber acts as a perfectly transmitting spatial filter. Furthermore, the monomode fiber propagates a pure mode of light without significant degradation of spatial coherence. However, the last advantage, like the preservation of polarization, is not crucial in the self-beating technique, which is the most popular method used in quasielastic light scattering. In Fig. 6.9.1, it is also interesting to note that placing a pinhole stop at the focal distance f of the lens is not exactly

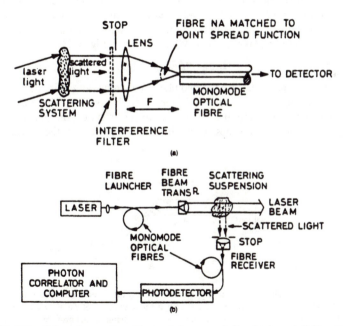

FIG. 6.9.1. Schematic operation of lens-ended monomode fiber for dynamic light scattering: (a) detail; (b) system for 90° scattering. (From R. G. W. Brown, *Appl. Optics* **26**, 4846, 1987).

equivalent to placing the monomode fiber at f, i.e., it is not helpful to put a pinhole (aperture stop) in front of the monomode fiber, especially if the aperture stop and the fiber are separated by some distance. As the core diameter of a monomode fiber is usually quite small, such a pinhole would, in any case, be difficult to insert, even immediately before the fiber. One may also view the fiber launcher in Fig. 6.9.1 as a spatial filter, and the fiber transmission optics as "symmetrical" to that of the fiber receiver, except that no field stop is to be used after the microlens focusing, since we want to preserve the power of the laser beam. For higher-power lasers, it is easier to avoid the use of fiber optics for the incident laser beam.

A single coherence area (A_{coh}), or speckle (Dainty, 1984), is defined by

$$A_{\text{coh}} = \lambda^2 R^2 / A_{\text{sca}}, \tag{6.9.1}$$

where A_{sca} is the cross-sectional area of scattering volume v. More specifically, the scattering volume is a three-dimensional source, and one or more of the three dimensions can change with the scattering angle θ. In Chapter 3, the coherence solid angle Ω_{coh} was derived for a rectangular scattering volume of fixed dimensions (Eq. (3.1.19)). In actuality, the scattering volume (SV = v) for most light scattering experiments has the optical geometry of a parallelepiped, as shown in Fig. 6.9.2. For an incident laser beam linearly polarized in the x-direction and propagating in the z-direction, and with the scattering plane confined in the $y-z$ plane (i.e., $\phi = 90°$ as shown in Fig. 6.9.3),

$$\Omega_{\text{coh}} = 4(\Delta\theta)_{\text{coh}}(\Delta\phi)_{\text{coh}}, \tag{6.9.2}$$

where $(\Delta\theta)_{\text{coh}}$ and $(\Delta\phi)_{\text{coh}}$ are the two planar coherence angles in the $y-z$ and $x-\mathbf{k}_s$ planes, respectively, and $\mathbf{k}_s$ is the scattering vector. The diffraction approach for computing the coherence angles is based on the fact that complete coherence vanishes when the relative phases of the wavefronts

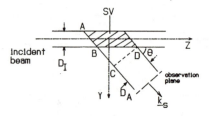

FIG. 6.9.2. A parallelepiped scattering volume defined by the intersection of the incident laser beam D_I and the detector entrance pupil D_A; SV is the scattering volume shown for the case $D_I < D_A$; θ is the scattering angle. The total relative phase change at the observer, due to fields originating from A and D, is $\Phi = AB + BC = \Delta_1 + \Delta_2 = D_I/\sin\theta + D_A/\tan\theta$. (After Dhadwal and Chu, 1989.)

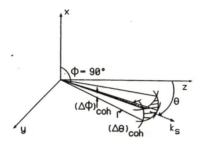

FIG. 6.9.3. Cartesian coordinate system centered in the scattering volume. Incident laser beam is polarized in the x-direction and propagates in the z-direction; θ, the scattering angle, lies in the y-z plane, since the angle between the direction of polarization and the scattering plane is $\phi = 90°$; the scattering vector $\mathbf{k}$ lies in the y-z plane, and $(\Delta\phi)_{\mathrm{coh}}$ is in the x-$\mathbf{k}$ plane. The coherence solid angle is given by $\Omega_{\mathrm{coh}} = 4(\Delta\theta)_{\mathrm{coh}}(\Delta\theta)_{\mathrm{coh}}$. (After Dhadwal and Chu, 1989.)

reaching the observer from any two points on the extended source change by $\pm\pi$ (see also Eq. (3.1.19)). By taking the incident laser beam to have a circular cross section of diameter D_I and for $D_\mathrm{I} < D_\mathrm{A}$ and $\phi = 90°$, the total relative phase Φ arising from two extremal points A and D, as shown in Fig. 6.9.2, is (after Dhadwal and Chu, 1989)

$$\Phi = \frac{2\pi}{\lambda}(\Delta_1 + \Delta_2), \tag{6.9.3}$$

where $\Delta_1 = D_\mathrm{I}/\sin\theta$ and $\Delta_2 = D_\mathrm{A}/\tan\theta$ with D_A being the diameter of the field stop. As Φ is a function of scattering angle θ,

$$\frac{d\Phi}{d\theta} = -\frac{2\pi}{\lambda}\frac{D_\mathrm{I}\cos\theta + D_\mathrm{A}}{\sin^2\theta} \tag{6.9.4}$$

Thus, with $\Delta\Phi$ varying by π,

$$(\Delta\theta)_{\mathrm{coh}} = \Delta\Phi\left(\frac{d\Phi}{d\theta}\right)^{-1} = \frac{\lambda\sin^2\theta}{2(D_\mathrm{A} + D_\mathrm{I}\cos\theta)}, \tag{6.9.5}$$

and similarly, in the x-$\mathbf{k}_s$ plane,

$$(\Delta\phi)_{\mathrm{coh}} = \lambda/2D_\mathrm{I}. \tag{6.9.6}$$

By combining Eqs. (6.9.5) and (6.9.6) with Eq. (6.9.2) (which is Eq. (3.1.19)), we get

$$\Omega_{\mathrm{coh}} = \frac{\lambda^2\sin^2\theta}{D_\mathrm{I}(D_\mathrm{A} + D_\mathrm{I}\cos\theta)}. \tag{6.9.7}$$

Equation (6.9.7) is *different* from Eq. (3.1.19).

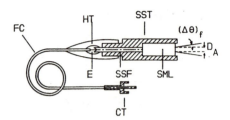

FIG. 6.9.4. Schematic of a typical fiber-optic detector probe. SST, machined piece of cylindrical stainless steel; SML, SELFOC microlens; SSF, stainless-steel or ceramic ferrule used for mounting the bare optical fiber; E, epoxy used for holding fiber in ferrule; HT, heat-shrink tubing; FC, fiber cable; CT, SMA type-II male connector. D_A and $(\Delta\theta)_f$ are the effective detector aperture and divergence angle, respectively, as defined in Eqs. (6.9.8) and (6.9.9). (After Dhadwal and Chu, 1989.)

In the Dhadwal–Chu design, the fiber-optic probe is made up of an optical fiber (single or multimode) and a SELFOC* graded index microlens, as shown in Fig. 6.9.4. The construction of such a probe is quite simple. For uninitiated readers who are not familiar with ray-tracing matrices, there are many optical companies which can construct such probes to specification. The design procedure is reproduced in Appendix 6.A. It is sufficient to mention that such fiber-optic probes are quite small. The $\frac{1}{4}$-pitch graded-index microlens and its outside stainless-steel housing for contact connection with the optical fiber measures typically ≈ 2 mm in diameter and 25 mm in length. The detector probe is characterized by an effective entrance pupil D_A and an angular uncertainty $(\Delta\theta)_f$ in air. The microlens has a quadratic refractive-index profile $n(r) = N_0[1 - (A/2)r^2]$ with A and N_0 being a refractive-index gradient constant and the refractive index of the microlens on the optical axis. We have, from Eqs. (6.A.11) and (6.A.12),

$$D_A = 2(NA)_f/N_0\sqrt{A}, \qquad (6.9.8)$$

$$(\Delta\theta)_f = (D_f/2)N_0\sqrt{A}, \qquad (6.9.9)$$

where D_f and $(NA)_f$ are the core diameter and the numerical aperture (in air) of the optical fiber, respectively.

The essential design requirement for efficient self-beating is that the divergence angle of the fiber probe must be smaller than $(\Delta\theta)_{coh}$. Again, it is not advisable to attempt to change $(NA)_f$ by using an aperture or pinhole in front of the graded-index microlens. In the Dhadwal–Chu approach, the optical fiber, for a nonimaged optical system, is used (1) as a field stop (which is determined by the core diameter) and (2) as a transmission line for light (so that

* A trademark of NSG America, Inc.

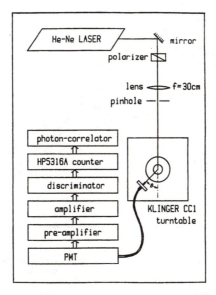

FIG. 6.9.5. Schematic diagram of a simplified light-scattering spectrometer using a microlens–fiber probe as the detection optics. (From H. S. Dhadwal, C. Wu and B. Chu. *Appl. Optics* **28**, 4199, 1989). The fiber was designated for 632.8-nm-wavelength operation and was mounted on a five-axis positioner.

the photodetector can be located at long distances from the scattering cell without appreciable signal loss), while the diameter of the microlens can act as a field stop. Thus, a combination of graded-index microlens and optical fiber replaces the entire optics and associated mechanical elements for the detection optics. Such a prealigned probe can become an integral part of the scattering cell. If the fiber-optic probe is mounted on a five-axis holder with θ, ϕ and x, y, z adjustments, as shown in Fig. 6.9.5, alignment of the light-scattering spectrometer can often be accomplished in about ten minutes.

A multiple-channel light-scattering photometer based on a combination of a fiber-optic device with an optical multichannel analyzer, as shown in Fig. 6.9.6, has been reported (Moser, 1973; Moser *et al.*, 1988). According to Moser, each fiber-optic channel is composed of 60 fibers of 250-μm diameter, each arranged in a rectangular cross section of 1.5×2.5 mm^2 (e.g., 6 fibers wide $\times$ 10 fibers high) towards the scattering volume, and in a row (i.e., in a straight slot 60 fibers long) towards the optical multichannel analyzer. The fiber bundles form a semicircle of 185-mm radius with the entrance surface of each fiber being perpendicular to the radius. The angular aperture per channel with respect to the center of the semicircle is $(1.5/92.5 =)$ 0.93° without focusing optics. Unfortunately, the large numerical aperture of the

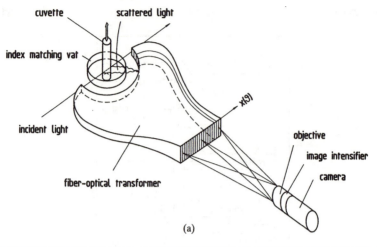

(a)

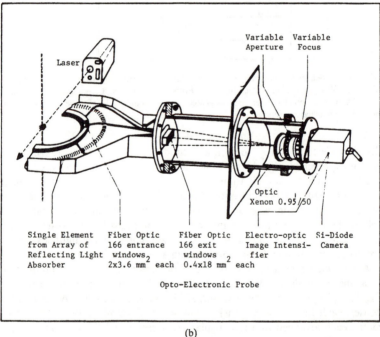

Single Element from Array of Reflecting Light Absorber

Fiber Optic 166 entrance windows 2x3.6 mm^2 each

Fiber Optic 166 exit windows 0.4x18 mm^2 each

Electro-optic Image Intensifier

Si-Diode Camera

Opto-Electronic Probe

(b)

FIG. 6.9.6. Schematic layout of the fast multichannel light-scattering photometer. (Reprinted with permission from Moser, *et al.*, ©1988, American Chemical Society). (a) Schematic diagram; (b) overall layout. (H. Versmold, private communication.) A more efficient way of detecting the scattered intensity from the fiber-optic transformer would be to fiber-optically couple the transformer to the intensifier and then transfer the intensifier output to the diode array or CCD chip by means of the much less efficient use of lens optics. However, the transformer output area then becomes more restrictive, as the larger-area intensifiers are very expensive. It should be emphasized that with the lens coupling between the fiber-optic transformer and the image intensifies, the intensity loss is of the order of 95%, an extremely inefficient method that should be avoided.

fiber remains operational. This fact can result in a low signal-to-noise ratio. Thus, it becomes essential to exclude stray light from the fiber and let the transformer see *only* the scattering volume, a very difficult experimental procedure.

The resolution also depends on the scattering volume, which is now defined by the intersection of the laser beam and the scattering medium, as the fibers, without the microlenses, collect all the light arriving within their aperture. Thus, such an optical detection system requires a different approach to the design of a scattering cell, i.e., the scattering volume is no longer defined by the intersection of the laser beam and the detection optics (field aperture).

By using either a photodiode array detector or a CCD detector (with or without an intensifier), the multichannel light-scattering photometer is capable of measuring the angular distribution of time-averaged scattered intensity almost simultaneously, and therefore can make time-resolved measurements. It would be interesting to replace some of the fiber-bundle channels with a microlens–fiber probe, enabling simultaneous measurements of the angular distribution of time-averaged scattered intensity and of time correlation functions in the time-resolved mode.

6.10. CORRELATOR COMPARISON

There are several commercial correlators on the market. Most dedicated correlators are able to *compute* the time correlation function from photoelectron pulses that have been standardized into TTL, NIM, or ECL pulses.

Further improvements are likely to be in convenience and miniaturization; the ability to compute continuously the correlation function over a fixed delay-time domain has essentially reached the limit of little or no signal loss. The main questions one should ask are:

(1) Is there any dead time (or loss of signal) in computing the time correlation function over the entire delay-time range? The answer should be no. One wants real-time operation over the entire delay-time range for the correlator.

(2) For variable or logarithmic spacings in the delay time, has the time correlation function been scaled to within the statistical error limits? The answer should be yes. Furthermore, it is important that the normalization be done properly so as not to produce a discontinuity from one delay-time value to another.

(3) How many data points are available over the measured delay-time range? In principle, only a few data points (say five) would be needed for each characteristic decay time separated by a factor of $\gtrsim 2$.

(4) What is the delay-time range? The broader the better. In practice, however, we need only a delay-time range that can cover the measured $g^{(1)}\tau$ from near 1 (say 99.5%) to near 0 (say 0.1%). Thus, the delay-time range depends on the nature of research problems of interest.

Table 6.10.1

Specifications of Two Commercial Correlators

BI-2030 (Brookhaven Instruments Corporation, Holtsville, New York)

Inputs[a] (single, dual)	TTL, ECL, NIM
Prescaling[b]	Input rates divisible by powers of 2 up to 128; independent prescaling on both inputs.
Multiplications[c]	$4 \times N$ bits in correlation modes
Overflows	Flashing indicator when 4 bits exceeded
Functions	Auto- and cross-correlation; multichannel scaling
Delay	Self-test mode
Times	100 ns to 990 ms
Duration	Total samples or total time
Data channels	72 standard; 136 or 264 optional; 36 hard bits/channel
Delay channels	6 more delayed; 1027 past last channel
Monitor channels	6 monitor channels record total samples, total count for both inputs, total overflows, reference intensity, space
Display modes	Scaled, auto, and baseline subtract
Data collection	Manual, auto, and remote control
Outputs	RS-232C (COM$_1$) serial; IEEE-488 parallel; LPT1 (Centronic type) parallel printer port
Software[d]	Correlator control program
BI-47	Multiple sample times (4 variable delay times)

ALV-5000 (ALV Laser Vertriebsgesellschaft mbH, Langen, Germany)

Inputs[a] single/dual	TTL (pulse width > 4 nsec)
Channel structure[b,c]	Delay times and the delay-time range fixed: 8×8 bit (channels 1–128), 16×16 bit (channel 129–256)

No. of Channels	$\Delta\tau$	Lag time	Multiplications
16	0.2 μsec	0.2–3.2 μsec	8 × 8
8	0.4 μsec	6.4 μsec	8 × 8
8	0.8 μsec	12.8 μsec	8 × 8
8	1.6 μsec	25.6 μsec	8 × 8
8	3.2 μsec	51.2 μsec	8 × 8
8	6.4 μsec	102.4 μsec	8 × 8
8	12.8 μsec	204.8 μsec	8 × 8
8	25.6 μsec	409.6 μsec	8 × 8
8	51.2 μsec	819.2 μsec	8 × 8
8	102.4 μsec	1.638 msec	8 × 8

(continued)

Table 6.10.1 (*continued*)

No. of Channels	$\Delta\tau$	Lag time	Multiplications
8	204.8 μsec	3.277 msec	8 × 8
8	409.6 μsec	6.544 msec	8 × 8
8	819.2 μsec	13.11 msec	8 × 8
8	1.638 msec	26.21 msec	8 × 8
8	3.277 msec	52.43 msec	8 × 8
8	6.544 msec	104.9 msec	8 × 8
8	13.11 msec	209.7 msec	8 × 8
8	26.21 msec	419.4 msec	8 × 8
8	52.43 msec	838.9 msec	8 × 8
8	104.9 msec	1.678 sec	8 × 8
8	209.7 msec	3.355 sec	8 × 8
8	419.4 msec	6.711 sec	8 × 8
8	838.9 msec	13.42 sec	8 × 8
8	1.678 sec	26.84 sec	8 × 8
8	3.355 sec	53.69 sec	16 × 16
8	6.711 sec	107.4 sec	16 × 16
8	13.42 sec	214.8 sec	16 × 16
8	26.84 sec	429.5 sec	16 × 16
8	53.69 sec	859.0 sec	16 × 16
8	107.4 sec	1718 sec	16 × 16
8	214.8 sec	3436 sec	16 × 16

Monitor channels	Separate direct data monitors for each sampling time; individual delayed data monitors for all channels beyond channel 6
Delay time	200-nsec initial delay time for single mode 400-nsec initial delay time for dual mode
Maximum count rates	≥ 100 MHz
Software[d]	Correlation control program.

(a) Input photoelectron pulses must satisfy the pulse specifications of the manufacturer in order for the correlator to function properly. The requirements are usually different for different correlators. The uninitiated reader should consult with the manufacturer to ensure that the specifications are met.

(b) Prescaling reduces the signal-to-noise ratio of the experiments and should be avoided whenever possible.

(c) 4 bit means 2^4; 8 bit means 2^8. Thus, an 8-bit correlator can take up to 256 counts per delay-time increment, while a 4-bit correlator can take up to 16 counts per delay-time increment. For short delay times, say $\Delta\tau = 10$ μsec, 10 counts per 10 μsec corresponds to an average of 10^6 counts per second, which is essentially the count-rate limit of the correlators. 8 bit becomes more useful for large delay-time increments. Then, scaling is used in any case. The important factors are how the pulses are scaled and how the baseline is normalized.

(d) Details of software operations are likely to change as the PCs or desktop computers become more powerful. Thus, the attractive features of the correlators are not listed.

In comparison with BI-2030, the ALV-5000 correlator according to specifications, can cover a broader delay-time range by a factor of ~10 in the long-delay-time domain, has taken care of the baseline fluctuations by using a more efficient normalization scheme, and is able to accept higher count rates. The delay times, however, are fixed as specified in Table 6.10.1. The ALV-5000 also has a shortest delay time of only 200 nsec. It has an edge in dynamic studies of viscous fluids, such as polymer melts and gels, with very broad frequency distributions. On the other hand, the BI-2030 has been tested thoroughly and can accommodate most time-correlation-function experiments. It also has a shorter delay time increment: 100 nsec. Thus, for particle sizing and for dynamic studies of polymer solutions, except in semidilute regimes, the BI-2030 is more than adequate.

Future correlators will be hardware–software hybrids, using PCs, or desktop computers as the base structure. Undoubtedly, shorter as well as longer delay times will be accessible. However, correlation lag times of a few thousand seconds (or even a couple of hundreds of seconds, as in the BI-2030) are about as long as one would need in most light scattering experiments.

Correlators manufactured by Malvern Instruments and Langley-Ford are similar in operation to the BI-2030.

APPENDIX 6.A. DESIGN OF FIBER-OPTIC PROBES FOR LIGHT SCATTERING (AFTER DHADWAL AND CHU, 1989)

A fiber-optic probe, whether to be used as a transmitter of incident optical radiation into the scattering cell or as a receiver of the scattered light from within the cell, must satisfy the same design criterion. A general fiber-optic probe comprises an optical fiber of the type used in light-wave communications and a SELFOC graded-index microlens. Careful matching of the two components gives the desired properties of the resulting probe, which can be used either in an imaging configuration or in a nonimaging mode. The latter is more appropriate for dynamic and/or static light scattering and is a special case of the former type.

Geometrical optics, in the form of ray tracing using transfer matrices (Gerrard and Burch, 1975), is employed as the design tool for describing the propagation of the laser light emanating from the tip of an optical fiber as it propagates through arbitrary stratified media. The propagating ray is defined by two parameters (Fig. 6.A.1): its height r at the point of intersection with a reference plane, and the angle $\dot{r}$ that a particular ray makes with the optic axis. In order to make the governing equations independent of the refractive index n of isotropic and homogeneous media, bounded by parallel planes perpendicular to the direction of propagation, a normalized angle $v = n\dot{r}$ is used

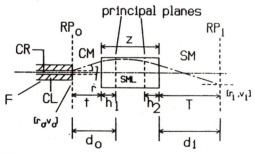

FIG. 6.A.1. Schematic of a fiber-optic probe. SML, SELFOC graded-index microlens; RP_o and RP_i, object and image reference planes; h_1 and h_2, principal planes of the microlens measured from the two end faces; CL and CR, cladding and core regions of the optical fiber (F); d_o and d_i, object and image distances measured from the principal planes; n_{cm}, refractive index of coupling medium; n_{sm}, refractive index of scattering medium. The short-dash–long-dash line indicates a typical ray path through the optical system. (After Dhadwal and Chu, 1989.)

to specify the optical direction cosine of a particular ray. With this notation the propagation of a ray through a series of P different media, bounded by parallel planes, can be expressed through the matrix relation

$$\begin{bmatrix} r_{p+1} \\ v_{p+1} \end{bmatrix} = [R_p][R_{p-1}]\cdots[R_1]\begin{bmatrix} r_1 \\ v_1 \end{bmatrix}, \qquad (6.A.1)$$

where $[R_1]$ is the ray transfer (or $ABCD$) matrix for the pth medium and $[r_1, v_1]$ and $[r_{p+1}, v_{p+1}]$ define the ray parameters in the input and output reference planes, respectively. With knowledge of the transfer matrices for different types of media, the imaging conditions for an arbitrary optical system can be derived.

Figure 6.9.7 shows a schematic of a general fiber-optic probe. The optical fiber is characterized by a core diameter D_f and a numerical aperture $(NA)_f$ in air. Versatility is introduced in the design stage by having an arbitrary coupling medium between the optical fiber and the microlens, and by allowing the output face of the microlens to be imbedded in an arbitrary scattering medium. The coupling and scattering media are assumed to be isotropic and homogeneous with refractive indices n_{cm} and n_{sm}, respectively. The $ABCD$ matrix for such a medium is given by

$$\begin{bmatrix} 1 & \hat{t} \\ 0 & 1 \end{bmatrix}, \qquad (6.A.2)$$

where $\hat{t}$ is the optical path length in the medium. The graded-index microlens, a distributed lenslike medium, has a radial index profile $n(r) = N_0(1.0 - \frac{1}{2}Ar^2)$.

The corresponding $ABCD$ matrix is

$$\begin{bmatrix} \cos(\sqrt{A}\,z) & \dfrac{1}{N_0\sqrt{A}}\sin(\sqrt{A}\,z) \\ -N_0\sqrt{A}\,\sin(\sqrt{A}\,z) & \cos(\sqrt{A}\,z) \end{bmatrix}, \tag{6.A.3}$$

where z is the length of the microlens in millimeters, A is a refractive index gradient constant in mm^{-2}, and N_0 is the refractive index of the microlens on the axis. Thus, for the general fiber-optic probe shown in Fig. 6.A.1 the output ray parameters $[r_1, v_1]$ can be related to the input ray $[r_0, v_0]$ through

$$\begin{bmatrix} r_1 \\ v_1 \end{bmatrix} = \begin{bmatrix} A & B \\ C & D \end{bmatrix}\begin{bmatrix} r_0 \\ v_0 \end{bmatrix}, \tag{6.A.4}$$

where

$$A^+ = \cos(\sqrt{A}\,z) - \hat{T}N_0\sqrt{A}\,\sin(\sqrt{A}\,z), \tag{6.A.5a}$$

$$\begin{aligned} B = (1/N_0\sqrt{A})\sin(\sqrt{A}\,z) + \hat{t}\cos(\sqrt{A}\,z) \\ + \hat{T}\cos(\sqrt{A}\,z) - \hat{T}\hat{t}N_0\sqrt{A}\,\sin(\sqrt{A}\,z), \end{aligned} \tag{6.A.5b}$$

$$C = -N_0\sqrt{A}\,\sin(\sqrt{A}\,z), \tag{6.A.5c}$$

$$D = \cos(\sqrt{A}\,z) - \hat{t}N_0\sqrt{A}\,\sin(\sqrt{A}\,z). \tag{6.A.5d}$$

In Eq. (6.A.5), $\hat{t} = t/n_{cm}$, $\hat{T} = T/n_{sm}$, where t and T are the distances of the object and image planes from the front and back surfaces of the microlens, respectively; the pitch of the microlens is given by $P = 2\pi/\sqrt{A}$, and the focal length of the microlens is $[N_0\sqrt{A}\sin(\sqrt{A}z)]^{-1}$.

The imaging condition is obtained by considering a point source in the object reference plane RP_0, that is, $r_0 = 0$. The image of a point source will be another point, that is, $r_i = 0$—requiring that $B = 0$. Reordering Eq. (6.A.5b) gives

$$\hat{T} = \frac{\hat{t}N_0\sqrt{A}\cos(\sqrt{A}\,z) + \sin(\sqrt{A}\,z)}{[\hat{t}N_0\sqrt{A}\sin(\sqrt{A}z) - \cos(\sqrt{A}\,z)]N_0\sqrt{A}}, \tag{6.A.6}$$

and the magnification $m(= r_1/r_0)$ is given by

$$m = \frac{-(N_0\sqrt{A})^2\sin(\sqrt{A}z)}{N_0\sqrt{A}\,\hat{t} - \cot(\sqrt{A}\,z)}. \tag{6.A.7}$$

Equations (6.A.6) and (6.A.7) can be used to determine the position and magnification of the image for any combination of optical fiber, microlens, coupling, and scattering medium. Note, that, by replacing t with $d_o - h_1$ and T with $d_i - h_2$, it can be easily shown that Eq. (6.A.6) is consistent with the usual lens formula associated with aspherical lenses. h_1 and h_2 are the positions of the principal planes of the microlens as measured from the front and back surfaces, respectively (Fig. 6.A.1); here $h_1 = h_2 = (N_0)\sqrt{A} \tan(\sqrt{A}z/2)$. However, a fiber-optic probe suitable for light scattering has a nonimaging configuration, that is, the image is at infinity ($T = \infty$). Under this condition Eq. (6.A.b) gives

$$\hat{t} = [N_0\sqrt{A} \tan(\sqrt{A}\,z)]^{-1}. \tag{6.A.8}$$

Substituting Eq. (6.A.8) into Eq. (6.A.5) and using Eq. (6.A.4), we can express the diameter D_A and divergence $(\Delta\theta)_f$ of the collimated beam at any plane parallel to the output face of the microlens as follows:

$$D_A = 2r_1 = 2\left\{r_0\cos(\sqrt{A}\,z) - r_0\hat{T}N_0\sqrt{A}\sin(\sqrt{A}\,z) + \frac{v_0}{N_0\sqrt{A}\sin(\sqrt{A}\,z)]}\right\} \tag{6.A.9}$$

and

$$(\Delta\theta)_f = \frac{v_1}{n_{sm}} = \frac{-r_0 N_0\sqrt{A}\sin(\sqrt{A}\,z)}{n_{sm}}. \tag{6.A.10}$$

Equations (6.A.9) and (6.A.10) can be simplified considerably if a quarter-pitch microlens is used ($z = 0.25P$): the equations reduce to

$$D_A = 2v_0/N_0\sqrt{A}, \tag{6.A.11}$$

$$(\Delta\theta)_f = r_0 N_0\sqrt{A}/n_{sm}. \tag{6.A.12}$$

Note that v_0 is equivalent to the numerical aperture of the optical fiber in the coupling medium, $r_0 = D_f/2$, and the minus sign in Eq. (6.A.12) has been dropped.

In this appendix we have outlined a generalized approach to the design of fiber-optic probes. The agreement between the predicted and measured characteristics, of course, depends upon the accuracy of the governing constants and input parameters. In particular, the optical fiber has been assumed to be a planar extended source whose far-field radiation pattern is adequately defined by its diameter and numerical aperture. This approximation is generally very good for multimode optical fibers, but for monomode fibers the $(NA)_f$, as expressed by $(n_1^2 - n_2^2)^{1/2}$, is an overestimate (here n_1 and n_2 are the refractive indices of the core and cladding of the optical fiber, respectively). However, the far-field radiation pattern is easily measured for a more precise estimate (Marcuse, 1981). Other configurations, using different values of z, may also be employed to design fiber-optic detector probes. On occasion it may be advantageous to use a $0.23P$ microlens, allowing the use of a coupling medium.

REFERENCES

Adam, M., Hamelin, A., and Berge, P. (1969). *Opt. Acta* **16**, 337.

Akins, D. L., Schwartz, S. E., and Moore, C. B. (1968). *Rev. Sci. Instrum.* **39**, 715.

Alfano, R. R. and Ockman, N. (1968). *J. Opt. Soc. Amer.* **58**, 90.

Auweter, H. and Horn, D. (1985). *J. Colloid Interface Sci.* **105**, 399.

Barr, F. H. and Eberhardt, E. H. (1965). "Research in the Development of an Improved Multiplier Phototube," ITT Final Rep., Contr. No. NASW-1038.

Bates, B., Conway, J. K., Courts, G. R., McKeith, C. D., and McKeith, N. E. (1971). *J. Phys. E* **4**, 899.

Belland, P. and Lecullier, J. C. (1980). *Appl. Opt.* **19**, 1946.

Benedek, G. and Greytak, T. (1965). *Proc. IEEE* **53**, 1623.

Bennett, Jr., W. R. and Kindlmann, P. J. (1962). *Rev. Sci. Instrum.* **33**, 601.

Berge, P. (1967). *Bull. Soc. Fr. Mineral, Cristallogr.* **90**, 508.

Biondi, M. A. (1956). *Rev. Sci. Instrum.* **27**, 36.

Birenbaum, L. and Scarl, D. B. (1973). *Appl. Opt.* **12**, 519.

Born, M. and Wolf, E. (1956). "Principles of Optics," 3rd (revised) ed., p. 312, Pergamon, London.

Brown, R. G. W. (1987). *Appl. Optics* **26**, 4846.

Brown, R. G. W. and Grant, R. S. (1987). *Rev. Sci. Instrum.* **56**, 928.

Brown, R. G. W., and Jackson, A. P. (1987). *J. Phys. E (Sci. Instrum.)* **20**, 1503.

Brown, R. G. W., Ridley, K. D., and Rarity, J. G. (1986). *Appl. Optics* **25**, 4122.

Brown, R. G. W., Jones, R., Rarity, J. G., and Ridley, K. D. (1987). *Appl. Optics* **26**, 2383.

Bruce, C. F. and Hill, R. M. (1961). *Aust. J. Phys.* **14**, 64.

Burger, H. C. and van Cittert, P. H. (1935). *Physica* (Utrecht) **2**, 87.

Burleigh, Instruments, Inc. Burleigh Park, Fishers, New York (1986). Fabry Perot Interferometry Bibliography, "over 200 selected technical references compiled by Burleigh. A convenient source, though not an endorsement."

Cannell, D. S. and Benedek, G. B. (1970). *Phys. Rev. Lett.* **25**, 1157.

Cannell, D. S., Lunacek, J. H., and Dubin, S. B. (1973). *Rev. Sci. Instrum.* **44**, 1651.

Chabbal, R. and Jacquinot, P. (1961). *Rev. Opt.* **40**, 157.

Chodil, G., Hearn, D., Jopson, R. C., Mark, H., Swift, C. D., and Anderson, K. A. (1965). *Rev. Sci. Instrum.* **36**, 394.

Chu, B. (1968). *J. Chem. Educ.* **45**, 224.

Chu, B. (1977). *Pure and Appl. Chem.* **49**, 941.

Chu, B. (1985). *Polymer J.* (Japan) **17**, 225.

Chu, B. and Wu, C. (1987). *Macromolecules* **20**, 93.

Chu, B., and Xu, R. (1988). *In* "OSA Proceedings on Photon Correlation Techniques and Applications," Vol. 1, pp. 137–146.

Chu, B., Wu, C., and Buck, W. (1988a) *Macromolecules* **21**, 397.

Chu, B., Xu, R., Maeda, T., and Dhadwal, H. S. (1988b). *Rev. Sci. Instrum.* **59**, 716.

Chu, B., Xu, R., and Nyeo, S. (1989). *Part. Part. Syst. Charact.* **6**, 34.

Cole, M. and Ryer, D. (1972). "Electro Optical Systems Design," pp. 16–19.

Connes, P. (1956). *Rev. Opt.* **35**, 37.

Connes, P. (1958). *J. Phys. Radium* **19**, 262.

Cooper, J. and Greig, J. R. (1963). *J. Sci. Instrum.* **40**, 433.

Cooper, V. G., Gupta, B. K., and May, A. D. (1972). *Appl. Opt.* **11**, 2265.

Cummins, H. Z., Knable, N., and Yeh, Y. (1964). *Phys. Rev. Lett.* **12**, 150.

Cummins, H. Z., Carlson, F. D., Herbert, T. J., and Woods, G. (1969). *Biophys. J.* **9**, 518.

Dainty, J. C., ed., (1984). "Laser Speckle and Related Phenomena," Springer-Verlag, Berlin.

Dhadwal, H. S. and Chu, B. (1987). *J. Colloid Interface Sci*, **115**, 561.

Dhadwal, H. S. and Chu, B. (1989). *Rev. Sci. Instrum.* **60**, 845.

Dhadwal, H. S. and Ross, D. A. (1980). *J. Colloid Interface Sci.* **76**, 478.

Dhadwal, H. S., Wu, C., and Chu, B. (1989). *Appl. Opt.* **28**, 4199.

Dressler, K. and Spitzer, L. (1967). *Rev. Sci. Instrum.* **38**, 436.

Dubin, S. B., Clark, N. A., and Benedek, G. B. (1971). *J. Chem. Phys.* **54**, 5158.

Durst, F., Melling, A., and Whitelaw, J. H. (1976). "Principles and Practice of Laser-Doppler Anemometry," Academic Press, New York, 405 pp.

Dyott, R. B. (1978). *Microwaves, Opt. and Acoust.* **2**, 13.

Eberhardt, E. H. (1967). *IEEE Trans. Nucl. Sci.* **NS-14** (2), 7.

Fleury, P. A. and Chiao, R. Y. (1966). *J. Acoust. Soc. Amer.* **39**, 751.

Foord, R., Jones, R., Oliver, C. J., and Pike, E. R. (1969). *Appl. Opt.* **8**, 1975.

Ford, Jr., N. C. and Benedek, G. B. (1965). *Phys. Rev. Lett.* **15**, 649.

Fork, R. L., Herriott, D. R., and Kogelnik, H. (1964). *Appl. Opt.* **3**, 1471.

Fray, S., Johnson, F. A., Jones, R., Kay, S., Oliver, C. J., Pike, E. R., Russell, J., Sennett, C., O'Shaughnessy, J., and Smith, C. (1969). *In* "Light Scattering Spectra of Solids" (G. B. Wright, ed.), pp. 139–150, Springer-Verlag, Berlin and New York.

Gadsden, M. (1965). *Appl. Opt.* **2**, 1446.

Gerrard, A. and Burch, J. M. (1975). "Introduction to Matrix Methods in Optics," Wiley, London.

Greig, J. R. and Cooper, J. (1968). *Appl. Opt.* **7**, 2166.

Haller, H. R., Destor, C., and Cannell, D. S. (1983). *Rev. Sci. Instrum.* **54**, 973.

Han, C. C.-C. and Yu, H. (1974). *J. Chem. Phys.* **61**, 2650.

Hariharan, P. and Sen, D. (1961). *J. Amer. Opt. Soc.* **51**, 398.

Harker, Y. D., Masso, J. D., and Edwards, D. F. (1969). *Appl. Opt.* **8**, 2563.

Hernandez, G. and Mills, O. A. (1973). *Appl. Opt.* **12**, 126.

Hicks, T. R., Reay, N. K., and Scaddan, R. J. (1974). *J. Phys. E* **7**, 27.

Hindle, P. H. and Reay, N. K. (1967). *J. Sci. Instrum.* **44**, 360.

Huglin, M. (1978). *Topics Current Chem.* **77**, 141.

Jackson, D. A. and Jones, J. D. C. (1986). *Optics and Laser Technology*, **18**, 243–252, 299–302.

Jackson, D. A. and Paul, D. M. (1969). *J. Phys. E* **2**, 1077.

Jackson, D. A. and Pike, E. R. (1968). *J. Phys. E* **1**, 394.

Jacquinot, P. (1960). *Rep. Progr. Phys.* **23**, 267.

Jacquinot, P. and Dufour, C. (1948). *J. Rech. Cent. Nat. Rech. Sci.* **2**, 91.

Jerde, R. L., Peterson, L. E., and Stein, W. (1967). *Rev. Sci. Instrum.* **38**, 1387.

Jonas, M. and Alon, Y. (1971). *Appl. Opt.* **10**, 2436.

Jones, R., Oliver, C. J., and Pike, E. R. (1971). *Appl. Opt.* **10**, 1673.

Kittel, C. (1968). "Introduction to Solid State Physics," 3rd ed., p. 247, Wiley, New York.

Klein, M. V. (1970). "Optics," Wiley, New York.

Knuhtsen, J., Olldag, E., and Buchhave, P. (1982). *J. Phys. E.* **15**, 1188.

Kogelnik, H. and Li, T. (1966). *Appl. Opt.* **5**, 1550.

Krall, H. R. (1967). *IEEE Trans. Nucl. Sci.* **NS-14**, 455.

Lai, C. C. and Chen, S. H. (1972). *Phys. Rev. Lett.* **29**, 401.

Lakes, R. S. and Poultney, S. K. (1971). *Appl. Opt.* **10**, 797.

Lan, K. H., Ostrowsky, N., and Sornette, D. (1986). *Phys. Rev. Lett.* **57**, 17.

Lao, Q. H., Schoen, P. E., and Chu, B. (1976). *Rev. Sci. Instrum.* **47**, 418.

Lastovka, J. B. (1967). Ph.D. thesis, MIT, Cambridge, Massachusetts.

Lastovka, J. B. and Benedek, G. B. (1966a). *In* "Physics of Quantum Electronics," Conf. Proc., San Juan, Puerto Rico, 1965 (P. L. Kelley, B. Lax, and P. E. Tannenwald, eds.), pp. 231–240, McGraw-Hill, New York.

Lastovka, J. B., and Benedek, G. B. (1966b). *Phys. Rev. Lett.* **17**, 1039.

Lecullier, J. C. and Chanin, G. (1976). *Infrared Phys.* **16**, 273.

Macfadyen, A. J. and Jennings, B. R. (1990). *Optics and Laser Technology* **22**, 175.

Marcuse, D. (1981). "Principles of Optical Fiber Measurements," Academic, London.

Marshall, L. (1971). *Laser Focus* 7, 26.

McLaren, R. A. and Stegeman, G. I. A. (1973). *Appl. Opt.* 12, 1396.

Morton, G. A. (1968). *Appl. Opt.* 7, 1.

Moser, H. O., Vorrichtung zur schnellen Messung der zeitlichen Änderung der Strahlungsintensität, German Patent Application DE 2338481 C2, June 28, 1973.

Moser, H. O., Fromhein, O., Hermann, F., and Versmold, H. (1988). *J. Phys. Chem.* 92, 6723.

Ohbayashi, K., Kohno, T., Fukino, Y., and Terui, G. (1986). *Rev. Sci. Instrum.* 57, 2983.

Oliver, C. J. and Pike, E. R. (1968). *J. Phys. D* 2, 1459.

Pao, Y. H. and Griffiths, J. E. (1967). *J. Chem. Phys.* 46, 1671.

Pao, Y. H., Zitter, R. N., and Griffiths, J. E. (1966). *J. Opt. Soc. Amer.* 56, 1133.

Peacock, N. J., Cooper, J., and Greig, J. R. (1964). *Proc. Phys. Soc. London* 83, 803.

Phelps, F. M., III (1965). *J. Opt. Soc. Amer.* 55, 293.

Poultney, S. K. (1972). *In* "Advances in Electronics and Electron Physics" (L. Marton, ed.), Vol. 31, Academic Press, New York and London.

Ramsay, J. V. (1962). *Appl. Opt.* 1, 411.

Ramsay, J. V. (1966). *Appl. Opt.* 5, 1297.

Ramsay, J. V. and Mugridge, E. G. V. (1962a). *J. Sci. Instrum.* 39, 636.

Ramsay, J. V. and Mugridge, E. G. V. (1962b). *Appl. Opt.* 1, 538.

Reisse, R., Creecy, R., and Poultney, S. K. (1973). *Rev. Sci. Instrum.* 44, 1666.

Robben, F. (1971). *Appl. Opt.* 10, 2560.

Robinson, W., Williams, J. and Lewis, T. (1971). *Appl. Opt.* 10, 2560.

Rodman, J. P. and Smith, H. J. (1963). *Appl. Opt.* 2, 181.

Rolfe, J. and Moore, S. E. (1970). *Appl. Opt.* 9, 63.

Ross, D. A., Dhadwal, H. S., and Dyott, R. B. (1978). *J. Colloid Interface Sci.*, 64, 533.

Sandercock, J. R. (1970). *Opt. Commun.* 2, 73.

Sandercock, J. R. (1971). *In* "Light Scattering in Solids" (M. Balkanski, ed.), Flammarion Sciences, Paris.

Schatzel, K. (1987). *Appl. Phys.* B42, 193.

Schurr, J. M. and Schmitz, K. S. (1973). *Biopolymers* 12, 1021.

Simic-Glavaski, B. and Jackson, D. A. (1970). *J. Phys. E* 3, 660.

Smeethe, M. J. and James, J. F. (1971). *J. Phys. E* 4, 429.

Tanaka, T. and Benedek, G. B. (1975). *Appl. Opt.* 14, 189.

Terui, G., Kohno, T., Fukino, Y., and Ohbayashi, K. (1986). *Jap. J. Appl. Phys.* 25, 1243.

Tolansky, S. and Bradley, D. J. (1960). *In* "Symposium on Interferometry," N. P. L. Symposium No. 11, pp. 375–386, H. M. Stationary Office, London.

Tuma, W. and Van der Hoeven, C. J. (1973). *J. Phys. E* 6, 169.

Uzgiris, E. E. (1972). *Rev. Sci. Instrum.* 43, 1383.

van der Ziel, A. "Noise in Solid State Devices," Wiley, 1986.

Vaughan, A. H. (1967). *Annu. Rev. Astron. Astrophys.* 5, 139.

Wada, A., Suda, N., Tsuda, T., and Soda, K. (1969). *J. Chem. Phys.* 50, 31.

Wada, A., Soda, K., Tanaka, T., and Suda, N. (1970). *Rev. Sci. Instrum.* 41, 845.

Wada, A., Tsuda, T., and Suda, N. (1972). *Jap. J. Appl. Phys.* 11, 266.

Ware, B. and Flygare, W. H. (1971). *Chem. Phys. Lett.* 12, 81.

Yau, W. W. (1990). *Chemtracts–Macromolecular Chem.* 1, 1.

Yeh, Y. (1969). *Appl. Opt.* 8, 1254.

Young, A. T. (1966). *Rev. Sci. Instrum.* 37, 1472.

Young, A. T. (1969). *Appl. Opt.* 8, 2431.

Young, A. T. and Schild, R. E. (1971). *Appl. Opt.* 10, 1668.

Zatzick, M. R. (1971). Application Note F1021, SSR Instruments Co., Santa Monica, California.

VII

METHODS OF DATA ANALYSIS

7.1. NATURE OF THE PROBLEM

For a polydisperse system of particles in suspension (or macromolecules in solution), the first-order electric field time correlation function $g^{(1)}(\tau)$ is related to the characteristics linewidth (Γ) distribution function $G(\Gamma)$ through a Laplace transform relation:

$$g^{(1)}(\tau) = \int_0^\infty G(\Gamma)e^{-\Gamma\tau}d\Gamma, \tag{7.1.1}$$

where τ denotes the delay time. $g^{(1)}(\tau)$ is measured and known. $G(\Gamma)$ is the unknown characteristic linewidth distribution function, which we want to determine. Solving for $G(\Gamma)$ from Eq. (7.1.1), either analytically or numerically, is known as Laplace inversion and is the main subject of this chapter. The recovery of $G(\Gamma)$ from $g^{(1)}(\tau)$ using data that contains noise is an ill-posed problem. A correlator has a finite delay-time increment $\Delta\tau$ and contains a finite number N of delay-time (τ) channels, whether equally or logarithmically spaced in delay time; e.g., a correlator can cover a finite range of delay times with $\tau_{min} \simeq \Delta\tau, \tau_{max}(\equiv \tau_N) = N\Delta\tau$ for an equally spaced delay-time correlator, implying that the integration limits of Eq. (7.1.1) cover from Γ_{min} to Γ_{max} instead of 0 to ∞. Therefore, $g^{(1)}(\tau)$ is bandwidth-limited. The above illustrations tell us that the measured data $g^{(1)}(\tau)$ contains noise *and* is bandwidth limited. What the correlator measures is in fact an *estimate* of the correlation function, i.e., it is a *time-averaged* time correlation function, measured using the usual optical arrangements that we have discussed in previous chapters (e.g., see Chapter 6).

The ill-conditioned nature of Laplace-transform inversion suggests that we should try to measure the intensity time correlation function $G^{(2)}(\tau)$ to a high level of precision (low noise) and over a broad range of delay time (decreasing the bandwidth limitation, which is governed by the *unknown* $G(\Gamma)$) before we attempt to perform the Laplace-transform inversion according to Eq. (7.1.1). The uninitiated reader may immediately wonder what (1) a high level of precision and (2) a broad range of delay time mean. The following discussions and Section 7.3 represent some guidelines to these two conditions.

The noise can come from the photon-counting statistics and the statistical nature of intensity fluctuations (Saleh and Cardoso, 1973). The photode-tection noise can be reduced by increasing the total number of counts per delay-time increment. As

$$G^{(2)}(\tau) = \langle n(t)n(t+\tau) \rangle = N_S \langle n(t) \rangle^2 (1 + \beta |g^{(1)}(\tau)|^2), \qquad (7.1.2)$$

the magnitude of $G^{(2)}(\tau)$ is closely related to the background $A (\equiv N_S \langle n(t) \rangle^2)$, i.e., to the total number of samples, N_S, and the mean number of counts per sample time, $\langle n(t) \rangle$. Thus, aside from $\beta |g^{(1)}(\tau)|^2$, the magnitude of the intensity correlation-function estimate can be increased by increasing the number of samples, N_S, i.e., by increasing the measurement time and/or by increasing the incident light intensity. Furthermore, as the photodetection process is uncorrelated among neighboring channels, there should be a sufficient number of channels in the measured intensity correlation function, allowing better curve fitting through the centroid of the noise fluctuations. However, in the *correlated* intensity fluctuations, the intensity-fluctuation noise can be reduced only by increasing the measurement time. In any case, we should recognize that the presence of noise cannot be avoided. Indeed, even computer-simulated intensity correlation data contains noise because of roundoff errors in the computer. We wish to obtain an estimate of $G(\Gamma)$, having measured an inevitably noisy and incomplete representation of the intensity time correlation function. Due to the ill-conditioning, the amount of information that can be recovered from the data is quite limited, i.e., small errors in the measurement of $G^{(2)}(\tau)$ can produce very large errors in the reconstruction of $G(\Gamma)$. Attempts to extract too much information result in physically unreasonable solutions. Therefore, a correct inversion of Eq. (7.1.1) must involve some means of limiting the amount of information requested from the problem. In recent years, much attention has been focused on this inversion problem (Bertero *et al.*, 1984, 1985; Refs. 18–29 of Chu *et al.*, 1985; Livesey *et al.* 1986; Vansco *et al.*, 1988; Dhadwal, 1989; Nyeo and Chu, 1989; Langowski and Byran, 1990).

7.2. A SCHEMATIC OUTLINE OF THE PROCEDURE

In an ill-conditioned problem, it is important to know that fitting the data to within experimental error limits does not necessarily yield the correct answer. A pictorial scheme by Livesey *et al.* 1986, as shown in Fig. 7.2.1, demonstrates the general concept. If we let the set $\mathscr{G}$ corresponding to all the $G(\Gamma)$ be represented by a rectangle (bounded by the outer solid lines), as shown in Fig. 7.2.1, then by means of Eq. (7.1.1), a subset (bounded by the dot-dashed lines) of $\mathscr{G}$ can be generated for which each $G(\Gamma)$ satisfies the condition that the computed $G^{(2)}(\tau)$ agrees with the measured estimates $\hat{d}_i$ on the data set $d_i (= \beta^{1/2}|g^{(1)}(\tau_i)|)$ to within experimental error limits. However, many of these $G(\Gamma)$ contain unphysical features, such as negative regions. The unphysical features are rejected from the subset of $\mathscr{G}$ yielding the feasible set

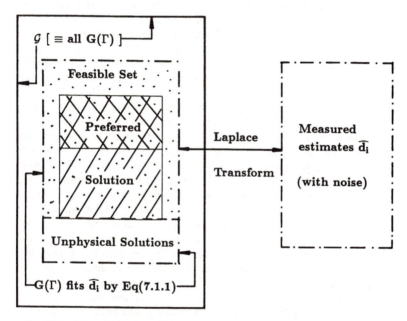

FIG. 7.2.1. Diagram showing the set of all $G(\Gamma)$ spectra (solid rectangle), the boundary of the subset $G(\Gamma)$ that fits $\hat{d}_i$ by Eq. (7.1.1) within experimental accuracy (dot-dash rectangle), and the boundary of the set of physically allowable $G(\Gamma)$ (dashed rectangle); the intersection of the two sets is the feasible set (stippled region). Inside the feasible set, the aim is to choose a preferred solution by means of various methods of data analysis, such as methods of regularization or maximum entropy (Livesey *et al.* 1986). The preferred solution can become more restrictive if additional information about the system can be introduced during the analysis as shown schematically by the double-hatched region.

(stippled region) in which every $G(\Gamma)$ yields a computed $G^{(2)}(\tau)$ in agreement with $\hat{d}_i$ and is physically allowable. The various methods of data analysis try to choose a "preferred" solution (hatched region). If additional boundary conditions can be incorporated into the data analysis procedure, a further reduced set of preferred solutions (double-hatched region) can be chosen.

7.3. EXPERIMENTAL CONSIDERATIONS

From the discussions in Sections 7.1 and 7.2, it is evident that the measured estimates $\hat{d}_i$ on the data d_i must be treated with care, e.g., $\hat{d}_i$ should cover a range of τ_i so that τ_{min} and τ_{max} represent the corresponding values of $G^{(2)}(\tau)$ with $G(\Gamma_{max}) \simeq 0$ and $G(\Gamma_{min}) \simeq 0$. In other words, we should cover $\hat{d}_i$ so that, as $\tau \to 0$, $d^2 = \beta|g^{(1)}(\tau)|^2 \to \beta$. So, we arbitrarily set $|g^{(1)}(\tau = \Delta\tau)|^2 \simeq$ 0.998. $\hat{d}_i$ must also cover a range of τ such that at τ_{max}, $|g^{(1)}(\tau_{max})|^2 \lesssim 0.005$, i.e., the agreement between the measured and the computed baseline should agree to within a few tenths of a percent in order to avoid significant distortion of the resultant $G(\Gamma)$ from the Laplace-transform inversion.

In Eq. (7.1.2), the normalization of $b_i(= G^{(2)}(\tau_i)\text{-}A)$ by $N_S \langle n(t) \rangle^2$ suggests a special importance for the baseline $A(\equiv N_S\langle n(t)\rangle^2)$, which depends on N_S and $\langle n(t) \rangle$ as well as on the delay time over which $G^{(2)}(\tau)$ is measured, since we want to set $G^{(2)}(\tau = \tau_{max})$ close to $N_S\langle n(t)\rangle^2$, i.e., to within 0.005β. The importance of the baseline has been mentioned by many authors (Oliver, 1981; Chu, 1983; Weiner and Tscharnuter, 1985) and treated in a more quantitative manner by Ruf (1989). The relationship between the level of statistical errors in a measured intensity autocorrelation function and the relative size distribution width (uncertainty) has been considered (Kojro, 1990). Errors in the baseline uncertainties play a more complex role than merely shifting $g^{(1)}(\tau)$. In addition to covering the proper τ-range and to recognizing the importance of the baseline before attempting a meaningful data analysis, the statistical errors due to the photodetection process and intensity fluctuations should be minimized by taking short batches of the net intensity correlation function and normalizing the (calculated) baseline for each batch separately (Oliver, 1979); and for long delay times, by considering the difference between $\langle n(t) \rangle$ and $\langle n(t + \tau) \rangle$, as has been discussed by Schatzel et al. (1988).

Finally, we must recognize the fact that Eq. (7.1.1) cannot be implemented unless we have good data, i.e., the sample solution must be properly clarified, so that spurious effects, such as contributions due to dust, are absent, and the instrument properly prepared, such that there is no stray light or after pulsing. There is a general saying, "garbage in, garbage out." The ill-conditioning of the Laplace inversion makes data taking especially crucial, as we need very precise data with some ideas on the error limits of our photon correlation measurements in order to reconstruct $G(\Gamma)$.

To summarize the experimental conditions for measuring the intensity time correlation function (ICF), we need to consider the following steps.

(1) Have a good sample preparation procedure, including clarification of the solution. The presence of foreign particles, such as dust particles, will be reflected in the ICF and consequently $G(\Gamma)$. Dust discrimination by electronic means is feasible, provided that one knows the geometry of the scattering volume and the size of the dust particles, i.e., the average transit time for the dust particle to pass through the scattering volume. However, it is advisable to start with a clean system in which only the particles (or macromolecules) of interest are present. As dust particles are usually large, even a small number can influence the scattering behavior of smaller particles at small scattering angles. For dust discrimination by electronic means, we monitor the scattered intensity over a short delay-time increment. While allowing for normal intensity fluctuations, one can set a threshold level indicating the entrance of large dust particles into the scattering volume. The correlator is then shut off electronically for a preset (transition) time before a resumption of operation. The problem with this type of dust discrimination is that it cannot be entirely objective, because we need to know the threshold level and the dust transit time.

(2) Test the correlator over the delay-time range of interest, making sure that afterpulsing and other electronics problems have been eliminated. Afterpulsing is a serious problem for certain types of photomultipliers operating at short delay-time increments ($\lesssim 100$ nsec).

(3) Measure the ICF over a broad delay-time range such that, e.g.,

(a) $d^2 = \beta|g^{(1)}(\tau)|^2 \to \beta$ as $\tau \to 0$, e.g., $|g^{(1)}(\tau = \Delta\tau)|^2 = 0.998$,
(b) $\hat{d}_i$ decay to A at $\tau = \tau_{max}$, e.g., $|g^{(1)}(\tau_{max})|^2 \lesssim 0.005$.

(4) Have extra channels with long delays to measure the baseline A experimentally. The measured baseline A should agree with the computed baseline $A = N_S \langle n(t) \rangle^2$ to within a few tenths of a percent.

(5) It is better to take short batches of the net ICF, each of which has been normalized by a separate (calculated) baseline, and then add the net ICFs together.

(6) For long delay times, the averages of $\langle n(t) \rangle$ and $\langle n(t + \tau) \rangle$ can be different. Thus, one should use $\langle n(t) \rangle \langle n(t + \tau) \rangle$, not $\langle n(t) \rangle^2$, in the baseline computation.

7.4. BRIEF OUTLINE OF CURRENT DATA ANALYSIS TECHNIQUES

We may subdivide current data analysis techniques into several operational categories:

(1) *Cumulant expansion:*

$$\ln|g^{(1)}(\tau)| = -\bar{\Gamma}\tau + \frac{1}{2!}\mu_2\tau^2 - \frac{1}{3!}\mu_3\tau^3 + \cdots, \qquad (7.4.1)$$

where

$$\bar{\Gamma} = \int_0^\infty \Gamma G(\Gamma) \, d\Gamma, \tag{7.4.2}$$

$$\mu_i = \int_0^\infty (\Gamma - \bar{\Gamma})^i G(\Gamma) \, d\Gamma. \tag{7.4.3}$$

The cumulant expansion (Koppel, 1972) is valid for small τ and sufficiently narrow $G(\Gamma)$. One should seldom use parameters beyond μ_3, because overfitting of data with many parameters in a power-series expansion will render all the parameters, including $\bar{\Gamma}$ and μ_2, less precise. This expansion has the advantage of getting information on $G(\Gamma)$ in terms of $\bar{\Gamma}$ and μ_2 without *a priori* knowledge of the form of $G(\Gamma)$, and is fairly reliable for a variance $\mu_2/\bar{\Gamma}^2 \lesssim 0.3$. The cumulant expansion algorithm is fast and easy to implement. One often uses its results as a starting point for more detailed data analysis.

(2) *Known $G(\Gamma)$:* The ill-conditioned nature of Laplace inversion is essentially removed if the form of $G(\Gamma)$ is known. Thus, for known $G(\Gamma)$, the problem is not so difficult and will not be discussed here. It is sufficient to say that one can achieve much better results on the parameters of the distribution function $G(\Gamma)$ if its form is known.

(3) *Double-exponential distribution:*

$$|g^{(1)}(\tau)| = A_1 \exp(-\Gamma_1 \tau) + A_2 \exp(-\Gamma_2 \tau), \tag{7.4.4}$$

where $A_1 + A_2 = 1$. With the three unknown parameters Γ_1, Γ_2, and A_1 (or A_2), this is a nonlinear least-squares fit to $|g^{(1)}(\tau)|$. If one knows $G(\Gamma)$ to be a bimodal distribution consisting of two δ-functions, Eq. (7.4.4) should be very easy to use. Without knowing the form of $G(\Gamma)$, it is still useful to fit fairly broad $G(\Gamma)$ distributions (say with $\mu_2/\bar{\Gamma}^2 \lesssim 0.5$). Then, the computed A_i and Γ_i values can be used to compute a meaningful average linewidth ($\bar{\Gamma}$) and variance ($\mu_2/\bar{\Gamma}^2$) with $\bar{\Gamma} = A_1 \Gamma_1 + A_2 \Gamma_2$ and $\mu_2/\bar{\Gamma}^2 = (A_1 \Gamma_1^2 + A_2 \Gamma_2^2)/(A_1 \Gamma_1 + A_2 \Gamma_2)^2 - 1$. The double-exponential form can fit $|g^{(1)}(\tau)|$ over a broader range of τ than the cumulant expansion for broader $|g^{(1)}(\tau)|$. However, its application remains limited.

(4) *Method of regularization:* This is a smoothing technique that tries to overcome the ill-posed nature of the Laplace-transform inversion. The regularized inversion of the Laplace integral equation requires little prior knowledge on the form of $G(\Gamma)$. The most tested and well-documented algorithm has been kindly provided by Provencher (1976, 1978, 1979, 1982a, b, c, 1983) to research workers upon request and is known as CONTIN. The serious worker in the field who wants to determine reliable $\bar{\Gamma}$ and μ_2 values should learn how to use CONTIN, which has a detailed set of instructions.

(5) *Method of maximum entropy:* The method of maximum entropy (MEM) provides the best possibility for future improvement in data analysis. It can proceed from an unbiased, objective viewpoint and permits the introduction of constraints that reflect the physics of the problem of interest.

Unfortunately, inversion of the Laplace integral equation cannot be understood on an elementary mathematical level. Thus, the following discussions (Sections 7.5–7.6) can be *skipped* by the less mathematically oriented reader. One should learn how to use the cumulant expansion, the double-exponential fit, and CONTIN, and watch out for the development of the MEM. In the analysis procedure, always try to generate simulated data sets with different noise levels and delay-time ranges that are comparable to (and broader than) the measured data set, and test the programs in order to gain confidence in their validity. *Objectivity is the key.* Normally, one should not try to get more than three parameters $(\bar{\Gamma}, \mu_2, \mu_3)$ from the fitting procedures.

Many data analysis techniques rely on minimizing the sum of the squares of the normalized residuals. We seek solutions to Eq. (7.1.1) that minimize some measure of the discrepancies between the discrete sample data points $\hat{d}_i$ and the corresponding values $\beta^{1/2}|g^{(1)}(\tau_i)|$ that can be calculated from Eq. (7.1.1), subject to the requirement that the solution be well behaved. The minimization of the L^2 norm of residuals (i.e., least-squares minimization) involves an iteration scheme that aims at descending the χ^2 surface to the minimum.

If a $G(\Gamma)$ is estimated by simply minimizing χ^2, the solution to Eq. (7.1.1) may be overfitted, resulting in erroneous $G(\Gamma)$ even when χ^2 can be fitted to within experimental error limits, as shown schematically by the dot-dashed rectangle in Fig. 7.2.1. The particular solution chosen could also depend on the computer algorithm or be biased by the choice of starting solution. ·

In order to remove the ill-conditioning, the correct inversion of Eq. (7.1.1) must involve some means to reduce the size of the feasible set, i.e., to reduce the amount of information requested from the problem; this has been done in a number of ways. The more popular ones, though by no means all, are summarized in the following sections. This summary with comments tries to inform the reader what has been accomplished. The cumulant method (Section 7.4.6), the multiexponential singular-value decomposition (Section 7.5), and the maximum-entropy formalism (Section 7.6) are presented in some detail for those readers interested in learning more.

7.4.1. EIGENVALUE DECOMPOSITION

Eigenvalue decomposition of the Laplace kernel coupled with reliable estimates of the noise in the data can be used to recover estimates of $G(\Gamma)$ (McWhirter and Pike, 1978). The eigenvalues are shown to decrease rapidly below the noise level of the data, at which point it becomes meaningless to extract further knowledge about the characteristic linewidth distribution $G(\Gamma)$. Unfortunately, it is difficult to formulate a quantitative description of the noise in the data. The discussions in Section 7.3 have treated some aspects of sources of noise in the measured data, and attempts have been made, at least in the first approximation, to incorporate the noise error in the baseline subtraction (Ruf, 1989). Nevertheless, it would be difficult to unambiguously determine the amount of noise in the measured data and so to determine the number of terms that should be retained in the eigenfunction expansion.

By limiting the amount of information in the beginning so that the support of the solution is restricted (between Γ_{min} and Γ_{max}, the upper and lower frequencies of the characteristic linewidth distribution) due to a prior knowledge of the spread of molecular sizes (Bertero et. al., 1982, 1984), Pike and his coworkers proposed to use a sufficiently small number N of characteristic linewidths in order to digitize $G(\Gamma)$ discretely and to ensure uniqueness of the solution.

Two schemes have been suggested by Pike and his coworkers to interpolate the shape of $G(\Gamma)$ between the few (N) exponentials permitted by the eigenvalue decomposition techique. Pike et al. (1983) proposed an interpolation formula that sets the coefficients of the eigenfunction expansion to zero beyond a cutoff. An alternative solution (Pike, 1981; Ostrowsky et al., 1981) is to phase-shift the transform coefficient.

Bertero et al. (1982), for example, have shown that the obtainable resolution is increased as the support ratio $\gamma \, (\equiv \Gamma_{max}/\Gamma_{min})$ is decreased. It should be noted that Γ_{min} and Γ_{max} refer to the bounds of the *true* distribution. More information on $G(\Gamma)$ can be recovered with the introduction of additional physical constraints. Nevertheless, the sharp truncation at Γ_{min} and Γ_{max} produces too much sensitivity to the position of $\bar{\Gamma} \, (\equiv \int \Gamma G(\Gamma) \, d\Gamma)$ within the interval from Γ_{min} to Γ_{max}, as well as some unphysical edge effects.

In a recent LS polydispersity analysis of molecular diffusion by Laplace-transform inversion in weighted species, Bertero et al. (1985) imposed a profile function $f_0(\Gamma)$ having the measured mean $\bar{\Gamma}$ and polydispersity index Q $(=\mu_2/\bar{\Gamma}^2)$ to localize the recovered solution and to improve its resolution. The profile function is large in the expected region of the solution and tapers off smoothly to zero on either side, so that the integration in Eq. (7.1.1) is no longer restricted to $[\Gamma_{min}, \Gamma_{max}]$ but is returned to $[0, \infty]$. The problem becomes

$$g^{(1)}(\tau) = \int_0^\infty e^{-\Gamma \tau} f_0(\Gamma) \phi(\Gamma) \, d\Gamma, \qquad (7.4.5)$$

where $G(\Gamma) = f_0(\Gamma)\phi(\Gamma)$. Bertero et al. (1985) have, for example, chosen a gamma distribution for f_0:

$$f_0(\Gamma) = \frac{B^B}{(B-1)!} \left(\frac{\Gamma}{\bar{\Gamma}}\right)^{B-1} e^{-B(\Gamma/\bar{\Gamma})} \qquad (7.4.6)$$

with $Q_{max} = (\mu_2/\bar{\Gamma}^2) = 1/B$. There is also a lower limit Q_{min}, which is determined empirically. The reader is referred to the original article for details.

7.4.2. SINGULAR-VALUE DECOMPOSITION

The singular-value decomposition technique (Golub and Reinsch, 1970; Hanson, 1971; Lawson and Hanson, 1974) provides a means of determining the information elements of the actual problem (including constraints) by ordering these elements in decreasing importance. It has been applied by Bertero et al. (for example, see 1982). The problem of how many terms to use before cutoff remains. This is, however, one of the simpler methods. So we shall discuss the technique in some detail later (see Section 7.5).

7.4.3. REGULARIZATION METHODS

Regularization methods aim at seeking solutions to Eq. (7.1.1) from a class of functions that make the inversion of the integral equation well behaved. These pseudosolutions generally correspond to solutions of minimum energy, nonnegativity, or some other reasonable physical constraints. A regularizer may be used to restrict the curvature of $G(\Gamma)$ in order to give a smoother distribution. By imposing nonnegativity on the $G(\Gamma)$ distribution and by choosing the degree of smoothing based on the Fisher test, Provencher (1976, 1978, 1979, 1982a,b,c, 1983) has made available an automated procedure that combines regularization with the eigenfunction analysis to yield a thoroughly tested solution to Eq. (7.1.1). The CONTIN program is well documented and widely used, and has been adapted by workers to IBM PC/AT microcomputers. It is also available through Brookhaven Instruments (New York) for the BI correlators (not an endorsement). A simpler version (Chu et al. 1985), based on the procedure by Miller (1970), is limited to mainly unimodal distributions. Provencher's CONTIN program should be one of the programs at the disposal of the reader who is interested in using quasielastic light scattering for polydispersity analysis.

7.4.4. METHOD OF MAXIMUM ENTROPY (MEM)

The method of maximum entropy has been applied to a variety of problems in image (or structural) analysis. Its application to the reconstruction of $G(\Gamma)$ based on the Shannon–Jaynes entropy provides the best introduction of a priori information for the structure of $G(\Gamma)$ and has been explored by several authors (Livesey et al. 1986; Vansco et al., 1988; Nyeo and Chu, 1989). A direct comparison of the present-day methods shows that CONTIN and the MEM yield comparable results (see Section 7.7). It is likely that with additional knowledge of $g^{(1)}(\tau)$, the MEM could be improved further and could yield somewhat higher resolving power than CONTIN (Langowski and Bryan, 1990).

7.4.5. EMPIRICAL FORMS FOR $G(\Gamma)$ OR $|g^{(1)}(\tau)|$

If for any reason, such as in a polymerization process, we know the form of $G(\Gamma)$, then the parameters of $G(\Gamma)$ can be determined more precisely because we can introduce the information directly into Eq. (7.1.1). However, if the calculated curve represents only an analytical *approximation* to the expected curve, the true solution may now be excluded from the feasible or allowed set of solutions. For example, if $G(\Gamma)$ is a truncated Gaussian or Pearson curve, we can generate $\beta|g^{(1)}(\tau_i)|^2$ in order to compare it with d_i^2. A better comparison can be made if β is known experimentally, instead of as a fitting parameter. The results with known $G(\Gamma)$ can still be elusive if there were too many parameters, as in a Pearson's curve (Chu *et al.*, 1979).

7.4.6. DIRECT INVERSION TECHNIQUES (CHU ET AL., 1985)

Direct inversion techniques, such as calculating the cumulants (Koppel, 1972) or the moments (Isenberg *et al.* 1973) of $G(\Gamma)$ are well suited to reasonably narrow, well-behaved characteristic linewidth distribution functions, but have grave difficulties in handling broad ($\mu_2/\bar{\Gamma}^2 \gtrsim 0.3$) or multimodal distributions in $G(\Gamma)$.

The following paragraphs are transferred with only three minor changes from Chu *et al.* (1985, pp. 264, 267–269, 278–287). $G(\Gamma)$ is described in terms of a moment expansion:

$$\ln|g^{(1)}(\tau)| = -\bar{\Gamma}\tau + \frac{1}{2!}\mu_2\tau^2 - \frac{1}{3!}\mu_3\tau^3 + \frac{1}{4!}(\mu_4 - 3\mu_2)\tau^4 - \cdots$$
$$= K_m(\Gamma)(-\tau)^m/m!, \tag{7.4.1}$$

where $K_m(\Gamma)$ is the mth cumulant. $K_1 = \mu_1 = 0$, $K_2 = \mu_2$, $K_3 = \mu_3$, and $K_4 = \mu_4 - 3\mu_2$. One can fit this directly to the measured quantity $\hat{b}_i$ ($= G^{(2)}(\tau_i) - A$), recognizing that a constant $(A\beta)^{1/2}$ will be added to Eq. (7.4.1). Note that $A\beta$ (not A) is an adjustable parameter that will be determined by the algorithm as presented here. This is a straightforward weighted linear least-squares problem, the weighting factors being necessary because the operation of taking the logarithm of the data has affected the (approximately) equal weighting of the data points. In fact, the data points are not equally significant. Jakeman *et al.* (1971) have derived expressions for the variance of the correlation function as a function of delay channel, but only in the limit of single-exponential decays. If we assume that the major source of statistical uncertainty is counting statistics, and that the errors in adjacent channels are uncorrelated, we conclude that the error in a particular channel ought to be approximately constant, since the first channel (with the maximum in the correlation function) rarely differs by more than 40% from the last. If we introduce a weighting of each data point by another factor corresponding to b_i^{-1}, additional complications may result, as the net value of the data points becomes very small near τ_{max}. Alternatively, we can use the net autocorrelation function directly (after subtracting the baseline) and fit to

$$\hat{b}_i = A\beta \exp\left[2\left(-\bar{\Gamma}\tau_i + \frac{1}{2!}\mu_2\tau_i^2 - \frac{1}{3!}\mu_3\tau_i^3 + \cdots\right)\right]. \tag{7.4.7}$$

We introduce $\hat{b}_i$ to distinguish the data vector of this problem from that of b_i. The least-squares problem is a nonlinear one, but has the advantage that the relative statistical weighting of the data is unaffected. For the implementation of the nonlinear least-squares algorithm we have in this case the parameter vector **P** defined by

$$P_1 = A\beta,$$
$$P_2 = \bar{\Gamma}, \tag{7.4.8}$$
$$P_{j,\,j>2} = K_j(\Gamma)/(j-1)!.$$

The χ^2 surface is defined by the value of the sum of the squares of the residuals with M being the number of delay channels (i.e., data points),

$$\sum_{i=1}^{M} [\hat{b}_i - f(\tau_i, \mathbf{P})]^2,$$

at the point located by the vector $\mathbf{P} = (P_1, P_2, \ldots, P_n)$, with $f(\tau_i; \mathbf{P})$ being the value of b_i calculated from the model at the point τ_i where the parameters of the model have the values given by the vector **P**. Therefore, an initial estimate of **P** is supplied to the routine (i.e., a starting point at which to evaluate the χ^2 surface), and an appropriate algorithm is used to choose the direction of descent (i.e., to find some suitable modification to **P** so as to reduce χ^2). Three methods are in common use. They are the gradient method, the Gauss–Newton method, and the Levenberg–Marquardt method. Interested readers should refer to Chu *et al.* (1985) and the original references for details.

Eq. (7.4.7) becomes

$$\hat{b}_i = P_1 \exp[2(-P_2\tau_i + P_3\tau_i^2 - P_4\tau_i^3 + \cdots)], \tag{7.4.9}$$

and the elements of the Jacobian matrix are

$$\begin{aligned}
J_{i1} &= -\exp[2(-P_2\tau_i + P_3\tau_i^2 - P_4\tau_i^3 + \cdots)] = -\exp(Z), \\
J_{i2} &= 2P_1\tau_i\exp(Z), \\
J_{i3} &= -2P_1\tau_i^2\exp(Z), \\
J_{i4} &= 2P_1\tau_i^3\exp(Z),
\end{aligned} \tag{7.4.10}$$

where the expansion in Eq. (7.4.7) can be truncated after two, three, or four terms, giving rise to second-, third-, or fourth-order cumulant fits. Starting estimates for the parameters can be provided directly from the data themselves:

$$P_1 = b_1,$$

$$P_2 = \frac{\ln(b_{10}/b_1)}{2(\tau_1 - \tau_{10})},$$

$$P_{n,\,n>2} = 0,$$

where the expression for P_2 is derived from Eq. (7.4.7) as $\tau \to 0$. The use of the first and tenth points to determine the initial slope helps to suppress the effect of statistical noise in the data, although the estimate need not be very close for the algorithm to converge. For wider or bimodal distributions, the cumulant approach does not usually describe the data well, and it is not often possible to recover accurately more than the first three cumulants. Nevertheless, the simplicity of the method and the lack of *a priori* assumptions about the distribution make it suitable as a first step in a more complicated analysis.

Table 7.4.1

Net Photocount Autocorrelation Function[a,b]

1.78550E+05	1.69120E+05	1.61470E+05	1.54000E+05	1.45600E+05
1.37810E+05	1.27660E+05	1.21880E+05	1.15850E+05	1.13540E+05
1.07800E+05	9.90700E+04	9.12500E+04	9.13900E+04	8.39700E+04
7.84200E+04	7.74600E+04	7.40100E+04	7.06600E+04	6.34500E+04
6.13300E+04	6.06300E+04	5.77000E+04	5.66500E+04	5.39500E+04
5.27200E+04	4.85700E+04	4.65200E+04	4.89500E+04	4.36200E+04
4.21800E+04	4.01800E+04	3.90300E+04	3.64300E+04	3.51400E+04
3.54600E+04	3.49200E+04	3.28300E+04	3.25900E+04	2.72100E+04
2.58700E+04	2.44100E+04	2.37500E+04	2.36600E+04	2.16000E+04
2.03300E+04	2.16000E+04	1.80600E+04	1.69900E+04	1.90300E+04
1.67800E+04	1.61000E+04	1.41900E+04	1.37100E+04	1.39500E+04
1.48700E+04	1.57000E+04	1.49300E+04	1.42000E+04	9.66010E+03
1.26300E+04	1.22100E+04	1.21900E+04	1.04600E+04	8.74010E+03
9.47010E+03	1.09100E+04	9.80010E+03	8.77010E+03	9.46010E+03
9.57010E+03	1.03000E+04	7.58010E+03	6.92010E+03	7.42010E+03
7.27010E+03	8.81010E+03	7.78010E+03	9.21010E+03	5.50010E+03
7.78010E+03	8.62010E+03	7.74010E+03	5.38010E+03	6.85010E+03
4.88010E+03	6.32010E+03	3.80010E+03	3.27010E+03	2.34010E+03
3.98010E+03	3.86010E+03			

(a) PMMA in MMA at $\theta = 35$ and $50°$. Measured baseline of 3,429,707 subtracted, $\Delta t = 12\ \mu\text{sec}$, clip level 1.

(b) $\sqrt{A\beta}\,g^{(1)}(\tau = \Delta t) = 1.78550 \times 10^5$; $\sqrt{A\beta}\,g^{(1)}(\tau = 92\,\Delta t) = 3.86010 \times 10^3$.

Table 7.4.2

Cumulant Analysis of PMMA Data[a,b,c]

χ^2	$\lambda^{(N)}$	P_j			
		$j=1$	$j=2$	$j=3$	$j=4$
1.595E+09	1.000E+01	1.786E+05	2.096E+03	0.000E−01	0.000E−01
1.122E+09	1.000E+01	1.792E+05	2.085E+03	3.135E+04	−5.243E+07
6.966E+08	5.000E+00	1.801E+05	2.072E+03	6.758E+04	−1.130E+08
4.743E+08	2.500E+00	1.810E+05	2.064E+03	9.677E+04	−1.652E+08
4.058E+08	1.250E+00	1.819E+05	2.065E+03	1.115E+05	−2.020E+08
3.741E+08	6.250E−01	1.829E+05	2.078E+03	1.166E+05	−2.325E+08
3.448E+08	3.125E−01	1.839E+05	2.099E+03	1.230E+05	−2.643E+08
3.227E+08	1.562E−01	1.850E+05	2.126E+03	1.418E+05	−2.867E+08
3.087E+08	7.812E−02	1.860E+05	2.154E+03	1.868E+05	−2.767E+08
2.951E+08	3.906E−02	1.868E+05	2.186E+03	2.731E+05	−2.182E+08
2.786E+08	1.953E−02	1.876E+05	2.226E+03	4.111E+05	−1.098E+08
2.636E+08	9.766E−03	1.885E+05	2.276E+03	5.883E+05	3.320E+07
2.557E+08	4.883E−03	1.894E+05	2.323E+03	7.550E+05	1.691E+08
2.536E+08	2.441E−03	1.899E+05	2.352E+03	8.573E+05	2.532E+08
2.534E+08	1.221E−03	1.901E+05	2.362E+03	8.940E+05	2.836E+08
2.534E+08	6.104E−04	1.902E+05	2.364E+03	9.010E+05	2.895E+08
2.534E+08	3.052E−04	1.902E+05	2.364E+03	9.017E+05	2.900E+08
2.534E+08	1.526E−04	1.902E+05	2.364E+03	9.017E+05	2.901E+08

(a) $[G_k^{(2)}(\tau) - A] = P_1 \exp[2(-P_2\tau + P_3\tau^2 - \cdots)]$. Third-order fit to 92 points beginning with first point.

(b) $\bar{\Gamma} = 2.36 \times 10^3 \sec^{-1}$; $\mu_2/\bar{\Gamma}^2 = 0.323$.

(c) The iterative sequence used for computing the P_j's was terminated when none of the parameters changed by more than 0.1% between successive iterations.

Results of the third-order nonlinear cumulant analysis for the PMMA data of Table 7.4.1 are shown in Table 7.4.2. The values of the P_j at each iteration are listed along with the value of the Marquardt parameter $\lambda^{(N)}$, where the $(N + 1)^{\text{th}}$ estimate is $\mathbf{P}(N + 1) = \mathbf{P}^{(N)} - [\lambda^{(N)}I + J^T J]^{-1} J^T F^{(N)}$ with $\lambda^{(N)}$, I, and $F^{(N)}$ being, respectively, a scalar quantity that may change with each iteration, the identity matrix, and the residual vector at the Nth estimate, i.e., $(F^{(N)})_i = b_i - f(\tau_i; \mathbf{P}^{(N)})$:

$$\chi^2 = \sum_{i=1}^{M} \{b_i - P_1 \exp[2(-P_2\tau_i + P_3\tau_i^2 - \cdots)]\}^2. \qquad (7.4.11)$$

The convergence criterion satisfied was that the parameters were stable to three significant digits on successive iterations, i.e., no parameter changed by more than 0.1%. This should not be taken to mean that the uncertainty in the parameters is 0.1%, but rather that the parameters are most likely within a few tenths of a percent of the values they will eventually converge to. The variance is defined as $\mu_2/\bar{\Gamma}^2$, and a value of 0.323 indicates a reasonably wide distribution for this sample.

7.5. MULTIEXPONENTIAL SINGULAR-VALUE DECOMPOSITION (MSVD)

The MSVD scheme is inferior to CONTIN, but it is simpler to implement and can be used reasonably successfully for fairly broad unimodal distributions.

In the MSVD approach, $G(\Gamma)$ is approximated by a weighted sum of Dirac delta functions:

$$G(\Gamma) = \sum_{j=1}^{n} P_j \delta(\Gamma - \Gamma_j). \qquad (7.5.1)$$

Linear methods fix the location of the δ-function (i.e., the Γ_j) and fit for best values of the P_j; nonlinear methods allow the Γ_j to float as well. The choice of the number of δ-functions n, and hence the number of adjustable parameters (n for the linear approximations, $2n$ for the nonlinear), depends upon the method of inversion that is to be used. It must be kept in mind that the amount of information available is limited. Unless the method of inversion includes an appropriate rank-reduction step, the range in $G(\Gamma)$ determines n, as the resolution is fixed by the noise on the data. The nonlinear methods rapidly become very time-consuming and are subject to convergence problems as n is increased. In practice it is found that the nonlinear model, where the location of the δ-functions is an adjustable parameter, is essentially limited to the determination of a double-exponential decay, i.e., $n = 2$. The nonlinear problem can be solved by standard nonlinear least-squares techniques. By writing our model as

$$G(\Gamma) = w\,\delta(\Gamma - \Gamma_1) + (1 - w)\,\delta(\Gamma - \Gamma_2), \qquad (7.5.2)$$

where w is a normalized weighting factor, we have, after appropriate substitution of Eq. (7.5.2) into Eq. (7.1.1),

$$b_i = P_1 \exp(-P_2\tau_i) + P_3 \exp(-P_4\tau_i), \qquad (7.5.3)$$

with the elements of the Jacobian given by

$$\begin{aligned}
J_{i1} &= -\exp(-P_2\tau_i), \\
J_{i2} &= P_1\tau_i\exp(-P_2\tau_i), \\
J_{i3} &= -\exp(P_4\tau_i), \\
J_{i4} &= P_3\tau_i\exp(-P_4\tau_i).
\end{aligned} \qquad (7.5.4)$$

Initial estimates of the parameters can be obtained from the results of the cumulant analysis; we have found it simpler and usually sufficient to use the starting values

$$P_1 = P_3 = b_1/2,$$
$$P_2 = \bar{\Gamma}/2,$$
$$P_4 = 2\bar{\Gamma},$$

where $\bar{\Gamma}$ is estimated from the initial slope of the correlation function as in the cumulant technique.

The linear least-squares minimization problem can be stated as

$$C\mathbf{P} \simeq \mathbf{b}, \tag{7.5.5}$$

where C is the $(M \times n)$ curvature matrix, $\mathbf{P}$ is the parameter vector of length n, and $\mathbf{b}$ is again the data vector of length M. M and n are, respectively, the number of data points in the net correlation function and the number of adjustable parameters of the model. For example, under the linear multiexponential approximation, the elements of C can be determined from comparison of Eqs. (7.1.1) and (7.5.1) with Eq. (7.5.5) to be

$$c_{ij} = e^{-\Gamma_j \tau_i} \tag{7.5.6}$$

where i and j are the row and column indices, respectively, and the parameters to be determined are the P_j, i.e., the weighting factors of the δ-functions (see Eq. (7.5.1)). The symbol $\simeq$ in Eq. (7.5.5) is intended to imply solution of the overdetermined set of equations subject to the least-squares criterion (minimization of the Euclidean norm of the residual vector $\|\mathbf{b} - C\mathbf{P}\|$, where the Euclidean norm of a vector $\mathbf{h}$ can be computed from $\|\mathbf{h}\| = \left(\sum_i h_i^2\right)^{1/2}$). Prior to entering the singular-value decomposition routine, we generally scale the columns of C to unit norm in order to improve the numerical stability of the inversion (see for example, Lawson and Hanson, 1974, pp. 185–188). The scaling transforms Eq. (7.5.5) to

$$A\mathbf{x} \simeq \mathbf{b}, \tag{7.5.7}$$

where

$$A = CH \tag{7.5.8}$$

and

$$x = H^{-1}\mathbf{P}, \tag{7.5.9}$$

with H being a diagonal matrix whose nonzero elements are the reciprocals of the norms of the corresponding column of C,

$$h_{ij} = \left(\sum_{j=1}^{n} c_{ij}^2\right)^{-1/2} \delta_{ij}. \tag{7.5.10}$$

The resulting problem (Eq. (7.5.7)) is passed to a singular-value decomposition algorithm (for example, subroutine SVDRS from Appendix C of Lawson and Hanson, 1974), which transforms Eq. (7.5.7) to

$$USV^{-1}\mathbf{x} = \mathbf{b}, \tag{7.5.11}$$

where U and V^{-1} are orthogonal matrices and S is a diagonal matrix whose nonzero elements are the (monotonically decreasing) singular values of the stated problem. The explicit procedure of finding this singular-value decomposition (i.e., determining the matrices U, S, and V^{-1}) involves the application of properly chosen orthogonal transformations to the matrix A such that Eq. (7.5.11) holds; details of the procedure may be found in Lawson and Hanson (1974) and will not be discussed further. It is precisely this decomposition that subroutine SVDRS of Lawson ancd Hanson's book accomplishes. Defining the new vectors $\mathbf{y}$ and $\mathbf{g}$ by

$$\mathbf{x} = V\mathbf{y} \tag{7.5.12}$$

and

$$\mathbf{g} = U^T\mathbf{b}, \tag{7.5.13}$$

we have

$$S\mathbf{y} = \mathbf{g}. \tag{7.5.14}$$

The singular-value routine will typically return the matrices V and S and the vector $U^T\mathbf{b}$. Since S is diagonal, we have immediately

$$\mathbf{y}_i = \mathbf{g}_i/s_i \tag{7.5.15}$$

with s_i being the ith singular value, S_{ij}.

Equation (7.5.15) represents the full-rank solution to the problem (7.5.7) and assumes that all singular values are nonzero, i.e., the problem was nonsingular. This is generally not the case, since Eq. (7.5.7) is often pseudorank-deficient in that not all n elements of $\mathbf{x}$ are recoverable as independent resolution elements, due to the data noise. A rank-reduction step is employed to limit the amount of information necessary to express the solution. If we know the problem to be of pseudorank k, we can retain the first k diagonal elements of S and ignore the rest, i.e., we compute the pseudoinverse of S, S^+, and place zeros in the locations of S^+ that correspond to near-zero singular values s_i. This is consistent with the viewpoint of information theory, since the squares of the singular values are the eigenvalues of the matrix A^*A (A^* is the adjoint of A) and thus are closely related to the fundamental elements of resolution transmitted through the integral operator of Eq. (7.1.1). Generally, k is unknown; we therefore define a set of "candidate solutions" $\mathbf{x}^{(k)}$ from

Eq. (7.5.12) as

$$\mathbf{x}^{(k)} = V\mathbf{y}^{(k)}, \tag{7.5.16}$$

where

$$\mathbf{y}^{(k)} = \begin{cases} g_i/s_i & \text{for} \quad 1 < i \le k, \\ 0 & \text{for} \quad k < i < n. \end{cases} \tag{7.5.17}$$

In terms of the original problem (Eq. (7.5.5)) we have a set of candidate solutions for the vector $\mathbf{P}$ defined by

$$\mathbf{P}^{(k)} = H\mathbf{x}^{(k)}. \tag{7.5.18}$$

Furthermore, this procedure defines a set of residual vectors $\mathbf{r}^{(k)}$ such that

$$\mathbf{r}^{(k)} = C\mathbf{P}^{(k)} - \mathbf{b}. \tag{7.5.19}$$

There are several criteria for the selection of the particular value of k that can be applied (see Chapters 25 and 26 of Lawson and Hanson, 1974). One that we have found useful is to examine a plot of $\ln\|\mathbf{r}^{(k)}\|$ vs $\ln\|\mathbf{P}^{(k)}\|$. We endeavor to select k such that $\|\mathbf{r}^{(k)}\|$ is sufficiently small without $\|\mathbf{P}^{(k)}\|$ getting too large. This is directly analogous to the "energy" constraint of some regularization methods. $\|\mathbf{h}\|$ denotes the L^2 norm of $h(x)$ on $[x_1, x_2]$ with $\|\mathbf{h}\|^2 = \int_{x_1}^{x_2} |h(x)|^2 \, dx$.

Subroutine SVDRS of Lawson and Hanson would seem to deal satisfactorily with possible numerical problems that might arise during computation. Alternatively the IMSL routine LSVDF or equivalent may be used; again we caution the amateur computer enthusiast against attempting to improve on these routines.

The results of the nonlinear double-exponential analysis are presented in Table 7.5.1 in a fashion similar to Table 7.4.2. Again, convergence of the Levenberg–Marquardt algorithm was judged from the estimated number of significant digits in each of the parameters, i.e., none of the parameters changed by more than 0.1% in successive iterations. We note that although the value of $\bar{\Gamma}$ agrees with that of the cumulant analysis, the variance ($= \mu_2/\bar{\Gamma}^2$) is significantly lower.

The linear problem, where the Γ_j are fixed, is often used to obtain more information about the shape of a well-behaved unimodal linewidth distribution function. In practice we have found it extremely difficult to obtain data with sufficient accuracy to warrant the extraction of enough parameters to describe a bimodal distribution without additional constraints. Indeed, our studies on generated data have shown that with this approach resolution of a bimodal distribution composed of two narrow peaks of equal height and $\Gamma_2/\Gamma_1 = 4$ is only marginally possible with just the numerical noise of single-precision (32-bit) computations present. In the linear problem one fixes the

Table 7.5.1

Double-Exponential Analysis of PMMA Data[a,b,c]

		P_j			
χ^2	$\lambda^{(N)}$	$j = 1$	$j = 2$	$j = 3$	$j = 4$
1.519E + 04	1.000E + 01	2.113E + 02	1.048E + 03	2.113E + 02	4.192E + 03
1.085E + 04	1.000E + 01	2.125E + 02	1.042E + 03	2.137E + 02	4.122E + 03
7.091E + 03	5.000E + 00	2.140E + 02	1.036E + 03	2.169E + 02	4.036E + 03
5.097E + 03	2.500E + 00	2.153E + 02	1.037E + 03	2.202E + 02	3.954E + 03
4.339E + 03	1.250E + 00	2.160E + 02	1.048E + 03	2.227E + 02	3.885E + 03
3.965E + 03	6.250E − 01	2.161E + 02	1.067E + 03	2.240E + 02	3.814E + 03
3.719E + 03	3.125E − 01	2.159E + 02	1.088E + 03	2.238E + 02	3.721E + 03
3.558E + 03	1.562E − 01	2.155E + 02	1.107E + 03	2.225E + 02	3.613E + 03
3.484E + 03	7.812E − 02	2.152E + 02	1.124E + 03	2.211E + 02	3.520E + 03
3.466E + 03	3.906E − 02	2.150E + 02	1.134E + 03	2.202E + 02	3.464E + 03
3.464E + 03	1.953E − 02	2.150E + 02	1.138E + 03	2.198E + 02	3.444E + 03
3.464E + 03	9.766E − 03	2.151E + 02	1.139E + 03	2.195E + 02	3.441E + 03
3.464E + 03	4.883E − 03	2.154E + 02	1.140E + 03	2.192E + 02	3.443E + 03
3.463E + 03	2.441E − 03	2.160E + 02	1.142E + 03	2.187E + 02	3.449E + 03
3.463E + 03	1.221E − 03	2.171E + 02	1.146E + 03	2.176E + 02	3.459E + 03
3.463E + 03	6.104E − 04	2.190E + 02	1.152E + 03	2.158E + 02	3.476E + 03
3.463E + 03	3.052E − 04	2.218E + 02	1.162E + 03	2.130E + 02	3.502E + 03
3.463E + 03	1.526E − 04	2.251E + 02	1.173E + 03	2.098E + 02	3.534E + 03
3.462E + 03	7.629E − 05	2.278E + 02	1.181E + 03	2.072E + 02	3.561E + 03
3.462E + 03	3.815E − 05	2.291E + 02	1.186E + 03	2.059E + 02	3.575E + 03
3.462E + 03	1.907E − 05	2.295E + 02	1.187E + 03	2.055E + 02	3.579E + 03
3.462E + 03	9.537E − 06	2.296E + 02	1.187E + 03	2.054E + 02	3.580E + 03

(a) $[G_k^{(2)}(\tau) - A]^{1/2} = P_1 \exp(-P_2\tau) + P_3 \exp(-P_4\tau)$. Fit to 92 points beginning with first point.

(b) $\bar{\Gamma} = 2.32 \times 10^3 \ \text{sec}^{-1}$; $\mu_2/\bar{\Gamma}^2 = 0.266$.

(c) The iterative sequence used for computing the P_j's was terminated when none of the parameters changed by more than 0.1% between successive iterations.

locations of the δ-functions, based again on estimates made from the results of the cumulant or double-exponential analysis. As a first approximation, the characteristic linewidth distribution can be considered to be Gaussian and the first two moments of the cumulant expansion used to estimate the linewidth range that contains some large percentage of the distribution. One then finds the relative contributions from each of the δ-functions using a linear least-squares algorithm. If the range of the distribution is not estimated correctly, the inversion will often indicate this by letting the weighting function cross the axis (negative points in the distribution indicating that the range was overestimated on that end of the distribution) or by failure to achieve a distribution that approaches the axis at the endpoints (indicative of under-estimation of the range). One must be careful to address the problem of the ill-conditioning properly, however, as the trends described above can be a result of overspecification of the problem rather than true indicators of the proper linewidth range.

The multiexponential problem with fixed δ-functions can be further modified by the introduction of *a priori* constraints such as nonnegativity, smoothness, etc. In these cases, the problem again becomes a nonlinear one, and the choice of minimization must be made carefully and with a full understanding of the problem of the ill-posedness, since the aim of the introduction of the constraints is to reduce the ill-conditioning, and one must be cognizant of how much improvement has been effected.

Substituting the model (7.5.1) into Eq. (7.1.1) gives

$$b_i = \sum_{j=1}^{n} P_j \exp(-\Gamma_j \tau_i). \qquad (7.5.20)$$

In this linear problem, the elements of C are as described in Eq. (7.5.6) above, and the vector **P** holds the weighting factors P_j.

It should be added here that the eigenvalue analysis of the Laplace transform by McWhirter and Pike (1978) showed that the Γ_j should be spaced logarithmically in Γ (i.e., linear spacing in $\ln \Gamma$) to maximize the transmission of information across Eq. (7.1.1). When the δ-functions are spaced unequally, e.g., logarithmically, one should be careful in viewing the results. We have implicitly assumed that the continuous distribution can be sufficiently well represented by the discrete model. It must be kept in mind that the model, and thus the results, are discrete. One cannot draw a line between the points (Γ_j, P_j) and expect it to be a picture of the continuous distribution. If it is desired to *estimate* the behavior of the (assumed to be approximately equivalent) continuous distribution, one must correct for the unequal spacing of the Γ_j. This can be done by plotting the points $(\Gamma_j, P_j/\Gamma_j)$ and drawing a continuous curve through them, as has been shown in the figures that show the multi-exponential analysis results (Fig. 7.5.1(a)–(d)), with dashed vertical lines drawn from the Γ-axis to the curve to represent the locations of the functions used in the model. It is imperative to keep in mind that although the output of the algorithm can be a large number of parameters (typically 20), this does not mean that 20 *independent* parameters have been recovered from the data. Indeed, the vector **P** is reconstructed from a limited number (k, typically 3 or 4) of basis functions.

Results of the singular-value decomposition analysis of the multiexponential approximation applied to the PMMA data are shown in Fig. 7.5.1(a)–(d). Initial estimates were obtained from the cumulant results and a little Kentucky windage. In Fig. 7.5.1(a) is shown the $k = 3$ solution to the problem where the range of $G(\Gamma)$ was overestimated. The zeros of $G(\Gamma)$ were interpolated and used as improved estimates of the range; this process was repeated

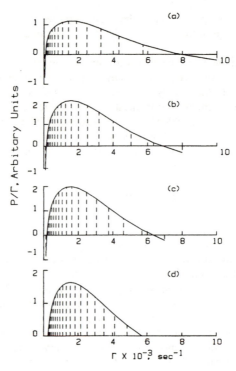

FIG. 7.5.1. Results of the linear multiexponential singular-value decomposition analysis of the PMMA data of Table 7.4.1. (a) $\Gamma_{min} = 50 \text{ sec}^{-1}$, $\Gamma_{max} = 10^4 \text{ sec}^{-1}$; (b) $\Gamma_{min} = 100 \text{ sec}^{-1}$, $\Gamma_{max} = 8000 \text{ sec}^{-1}$; (c) $\Gamma_{min} = 135 \text{ sec}^{-1}$, $\Gamma_{max} = 7000 \text{ sec}^{-1}$; (d) $\Gamma_{min} = 260 \text{ sec}^{-1}$, $\Gamma_{max} = 5700 \text{ sec}^{-1}$. For each range the $k = 3$ solution is shown; the $k = 4$ solution exhibited similar behavior. The values of Γ_{min} and Γ_{max} in (b) were estimated from the axis intercepts of (a), and similarly (c) from (b). (d) was the first solution satisfying positivity, and occurred several iterations after (c).

(Fig. 7.5.1(b) and (c)) until the resulting linewidth distribution was nonnegative (Fig. 7.5.1(d)). At each analysis, we required that the solution we felt to be the correct one for the range being investigated have a $k + 1$ solution that exhibited similar behavior. If the k and $k + 1$ solutions were not similar, we decided that the results did not indicate a true solution. This requirement is overstrict, but is the only one we have found that enables us to detect the occasional misleading results. Values of $\bar{\Gamma}$ and $\mu_2/\bar{\Gamma}^2$ computed from the distribution of Fig. 7.5.1(d) are shown in Table 7.5.2 along with the values of the P_j. We note that $\bar{\Gamma}$ again agrees with the previous results; the variance agrees well with that determined by the double-exponential method.

Table 7.5.2

Multiexponential Analysis of PMMA Data[a,b]

j	Γ_j	P_j	P_j/Γ_j
1	2.60000E+02	3.19001E−01	1.22693E−03
2	3.05877E+02	4.45129E+00	1.45525E−02
3	3.59849E+02	9.44640E+00	2.62510E−02
4	4.23345E+02	1.54904E+01	3.65906E−02
5	4.98045E+02	2.28029E+01	4.57848E−02
6	5.85925E+02	3.16337E+01	5.39894E−02
7	6.89312E+02	4.22522E+01	6.12961E−02
8	8.10942E+02	5.49178E+01	6.77210E−02
9	9.54033E+02	6.98258E+01	7.31901E−02
10	1.12237E+03	8.70107E+01	7.75239E−02
11	1.32042E+03	1.06200E+02	8.04292E−02
12	1.55341E+03	1.26616E+02	8.15085E−02
13	1.82750E+03	1.46751E+02	8.03014E−02
14	2.14997E+03	1.64188E+02	7.63677E−02
15	2.52933E+03	1.75555E+02	6.94078E−02
16	2.97564E+03	1.76722E+02	5.93895E−02
17	3.50069E+03	1.63260E+02	4.66364E−02
18	4.11839E+03	1.31107E+02	3.18346E−02
19	4.84508E+03	7.72596E+01	1.59460E−02
20	5.70000E+03	2.98877E+01	5.24346E−05

(a) $G(\Gamma) = \sum P_j \delta(\Gamma - \Gamma_j)$. Fit to 92 points beginning with first. Parameters of the $k = 3$ candidate solution.

(b) $\bar{\Gamma} = 2.30 \times 10^3 \sec^{-1}$; $\mu_2/\bar{\Gamma}^2 = 0.262$.

7.6. THE MAXIMUM-ENTROPY FORMALISM

7.6.1. INTRODUCTION

The following description is a slightly abbreviated version of the article by Nyeo and Chu (1989, pp. 3999–4004, 4007–4009).*

The first application of the principle of entropy maximization was made by Jaynes (1957a,b), who was concerned with providing a firm basis for the procedures of statistical mechanics, where one induces by the principle a certain probability structure of the states of some physical system, and then calculates therefrom the averages of the various thermodynamic quantities. The induced probability distribution of the system is the only one that has maximal entropy and corresponds to minimum assumptions about the system.

** Reprinted with permission from *Macromolecules*. Copyright 1989, American Chemical Society.

To describe a statistical-mechanical system in which inductive probabilities are useful and for which given information is insufficient to permit deductive inferences, information theory can be used to provide a useful quantitative measure of the missing information in a probability distribution.

A reasonable measure of missing information in a probability assignment $(p_1, p_2, \ldots, p_N)$ was first introduced by Shannon (1948) and has the form

$$S = -\sum_{i=1}^{N} p_i \log p_i, \tag{7.6.1}$$

with log being ln. Equation (7.6.1) has the maximal solution $p_i = $ constant, and it applies only to cases in which all events are equally likely to occur prior to the observation being made. This so-called entropy function was subsequently generalized by Kullback (1959) to

$$S = -\sum_{i=1}^{N} p_i \log \frac{p_i}{g_i}, \tag{7.6.2}$$

where g_i is a prior probability assignment, which plays the role of an initial model of the distribution of a physical system before any measurements are made. The generalization includes an important fact: that prior probabilities are not necessarily equal. They are constructed on the basis of all available theoretical information about the system. Prior information can lead to improvements and more reliability in estimating the parameters of a system. However, the solution then becomes more biased towards the prior information (Toutenburg, 1982).

The corresponding entropy of a continuous univariate probability distribution $P(x)$ of a mechanical system is defined by

$$S = -\int_{\Omega} P(x) \log \frac{P(x)}{m(x)} dx. \tag{7.6.3}$$

Here $m(x)$ is a prior probability distribution, which has the same transformation property as $P(x)$ and the same meanings as g_i in Eq. (7.6.2), and Ω denotes the domain of the distributions $P(x)$ and $m(x)$. The entropy functional S is an invariant with respect to the transformation of the variable x. If we express Eq. (7.6.3) in terms of another variable y, $y = y(x)$, then we must have

$$S = -\int_{\Omega} P(x) \log \frac{P(x)}{m(x)} dx = -\int_{\Omega^*} Q(y) \log \frac{Q(y)}{Q^0(y)} dy, \tag{7.6.4}$$

where Ω^* denotes the domain of the posterior distribution $Q(y)$ and the prior distribution $Q^0(y)$ of the system described in terms of the variable y. This invariance property provides a useful relation between distributions of two variables that we can use to describe a physical system. In photon correlation spectroscopy (PCS), for example, a characteristic linewidth distribution $G(\Gamma)$

or a relaxation-time distribution $H(t)$ can be used to describe a polymer system, where the characteristic linewidth Γ and the relaxation time t are related by $\Gamma = 1/t$. In this case, by the invariance definition of the entropy of the system, we should identify $P(x)$ and $Q(y)$ in Eq. (7.6.4) with $G(\Gamma)$ and $H(t)$, respectively, and introduce their corresponding prior distributions.

It should be realized that the use of an entropy expression for inference includes the introduction of a prior probability distribution, in which we summarize the prior information on the parameters of a system. Therefore, any sensible subsequent estimation of parameters is based on some problem-dependent choice of a prior distribution. This feature is another strength of the MEM, provided that the prior probability distribution is chosen properly. Clearly, some choices may bias the results of the estimation process so as to favor parameter values for which the prior distribution is relatively large. We note that a prior distribution $m(x)$ is uniform, or rectangular, if x is the canonical or natural variable of a system. In this instance, the system of interest can be described unambiguously by the parameter x. However, there are cases, referred to as noninformative, in which the scale of the variables of such systems is unknown; then the prior distributions should not be uniform (Jeffreys, 1961; Jaynes, 1968; Tarantola and Valette, 1982). For such cases, Jeffreys (1961) proposed a scheme that requires that $m(x)$ be invariant with respect to *powers* in x. Namely, we should choose

$$m(x) \propto x^{-1}, \tag{7.6.5}$$

so that for $y = x^n$,

$$Q^0(y) \propto y^{-1}. \tag{7.6.6}$$

This proposition was subsequently substantiated by Jaynes (1968) by a group invariance-principle argument. Such an assignment is essential if the use of the maximum-entropy formalism is to lead to objective and consistent solutions. PCS provides a typical example where such an assignment is needed (see Section 7.6.2).

According to the formalism, we should introduce into the optimization problem constraints, such as data, whose effect is to introduce dependency among the parameters of the system, and then maximize Eq. (7.6.2) or Eq. (7.6.3) subject to these constraints. This maximization then leads to the most probable distribution, which satisfies the constraints and corresponds to minimum assumptions about the system.

For general mathematically ill-posed inverse problems, Jaynes (1984) suggests that they should be treated as problems of inference and that the principle of maximum entropy can provide the least biased estimates on the basis of the available information about the problems. Such an approach is clearly conceptually distinct from the conventional differential-operator regularization approach. The major difference is that in the maximum-entropy (regularization) formalism, correlations in the maximal-entropy solution are dictated purely by data and by the nature of a problem, while in the differential-operator formalism, correlations are also effected by our choice of the order of an operator. In the next subsection, we shall describe how the maximum-entropy formalism can be applied to analyzing PCS data.

7.6.2. APPLICATION OF THE MEM TO PCS

Although the maximum-entropy formalism is usually applied to inducing a probability distribution on the basis of partial information, it can also be used (Levine, 1980; Jaynes, 1984) as an inversion procedure for making inductive inferences on the best estimates of general variables, or on the least-biased solutions of inverse problems. Thus, in this section we describe how the maximum-entropy formalism can be applied to PCS.

To get an estimate $\hat{G}(\Gamma)$ of the true characteristic linewidth distribution $G(\Gamma)$ from the data $d_i = \beta^{1/2}|g_{\exp}^{(1)}(\tau_i)|$, we first consider the discretized relation between $\hat{G}(\Gamma_j)$ and d_i:

$$\hat{d}_i = \sum_{j=1}^{N} K_{ij}\hat{G}(\Gamma_j)\Delta, \qquad i = 1,\ldots,M. \qquad (7.6.7)$$

Here $K_{ij} = \exp(-\Gamma_j\tau_i)$ is the curvature matrix, M is the number of data, N is the number of function values $\hat{G}(\Gamma_j)$ used in the discretization approximation (7.6.7), N is similar to n in Eq. (7.5.1), and Δ is a constant-Γ spacing. Then we expand $\hat{G}(\Gamma)$ in the set of random variables $\mathbf{X} = \{X_j\}$ comprising all possible solutions (Levine, 1980; Steenstrup, 1985):

$$\hat{G}(\Gamma_j) = \int X_j P(\mathbf{X})\, d\mathbf{X}, \int P(\mathbf{X})\, d\mathbf{X} = 1, \qquad (7.6.8)$$

where $P(\mathbf{X})$ is a multivariate probability distribution. We note that by defining $\hat{G}(\Gamma_j)$ as averages we get the best estimates in the sense that $\hat{G}(\Gamma_j)$ have minimal deviations from the true values $G(\Gamma_j)$. Substituting the first equation in (7.6.8) into (7.6.7) then gives

$$\hat{d}_i = \int \Phi_i(\mathbf{X}) P(\mathbf{X})\, d\mathbf{X}, \qquad (7.6.9)$$

where $\Phi_i(\mathbf{X}) = \sum_j K_{ij}X_j$. By the philosophy of the maximum-entropy formalism, we can now set up a Lagrangian for the probability distribution as follows:

$$S = -\int P(\mathbf{X})\log\frac{P(\mathbf{X})}{m(\mathbf{X})}d\mathbf{X} - \mu\int P(\mathbf{X})\, d\mathbf{X} - \sum_{i=1}^{M}\lambda_i\int\Phi_i(\mathbf{X})P(\mathbf{X})\, d\mathbf{X}. \quad (7.6.10)$$

Here, as usual, $m(\mathbf{X})$ is a prior distribution, and μ and λ_i are Lagrange multipliers, whose values are fixed by the conditions (7.6.8). Maximizing (7.6.10) with respect to $P(\mathbf{X})$ then leads to the Euler–Lagrange equation (Jaynes, 1957; Jaynes, 1957; Levine and Tribus, 1979; Levine, 1980)

$$P(\mathbf{X}) = m(\mathbf{X}) \exp\left[-1 - \mu - \sum_{i=1}^{M} \lambda_i \Phi_i(\mathbf{X}) \right] \qquad (7.6.11)$$

or the partition function

$$Z\left[\alpha_j = \sum_i \lambda_i K_{ij} \right] = \exp(1 + \mu) = \int m(\mathbf{X}) \exp\left[-\sum_{j=1}^{N} \alpha_j X_j \right] d\mathbf{X}. \qquad (7.6.12)$$

Equation (7.6.12) is in fact a general expression for an inverse problem. For an inverse problem, we only have to define a prior distribution $m(\mathbf{X})$ and a kernel K_{ij}. By using Eq. (7.6.12), we can calculate $\hat{G}(\Gamma_j)$ and $\hat{d}_i$ by Eq. (7.6.8) and (7.6.9), respectively. But we should remember to relate the experimental data d_i with the data estimates $\hat{d}_i$ by some statistic criterion, such as a χ^2 constraint. The χ^2 condition that we shall use is characterized by

$$\chi^2 = \left\| \frac{\hat{d}_i - d_i}{\sigma_i} \right\|^2 = \sum_{i=1}^{M} \left(\frac{\hat{d}_i - d_i}{\sigma_i} \right)^2 = M, \qquad (7.6.13)$$

which simply indicates that the estimated or predicted and the observed spreads in the observations are, on the average, equal for each channel value.

Equally important is to provide an estimate of the error-squared deviations σ_i^2 of the data d_i. *A data analysis method without some sort of error estimate is useless.* Since a general analytic expression for σ_i^2, which includes all possible experimental conditions, is not accessible, an approximation is necessary. We shall *assume* that the squared deviations of the correlation have a Poisson profile, so that the deviations of d_i are given by (Bartlett, 1947; Kendall and Stuart, 1968)

$$\sigma_i^2 \approx \frac{1 + d_i^2}{4 A d_i^2}. \qquad (7.6.14)$$

Equation (7.6.14) exhibits a reasonable feature of a correlation function: that data at longer delay times have larger σ_i^2. The form of σ_i^2 can be pivotal in the effective use of the MEM approach. This is a point often overlooked by investigators who want to improve the resolving power of the MEM. (See Appendix 7.B.)

With some prescribed $m(\mathbf{X})$, the Laplace inversion problem now consists of solving Eqs. (7.6.9), (7.6.12), and (7.6.13) for the parameters $\{\lambda_i\}$ or $\{\alpha_i\}$. This problem in its present form is numerically inefficient (Alhassid *et al.*, 1978). So instead of solving these equations, we shall use an equivalent but simpler maximum-entropy formalism (Levine, 1980; Steenstrup, 1985). Since a characteristic linewidth distribution is semipositive and bounded, we can take it as a probability density. Thus, we can define

$$S = - \sum_{j=1}^{N} p_j \ln \frac{p_j}{m_j}, \qquad p_j = \frac{\hat{G}(\Gamma_j)\Delta}{\sum_j \hat{G}(\Gamma_j)\Delta} \qquad (7.6.15)$$

as the entropy of our system, where p_j is the representative fraction of particles (or macromolecules) with characteristic linewidth in the interval $(\Gamma_j - \Delta/2,$ $\Gamma_j + \Delta/2)$. Here Δ is a constant spacing and Γ_j are equally spaced. However, in this case, the prior model distribution $\{m_j\}$ is not uniform. As mentioned in Section 7.6.1, the scale of the parameter Γ is *a priori* unknown. So we should choose $m_j \propto 1/\Gamma_j$, which is an objective choice. We can take either a characteristic linewidth Γ or a relaxation (or decay) time t $(=1/\Gamma)$ as a parameter for describing a system. The objective choice for the prior distribution is then either $m_j \propto 1/\Gamma_j$ or $m_j^0 \propto 1/t_j$ (cf. Eqs. (7.6.5) and (7.6.6)). Otherwise, the choice of $m_j = \text{constant}$ would lead to $m_j^0 \propto 1/(t_j)^2$ and not to $m_j^0 = \text{constant}$, and vice versa. If we now describe the problem in terms of $\ln \Gamma$ or $\ln t$, we can use objectively a uniform prior distribution. This choice also makes the description of a characteristic linewidth distribution more effective, so that very wide distributions can be better specified. Accordingly, we should define the fractions of particle sizes in the interval $(\ln \Gamma_j - \Delta/2, \ln \Gamma_j + \Delta/2)$ as $F_j = \hat{G}(\Gamma_j)\Gamma_j\Delta$, where $\Delta = (\ln \Gamma_N - \ln \Gamma_1)/(N - 1)$ is a spacing in $\ln \Gamma$, and N is the number of logarithmically spaced Γ_j-values satisfying $\ln \Gamma_{j+1} = \ln \Gamma_j + \Delta$. Normally, to describe very broad and multimodal distributions, we may take any value of N between 30 and 100. (In our program $N = 81$ is used.) The lower and upper bounds Γ_1 and Γ_N are chosen such that $\hat{G}(\Gamma_1)$ and $\hat{G}(\Gamma_N)$ are negligibly small. It should be emphasized that in order to define $\hat{G}(\Gamma)$, sample times should be chosen such that the relations $\Gamma_N = 1/\tau_1$ and $\Gamma_1 = 0.01/\tau_M$ are approximately satisfied (Provencher, 1979), where τ_1 and τ_M are the delay times of the first and the last channel, respectively. We can now use F_j as our "probabilities" and a uniform prior distribution to define an entropy expression. In this case the Laplace transform simply reads

$$\hat{d}_i = \sum_{j=1}^{N} F_j K_{ij}, \qquad K_{ij} = \exp(-\Gamma_j \tau_i), \qquad (7.6.16)$$

where K_{ij} is the curvature matrix (kernel). We should note that since in PCS characteristic linewidth distributions are only normalized to $\sqrt{\beta}$, i.e., $\sum_j F_j = \sqrt{\beta}$, the probabilities should be $p_j = F_j/\sum_j F_j$. But the entropy depends only on the form of $\{F_j\}$ and not on its normalization; thus it is computationally advantageous to use the following definition:

$$S = - \sum_{j=1}^{N} F_j \ln \frac{F_j}{b} = - \sum_{j=1}^{N} \left(F_j \ln \frac{F_j}{A_0} - F_j \right), \qquad (7.6.17)$$

where b and A_0 ($= b/e$, $e = 2.718\ldots$) are constants with A_0 being a default or predetermined value (Burch *et al.*, 1983; Gull and Skilling, 1984), which in the absence of data constraint is the maximal solution of Eq. (7.6.17). Here we choose $A_0 = \sqrt{\beta}/N$, so that F_j and A_0 are normalized to $\sqrt{\beta}$. The inverse problem now amounts to maximizing Eq. (7.6.17) subject to the χ^2 constraint (7.6.13) and the transform (7.6.16).

7.6.3. THE MAXIMAL-ENTROPY SOLUTION

In this section, we shall briefly describe a quadratic model approximation for solving the nonlinear maximum-entropy problem and how the solution can be attained.

The problem can be stated as follows: maximize the objective function with respect to the distribution $\{F_j\}$,

$$S = - \sum_{j=1}^{N} \left(F_j \ln \frac{F_j}{A_0} - F_j \right) \qquad (7.6.17)$$

subject to the χ^2 constraint

$$\chi^2 = \sum_{i=1}^{M} \left(\frac{\hat{d}_i - d_i}{\sigma_i} \right)^2 = M, \qquad (7.6.13)$$

where $\hat{d}_i = \sqrt{\beta} |g^{(1)}(\tau_i)|$ are the estimates on the data d_i and are related to F_j by Eq. (7.6.16). Equations (7.6.13), (7.6.16), and (7.6.17) describe a nonlinearly equality-constrained optimization problem and can be solved by several standard approaches (Wismer and Chattergy, 1978; Harley, 1986; Freeman, 1986; Beightler *et al.*, 1979). For instructional purposes, we briefly describe a quadratic model approximation method described by Burch *et al.* (1983) (see also Skilling and Bryan, 1984; Skilling and Gull, 1985).

The quadratic model approximation approach amounts to first approximating Eqs. (7.6.13) and (7.6.17) as

$$\hat{S} = S_0 + \sum_{j=1}^{N} \frac{\partial S}{\partial F_j} \delta F_j + \frac{1}{2} \sum_{j,k}^{N} \frac{\partial^2 S}{\partial F_j \partial F_k} \delta F_j \delta F_k, \qquad (7.6.18)$$

$$\hat{\chi}^2 = \chi_0^2 + \sum_{j=1}^{N} \frac{\partial \chi^2}{\partial F_j} \delta F_j + \frac{1}{2} \sum_{j,k}^{N} \frac{\partial^2 \chi^2}{\partial F_j \partial F_k} \delta F_j \delta F_k, \qquad (7.6.19)$$

where δF_j are the corrections to $F_j^0 = A_0$ (the default value), at which S_0, χ_0^2, and the derivatives of S and χ^2 are evaluated. Note that the χ^2 expansion terminates at the second derivative and is therefore exact. However, since the entropy function is highly nonlinear, the quadratic expansion of it is only approximately valid locally about $F_j^0 = A$; that is, the quadratic term in

Eqs. (7.6.18) must be smaller than its preceding terms. It is necessary to set an upper bound on the quadratic term:

$$-\sum_{j,k}^{N} \frac{\partial^2 S}{\partial F_j \partial F_k} \delta F_j \delta F_k \leq 0.1 \sum_{j}^{N} F_j^0, \qquad (7.6.20)$$

which is chosen by practical experience.

For nonlinear models, it is necessary to solve for the solution iteratively by calculating the corrections δF_j to the preceding approximate distribution. Thus, in the first iteration we calculate

$$F_j = F_j^0 + \delta F_j \qquad (7.6.21)$$

and use them as the second estimates at which a new quadratic model approximation can be constructed. Then we define δF_j in terms of several search directions e^μ (normally three is sufficient) in which we make the expansion:

$$\delta F_j = \sum_{\mu=1}^{3} x_\mu e_j^\mu. \qquad (7.6.22)$$

To achieve efficiently the maximization of $\hat{S}$ under the χ^2 (7.6.13), the following simple unit vectors are used (Burch et al., 1983):

$$e_j^1 \propto \frac{F_j \partial \chi^2}{\partial F_j}, \qquad (7.6.23)$$

$$e_j^2 \propto F_j \left(\alpha_1 \frac{\partial S}{\partial F_j} - \alpha_2 \frac{\partial \chi^2}{\partial F_j} \right), \qquad (7.6.24)$$

$$e_j^3 \propto \sum_{k=1}^{N} F_j \frac{\partial^2 \chi^2}{\partial F_j \partial F_k} e_k^2, \qquad (7.6.25)$$

where α_1 and α_2 are given by

$$\alpha_1 = \left[\sum_j F_j \left(\frac{\partial S}{\partial F_j} \right)^2 \right]^{-1/2}, \qquad \alpha_2 = \left[\sum_j F_j \left(\frac{\partial \chi^2}{\partial F_j} \right)^2 \right]^{-1/2}, \qquad (7.6.26)$$

so that the first and second quantities in Eq. (7.6.24) are normalized with respect to the metric tensor $g_{jk} = \delta_{jk}/F_j$ (δ_{jk} being the Kronecker delta). The metric tensor g_{jk}, which is minus the second derivative of the entropy function, $-\partial^2 S/\partial F_j \partial F_k$, provides a local measure of the magnitude and direction of the corrections (cf. Eq. (7.6.20)) about some map $\{F_j\}$. Thus, it is effective to use this metric tensor for defining local quantities. For instance, the unit vectors (7.6.23)–(7.6.25) are normalized according to

$$\|e^\mu\|^2 = \sum_{j,k}^{N} g_{jk} e_j^\mu e_k^\mu = 1, \qquad \mu = 1, 2, 3. \qquad (7.6.27)$$

Substituting the three vectors into Eqs. (7.6.18) and (7.6.19), we can rewrite them in terms of $x_\mu (\mu = 1, 2, 3)$, respectively, as

$$\hat{S} = S_0 + \sum_{\mu=1}^{3} S_\mu x_\mu - \frac{1}{2} \sum_{\mu,\nu}^{3} h_{\mu\nu} x_\mu x_\nu, \qquad (7.6.28)$$

$$\hat{\chi}^2 = \chi_0^2 + \sum_{\mu=1}^{3} C_\mu x_\mu + \frac{1}{2} \sum_{\mu,\nu}^{3} t_{\mu\nu} x_\mu x_\nu, \qquad (7.6.29)$$

where

$$S_\mu = \sum_{j=1}^{N} \frac{\partial S}{\partial F_j} e_j^\mu, \qquad h_{\mu\nu} = -\sum_{j,k}^{N} \frac{\partial^2 S}{\partial F_j \partial F_k} e_j^\mu e_k^\nu, \qquad (7.6.30)$$

$$C_\mu = \sum_{j=1}^{N} \frac{\partial \chi^2}{\partial F_j} e_j^\mu, \qquad t_{\mu\nu} = \sum_{j,k}^{N} \frac{\partial^2 \chi^2}{\partial F_j \partial F_k} e_j^\mu e_k^\nu. \qquad (7.6.31)$$

The distance constraint (7.6.20) reduces simply to

$$l = \sum_{\mu,\nu}^{3} h_{\mu\nu} x_\mu x_\nu \leq 0.1 \sum_{j}^{N} F_j^0. \qquad (7.6.32)$$

A Lagrangian can now be set up: $L(x) = \alpha S - \chi^2 - \beta l$, with α and β being positive parameters. (We note that we can use instead the Lagrangian $L'(x) = S - \alpha'\chi^2 - \beta'l$, with positive parameters α' and β', which amounts simply to the rescalings $L(x) \to \alpha L'(x)$, $\alpha' = 1/\alpha$, $\beta' = \beta/\alpha$. Thus, here the use of $L(x)$ is purely for convenience.) Maximizing $L(x)$ leads to a set of coupled equations for x_μ. For computational efficiency, we first decouple the components x_μ by transforming them (Beightler *et al.*, 1979; Birkhoff and MacLane, 1965) to an orthogonal set $\{y_\mu\}$, so that the matrices or tensors $h_{\mu\nu}$ and $t_{\mu\nu}$ become diagonal. By so doing, the Lagrangian now reads

$$L(y) = \alpha S_0 - \chi_0^2 + \sum_{\mu=1}^{3} (\alpha \tilde{S}_\mu - \tilde{C}_\mu) y_\mu$$
$$- \frac{1}{2} \sum_{\mu,\nu}^{3} ((\alpha + 2\beta) \delta_{\mu\nu} + \Lambda_{\mu\mu})) y_\mu y_\nu, \qquad (7.6.33)$$

where tildes denote quantities that are orthogonally transformed, $\delta_{\mu\nu}$ is a Kronecker delta, and $\Lambda_{\mu\mu} (\mu = 1, 2, 3)$ are the diagonal elements of the diagonalized matrix of $t_{\mu\nu}$. Maximization of this Lagrangian can be carried out in many ways. For example, the fact that Eq. (7.6.33) is stationary with respect to the variations of y_μ leads to the Euler–Lagrange equations for y_μ,

$$y_\mu = (\alpha \tilde{S}_\mu - \tilde{C}_\mu)(\alpha + 2\beta + \Lambda_{\mu\mu})^{-1}. \qquad (7.6.34)$$

Equation (7.6.34) gives infinitely many maximal-entropy solutions, which are classified by the parameters α and β. Of all the possible solutions we want to find one that satisfies our statistical criterion and the distance constraint (7.6.32), which is now given by $l = \sum_\mu y_\mu y_\mu$. But since at each iteration the minimum attainable χ^2-value is (cf. Eq. (7.6.33))

$$\chi^2_{\min} = \chi^2_0 - \frac{1}{2} \sum_{\mu=1}^{3} \frac{C_\mu C_\mu}{\Lambda_{\mu\mu}}, \qquad (7.6.35)$$

which can be larger than our required value M, a slightly higher χ^2-value is imposed to provide some flexibility in the iterative procedure:

$$\chi^2_* = \chi^2_0 - \frac{1}{3} \sum_{\mu=1}^{3} \frac{C_\mu C_\mu}{\Lambda_{\mu\mu}}. \qquad (7.6.36)$$

Thus, the refined χ^2-value in the procedure is given by

$$\tilde{\chi}^2 = \text{maximum}(\chi^2_0, \chi^2 = M). \qquad (7.6.37)$$

How the solution is chosen is briefly described as follows. At each iteration we first set $\beta = 0$ and find the largest α-value such that the required χ^2-value (7.6.37) is attainable. Then the parameter β is increased until the subsequent solution y_μ obeys the distance constraint. If the constraint is not satisfied, the $\tilde{\chi}^2$-value is increased to χ^2_0. The corrections δF_j are then obtained by backward substitution.

The required solution $\{F_j\}$ is the one that satisfies some terminating criteria, e.g.,

$$\max_i |\hat{d}_i^{(n+1)} - \hat{d}_i^{(n)}| \leq \delta_i \qquad (7.6.38)$$

and

$$\max_j |F_j^{(n+1)} - F_j^{(n)}| \leq \delta_j. \qquad (7.6.39)$$

Alternatively, the maximal-entropy solution can be selected by using the following criterion:

$$\sum_{j=1}^{N} F_j \left(\alpha_1 \frac{\partial S}{\partial F_j} - \alpha_2 \frac{\partial \chi^2}{\partial F_j} \right)^2 < 2 \times 10^{-3}, \qquad (7.6.40)$$

where α_1 and α_2 are given in (7.6.26). Equation (7.6.40) is a measure of the degree of entropy maximization, where the tolerance value of 2×10^{-3} is determined on practical grounds. We note that a zero tolerance value corresponds to the ideal maximal solution. In addition, the use of quadratic model approximation unavoidably loses the positivity property of the solution in the iterative procedure. Thus, it is necessary to reset any negative values in each iteration to a small but positive number.

For a set of 136 data points, about 6 seconds is needed for an iteration by using a microcomputer with a 12-MHz 80286 math coprocessor and Microsoft FORTRAN and about 100 iterations are required. By comparison, one second per iteration is needed if an IBM PS/2 model 80 microcomputer is used.

In Section 7.6, we have pedagogically (1) introduced the maximum-entropy formalism, (2) outlined its use in estimating solutions to inverse problems in general and to the Laplace inverse problem in PCS in particular, and (3) described a quadratic model approach for solving the nonlinearly constrained maximum-entropy optimization problem in PCS, and specified how its solution was achieved. The reliability of the formalism has been tested successfully by using several sets of numerically simulated time-correlation-function data corresponding to known characteristic linewidth distributions.

The main thrust of using the maximum-entropy (MEM) formalism lies in its proven objectivity. In this formalism, information in the maximum-entropy solution is dictated purely by the available data, and not by an arbitrarily introduced differential operator such as one may find in the regularization formalism. The maximum-entropy formalism requires introduction of a prior probability distribution in which we can summarize our prior information. For example, any theoretical information about the physical problem of interest can be taken into account here. Such an introduction depends on the nature of the problem. For most problems, the corresponding prior distributions are uniform. In the Laplace inverse problem in PCS, however, the prior distribution is not uniform but $\propto 1/\Gamma$ or $\propto 1/t$, depending on whether the characteristic linewidth distribution $G(\Gamma)$ or the relaxation-time distribution $H(t)$ is used for describing the system of interest. This choice of prior distribution leads to adopting a $\log \Gamma$ (or $\log t$) description of $G(\Gamma)$ (or $H(t)$) and hence to specifying very broad $G(\Gamma)$ (or $H(t)$) more effectively.

We have described one method for solving the nonlinear maximum-entropy problem, namely the quadratic model approximation (7.6.3). The quadratic model approximation yields an acceptable solution, which is, unfortunately, not positive definite. Better approaches to solving the MEM problem could undoubtedly be devised (e.g., see Section 7.8). It should, however, be emphasized that knowledge of the error deviations of the experimental data can substantially improve the solution for $G(\Gamma)$, including the resolution of $G(\Gamma)$. This error knowledge on the experimental data depends on subjective evaluation as well as systematic errors, which may be difficult to estimate. Preliminary analysis methods used for estimating or refining error deviations of measured data may introduce unintended bias into our $G(\Gamma)$ reconstruction. It remains to be seen what preliminary analysis method can best provide the least biased data deviations. As the analysis must necessarily depend on the physical nature of the experiment, a generalized approach may be difficult to achieve.

In addition to the noise in the measured data, we have alleviated the bandwidth limitation on the experimentally measured time correlation function by insisting on predetermining the value of β and by letting the first-channel value of the correlation function, $|g^{(1)}(\Delta t)|^2$, be greater than 0.998 and the last-channel value, $|g^{(1)}(\text{last channel})|^2$, be less than 0.005.

7.7. COMPARISON OF THE MEM WITH CONTIN

In our comparison of the MEM and CONTIN algorithms with simulated data, we have emphasized their utility for unimodal $G(\Gamma)$ and the limitations related to negatively skewed $G(\Gamma)$ distributions and to bimodal $G(\Gamma)$ distributions that are not delta functions. Finally, we demonstrated the application of the MEM to experimental data and showed its utility to be comparable to that of the CONTIN algorithm (Nyeo and Chu, 1989).

In making an actual comparison of the CONTIN and MEM members using experimentally measured photon correlation data, it should be noted that with normal unimodal characteristic linewidth distributions $G(\Gamma)$, both methods yield very good results. For skewed $G(\Gamma)$, the capability of CONTIN and the MEM excels, as shown in Table 7.7.2 for the methods listed in Table 7.7.1. In Table 7.7.2, we used a lognormal distribution (at constant K) for the simulation

$$G(\Gamma) = \frac{1}{\beta\sqrt{\pi}\Gamma} \exp\left[-\frac{1}{\beta^2}\left(\ln\frac{\Gamma}{\Gamma_0}\right)^2 \right], \tag{7.7.1}$$

where $\beta = \sqrt{2\ln(\sigma + 1)}$ and $\Gamma_0 = \bar{\Gamma}(\sigma + 1)^{1/2}$ with $\sigma\ (=\mu_2/\bar{\Gamma}^2)$ being the

Table 7.7.1

Methods of Correlation-Function Profile Analysis

Abbreviation	Method	Reference
CUMFIT	Cumulant expansion	Koppel (1972)
DEXP	Double exponential (nonlinear)	
EXPSAM	Exponential sampling	McWhirter and Pike (1978)
NNLS	Nonnegative least squares	Lawson and Hanson (1974)
MSVD	Multiexponential singular-value decomposition	Ford and Chu (1983)
RILIE	Regularized inversion of Laplace integral equation	Chu et al. (1985)
CONTIN	Constrained regularization method developed by Provencher	Provencher (1982a,b)
MEM	Method of maximum entropy	Nyeo and Chu (1989)

Table 7.7.2

Comparison of Different Methods of Data Analysis

	$\bar{\Gamma}$, rad/sec	σ	Skewness	Kurtosis
Input[a]	750	0.5	2.475	15.563
CUMFIT[b]:				
1st	645			
2nd	713	0.235		
3rd	736	0.349	0.64	
4th	744	0.413	1.10	3.73
DEXP	727	0.223	0.167	
EXPSAM	743	0.407	1.012	
NNLS	749	0.476	1.81	6.92
MSVD:				
4 eigenvalues	731	0.354		
5 eigenvalues	745	0.436		
RILIE	729	0.341		
CONTIN	749	0.476	1.90	7.55
MEM	750	0.488	2.05	

variance. $\sigma = 0.5$ represents a reasonably broad $G(\Gamma)$ distribution such as one encounters in a typical polydisperse synthetic polymer solution.

The simulated field correlation function $|g^{(1)}(j\Delta\tau)|$ (at constant K) has the form

$$|g^{(1)}(j\Delta\tau)| = \int_{\Gamma_{\min}}^{\Gamma_{\max}} G(\Gamma)e^{-\Gamma j\Delta\tau}\,d\Gamma, \qquad (7.7.2)$$

where $\Delta\tau$ is the delay time increment, $\Gamma_{\min} = \Gamma_0(\sigma + 1)^{-1}e^{-3\beta}$, and $\Gamma_{\max} = \Gamma_0(\sigma + 1)^{-1}e^{3\beta}$. The value of $|g^{(1)}(j\Delta\tau)|$ was computed using 1000 segment histograms with $|g^{(1)}(j\Delta\tau)| \simeq \sum_{i=1} \Delta\Gamma\, G(\Gamma_i)e^{-\Gamma_i j\Delta\tau}$ and $\Delta\Gamma = (\Gamma_{\max} - \Gamma_{\min})/1000$. The range of $g^{(1)}(j\Delta\tau)$ was constructed so that $g^{(1)}(\Delta\tau)$ was within $\approx 0.2\%$ of 1 and $g^{(1)}$(last channel) was within $\approx 0.5\%$ of the background (A). The total number of channels, $N(\equiv 136)$, was partitioned into four segments such that the effective delay time increments were $\Delta\tau$ for the first 32 channels, $8(=2^3)\Delta\tau$ for the second 32 channels. $32(=2^5)\Delta\tau$ for the third 32 channels, and $128(=2^7)\Delta\tau$ for the last 40 channels. A 0.1% random noise (RN), representing close to the optimal experimental signal-to-noise ratio achievable, was also introduced into the simulated $|g^{(1)}(j\Delta\tau)|$ such that

$$|g^{(1)}(j\Delta\tau, \text{Eq. (7.6.42) with noise})|$$
$$= [|g^{(1)}(j\Delta\tau, \text{Eq. (7.6.42)})|^2 + RN]^{1/2}. \qquad (7.7.3)$$

A comparison of two polystyrene species ($M_w = 4.48 \times 10^6$ g/mole, $M_w M_n = 1.14$ at 4.7×10^{-5} g/g, and $M_w = 2.33 \times 10^5$ g/mole, $M_w/M_n = 1.06$ at 9.5×10^{-4} g/g) in benzene at 25°C showed comparable findings by CONTIN and the MEM (Chu *et al.*, 1989). Thus, at this time, even with our present algorithm, we can claim the MEM to be at least comparable with CONTIN. With the improved MEM (Appendix 7.B), the resolution seems to exceed that of CONTIN. Further improvements (Section 7.8) will continue as we do not yet have a full understanding of the intricate nature of MEM.

7.8. MAXIMIM-ENTROPY METHOD USING A BAYESIAN ESTIMATE FOR THE REGULARIZATION PARAMETER

In the introduction to Section 7.4, the reader was reminded that Section 7.6 (on the maximum-entropy formalism) could be *skipped* by the less mathematically oriented reader. Sections 7.5 and 7.6 were included in this chapter in order to let interested readers see some aspects of the development of *one* approach to solving the nonlinear maximum-entropy problem. The details for using the MEM approach are complex. For example, by merely changing the form of σ_i^2 from Eq. (7.6.14) to Eq. (7.B.1), we can improve the resolution of MEM as presented in Section 7.6, which, nevertheless, still contains many undesirable features. These could include the problems that the MEM solution in Section 7.6 is very sensitive to baseline fluctuations, and that the $G(\Gamma)$ estimates are based on $|g^{(1)}(\tau)|$, which is only indirectly related to the measured intensity correlation function $G^{(2)}(\tau)$.

A new approach to the maximum-entropy analysis has been developed by using a Bayesian estimate for the regularization parameter. As first suggested by Gull (1989), who showed that the criterion to use χ^2 (Eq. (7.6.13)) tends to underfit the data, Langowski and Bryan (1990) and Bryan (1990b) reported a new computer program (MEXDLS) specifically designed for dynamic light scattering (DLS) data. With this algorithm (Bryan, 1990a,b), which permitted estimates of the baseline and of the variance of the calculated spectra, a better resolution than CONTIN could be achieved. Thus, an application of Bayesian statistics to the analysis of intensity time correlation functions seems to be a better approach for retrieving information from multimodal characteristic linewidth distributions. Interested readers should refer to the references for details. It should be noted that there are many volumes of *Maximum-Entropy and Bayesian Methods* in existence, with the latest two volumes being those edited by Skilling (1989) and by Fougere (1990). Thus, MEM is one of those established methods from another field. It is not as easy a task to transfer the MEM over to dynamic light scattering as one might expect, especially for the less mathematically oriented scientist. The reader should try to obtain well-tested algorithms rather than trying to develop what is already known.

Commercial programs for MEM are available. The reader is advised to obtain detailed information on the mathematical formalism and source code to understand what is in the program and at least what the boundary conditions are. Further improvements on MEM, especially those that take into account specific conditions related to experimental data, are likely to be forthcoming. The use of Laplace inversion as a black box could be a dangerous endeavor. Therefore, the reader should at least try to use simulated data to find out whether the *expected* solutions are reasonable.

7.9. PARTICLE POLYDISPERSITY ANALYSIS IN REAL SPACE

So far we have emphasized the Laplace inversion of the time correlation function from dynamic light scattering (Sections 7.1–7.8). An approximate computation of the distance function $p(r)$ is possible for light scattering intensity data from nonspherical particles in the range $1.0 \leq m \leq 1.3$ and $2.0 \leq \alpha \leq 12.0$ with m and $\alpha(=ka)$ being the relative refractive index and the size parameter, respectively. $k(=2\pi/\lambda)$ and a are the wave number (nm^{-1}) and the major semiaxis of the spheroid (nm), respectively. Figure 7.9.1 shows a

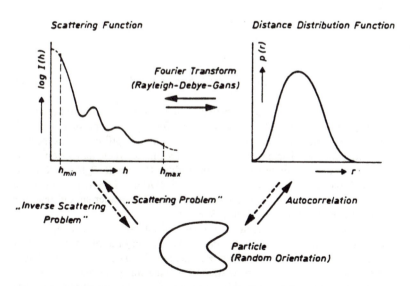

FIG. 7.9.1.   Schematic representation of the correlation between a particle, its observable experimental scattering data, and its distance distribution function. (After Glatter and Hofer, 1988a).

schematic representation of the correlation between a particle, its observable experimental data $I(h)$, and its distance distribution function $p(r)$ with $\mathbf{h} (\equiv \mathbf{q} \equiv \mathbf{K})$ being the scattering vector.

The interpretation of elastic light scattering data in real space for non-spherical and inhomogeneous monodisperse systems (Glatter and Hofer, 1988a) and for the determination of size distributions of polydisperse systems (Glatter and Hofer, 1988b) has been described in some detail.

The inverse scattering problem for $p(r)$ as illustrated in Fig. 7.9.1 has the form

$$p(r) = \frac{1}{2\pi^2} \int_0^\infty I(h)hr \sin hr \, dh. \tag{7.9.1}$$

The same truncation problem related to the finite range of h and unavoidable presence of experimental noise are present in the measured time-averaged scattered intensity $I(h)$. Thus, inversion of $I(h)$ to get $p(r)$ is an ill-posed problem. Similarly, $I(h)$ and the size distributions are related by the integrals

$$I(h) = \int_0^\infty D_n(R)\phi(h, m, R) \, dR, \tag{7.9.2}$$

$$I(h) = \int_0^\infty D_m(R)\phi(h, m, R) \, \phi(0, m, R)^{-1/2} \, dR, \tag{7.9.3}$$

and

$$I(h) = \int_0^\infty D_i(R)\phi(h, m, R)\phi(0, m, R)^{-1} \, dR$$

or

$$I(h) = \int_0^\infty D_i(R)\phi_0(h, m, R) \, dR, \tag{7.9.4}$$

where $D_m(R)$ and $D_n(R)$ are, respectively, the mass distribution and the number distribution. $\phi(h, m, R)$ is the scattering function of a particle of size R and relative refractive index m. $\phi_0(h, m, R) = \phi(h, m, R)\phi(0, m, R)^{-1}$ is the normalized $\phi(h, m, R)$. The parameter can be any characteristic dimension of the particle. In the case of spheres, R is the radius. $D_i(R) \simeq R^3 D_m(R) \simeq R^6 D_n(R)$ in the Rayleigh–Debye–Gans limit where $2kR(m-1) \ll 1$, i.e., for small particles with very low relative refractive index. The reader should read the original work (Glatter and Hofer, 1988a,b).

For polydisperse colloids in the presence of interactions, the coupling between scattering power and interactions plays a very important role. Pusey *et al.* (1982) and Vrij (1982) have considered a hard-sphere mixture based on the Percus–Yevick approximation. The hard-sphere formalism has been extended to polydisperse colloidal particles of spherosymmetrical arbitrary interaction potential (Licinio and Delaye, 1989).

APPENDIX 7.A. (RUF, 1989), RELATION BETWEEN BASELINE UNCERTAINTIES AND $\hat{d}_i^2$

If we let the measured normalized intensity time correlation function be $\hat{G}^{(2)}(\tau)/\hat{A}$, where the accent $\hat{}$ denotes an estimate of the true value, $\hat{G}^{(2)}(\tau) = G^{(2)}(\tau_i) + E(\tau_i)$ with E being an error function, and $\hat{A} = N_s[\hat{n}(t)]^2$, then

$$\frac{\hat{G}^{(2)}(\tau_i)}{\hat{A}} (\equiv 1 + \hat{d}_i^2) = \frac{G^{(2)}(\tau_i)}{A} - \frac{G^{(2)}(\tau_i)\Delta A}{A\hat{A}} + \frac{E(\tau_i)}{\hat{A}}, \qquad (7.A.1)$$

where the noise term $E(\tau_i)/\hat{A}$ is the normalized sum of statistical errors for correlator channel i. More importantly, with $\Delta A = \hat{A} - A$ and $A/\hat{A} = 1 - (\Delta A/\hat{A})$, the term $[G^2(\tau_i)/A](\Delta A/\hat{A})$ represents the deviation of the normalized intensity correlation function $G^{(2)}(\tau_i)/A$ ($\equiv 1 + d_i^2$) due to errors in the baseline $\Delta A/\hat{A}$. Thus, a normalization error in the baseline produces a constant shift in $G^{(2)}(\tau_i)/A$. However, in the Laplace transform, we measure $G^{(2)}(\tau_i)$, but we need $|g^{(1)}(\tau)|$ such that

$$\begin{aligned}
\hat{d}_i^2 &\equiv \frac{\hat{G}^{(2)}(\tau_i)}{\hat{A}} - 1 \\
&= \frac{G^{(2)}(\tau_i)}{A} - \frac{G^{(2)}(\tau_i)}{A}\frac{\Delta A}{\hat{A}} + \frac{E(\tau_i)}{\hat{A}} - 1 \\
&= \beta|g^{(1)}(\tau_i)|^2 - \frac{\Delta A}{\hat{A}}(1 + \beta|g^{(1)}(\tau_i)|^2) + \frac{E(\tau_i)}{\hat{A}} \qquad (7.A.2) \\
&= \left(1 - \frac{\Delta A}{\hat{A}}\right)(\beta|g^{(1)}(\tau_i)|^2) - \frac{\Delta A}{\hat{A}} + \frac{E(\tau_i)}{\hat{A}} \\
&= \left(1 - \frac{\Delta A}{\hat{A}}\right)d_i^2 - \frac{\Delta A}{\hat{A}} + \frac{E(\tau_i)}{\hat{A}}.
\end{aligned}$$

Thus, in using $\hat{d}_i$, the normalization errors $\Delta A/\hat{A}$ due to baseline uncertainties play a more complex role than merely shifting the first-order electric field correlation function.

APPENDIX 7.B. ANOTHER PROPOSED FORM FOR σ_i^2

For a Poisson profile,

$$\sigma_i^2 = \frac{1 + d_i^2}{4Ad_i^2}.$$

On closer examination, PCS is seen to be a measure of deviations of Poisson distributions. Thus, statistical fluctuations should be included in the error deviations. For example, the statistical error can be calculated by considering the signals, which obey a Gaussian distribution. Then σ_i^2 has two features. It is a constant according to the Poisson distribution, but it fluctuates at each data point when compared with the measured data. One proposed form is

$$\sigma_i^2 = \frac{1}{4A} d_i^{0.05}, \tag{7.B.1}$$

which is by no means unique. Its use instead of Eq. (7.6.14) improves the resolution of the MEM. Preliminary simulated tests show that the improved MEM can resolve two narrow (δ) bimodal distributions with $\bar{\Gamma}_1/\bar{\Gamma}_2 \lesssim 2$. Further improvements on the form of σ_i^2 are possible and should be forthcoming.

REFERENCES

Alhassid, Y., Agmon, N., and Levine, R. D. (1978). *Chem. Phys. Lett.* **53**, 22.

Bartlett, M. S. (1947). *Biometrics* **3**, 39.

Beightler, C. S., Phillips, D. T., and Wilde, D. J. (1979). *Foundations of Optimization*, 2nd ed., Prentice-Hall, Englewood Cliffs, N. J.

Bertero, M., Boccaci, P., and Pike, E. R. (1982). *Proc. R. Soc. London Ser. A* **383**, 15.

Bertero, M., Boccaci, P., and Pike, E. R. (1984). *Proc. R. Soc. London Ser. A* **393**, 51.

Bertero, M., Brianzi, P., Pike, E. R., de Villiers, G., Lan, K. H., and Ostrowsky, N. (1985). *J. Chem. Phys.* **82**, 1551.

Birkhoff, G. and MacLane, S. (1965). "Survey of Modern Algebra" 3rd ed., MacMillan, New York.

Burch, S. F., Gull, S. F., and Skilling, J. (1983). *J. Comp. Vis. Graph. Image Proc.* **23**, 113.

Bryan, R. (1990a). *Eur. Biophys. J.* **18**, 165.

Bryan, R. (1990b). *In* "Maximum-Entropy and Bayesian Methods" (P. F. Fougere, ed.), pp. 221–232, Dordrecht: Klewer Academic, Hingham, MA.

Chu, B. (1983). *In* "The Application of Laser Light Scattering to the Study of Biological Motion" (J. C. Earnshaw and M. W. Steer, eds.), NATO ASI Vol. A59, pp. 53–76, Plenum Press, New York and London.

Chu, B., Gulari, Es., and Gulari, Er. (1979). *Phys. Scr.* **19**, 476.

Chu, B., Ford, J. R., and Dhadwal, H. S. (1985). *In* Methods of Enzymology, Vol. 117 (S. P. Colowick and N. O. Kaplan, eds.), Chapter 15, pp. 256–297, Academic Press, New York.

Chu, B., Xu, R., and Nyeo, S.-L. (1989). *Part. Part. Syst. Charac.* **6**, 34.

Dahneke, B. E., ed., (1983). Essentials of size distribution measurement. *In* "Measurement of Suspended Particles by Quasi-elastic Light Scattering," pp. 81–252, Wiley, New York.

Dhadwal, H. S. (1989). *Part. Part. Syst. Charac.* **6**, 28.

Ford, J. R. and Chu, B. (1983). *In* "Proceedings of the 5th International Conference on Photon Correlation Techniques in Fluid Mechanics" (E. O. Schulz-DuBois, ed.), Springer Ser. Opt. Sci., p. 303, Springer-Verlag, Berlin and New York.

Fougere, P. F. (ed.) (1990). "Maximum-Entropy and Bayesian Methods," Dartmouth, Dordrecht: Klewer Academic, Hingham, MA.

Freeman, T. L. (1986). *In* "Numerical Algorithms" (J. L. Mohamed and J. Walsh, eds.), Oxford University Press, Oxford.

Glatter, O. and Hofer, M. (1988a). *J. Colloid and Interface Sci.* **122**, 484.

Glatter, O. and Hofer, M. (1988b). *J. Colloid and Interface Sci.* **122**, 496.

Golub, G. H. and Reinsch, C. (1970). *Numer. Math.* **14**, 403.

Gull, S. F. (1989). *In* "Maximum Entropy and Bayesian Methods," (J. Skilling, ed.), pp. 53–71, Dordrecht: Klewer Academic, Hingham, MA.

Gull, S. F. and Skilling, J. (1984). *IEE Proc.* **131**, 646.

Hanson, R. J. (1971). *SIAM J. Numer. Anal.* **8**, 616.

Harley, P. J. (1986). *In* "Numerical Algorithms" (J. L. Mohamed and J. Walsh, eds.), Oxford University Press, Oxford.

Hobson, A. (1971). "Concepts in Statistical Mechanics," Gordon & Breach Science, New York.

Isenberg, I., Dyson, R. D., and Hanson, R. (1973). *Biophys. J.* **13**, 1090.

Jakeman, E., Pike, E. R., and Swain, S. (1971). *J. Phys. A* **4**, 517.

Jaynes, E. T. (1957a). *Phys. Rev.* **106**, 620.

Jaynes, E. T. (1957b). *Phys. Rev.* **108**, 171.

Jaynes, E. T. (1968). *IEEE Trans. on Systems Science and Cybernetics* **SSC-4**, 227.

Jaynes, E. T. (1984). *In* "Inverse Problems" (D. W. McLaughlin, ed.), American Mathematical Society, Providence.

Jeffreys, H. (1961). "Theory of Probability," 3rd ed., Oxford University Press, London.

Katz, A. (1967). "Principles of Statistical Mechanics: The Information Theory Approach," W. H. Freeman, San Francisco and London.

Kendall, M. G. and Stuart, A. (1968). "The Advanced Theory of Statistics, Vol. 3: Design and Analysis, and Time-Series," 2nd ed., Hafner, New York.

Kojro, Z. (1990). *J. Phys. A: Math Gen.* **23**, 1363.

Koppel, D. E. (1972). *J. Chem. Phys.* **57**, 4814.

Kullback, S. (1959). "Information Theory and Statistics," Wiley, New York.

Landsberg, P. T. (1978). "Thermodynamics and Statistical Mechanics," Oxford University Press, Oxford.

Langowski, J., and Bryan (1990). *Prog. Colloid and Polym. Sci.* **81**, 269.

Lawson, C. L. and Hanson, R. J. (1974). "Solving Least Squares Problems," Prentice-Hall, Englewood Cliffs, N. J.

Levine, R. D. (1980). *J. Phys.* **A13**, 91.

Levine, R. D and Tribus, M., eds. (1979). "The Maximum Entropy Formalism," conference held at MIT May 2–4, 1978, MIT Press, Cambridge, Mass.

Licinio, P. and Delaye, M. (1989). *J. Colloid and Interface Sci.* **132**, 1.

Livesey, A. K., Licinio, P., and Delaye, M. (1986). *J. Chem. Phys.* **84**, 5102.

McWhirter, J. G. and Pike, E. R. (1978). *J. Phys. A; Math. Gen.* **11**, 1729.

Miller, K. (1970). *SIAM J. Math. Anal.* **1**, 52.

Nyeo, S.-L. and Chu, B. (1989). *Macromolecules* **22**, 3998.

Oliver, C. J. (1979). *J. Phys. A: Math. Gen.* **12**, 519.

Oliver, C. J. (1981). *In* "Scattering Techniques Applied to Supramolecular and Nonequilibrium Systems" (S.-H. Chen, B. Chu, and R. Nossal, eds.), NATO ASI Vol. 373, pp. 121–160, Plenum, New York and London.

Ostrowsky, N., Dornette, D., Parker, P., and Pike, E. R. (1981). *Opt. Acta* **28**, 1059.

Pike, E. R. (1981). *In* "Scattering Techniques Applied to Supramolecular and Non-equilibrium Systems" (S.-H. Chen, B. Chu and R. Nossal, eds.), p. 179, Plenum, New York.

Pike, E. R., Watson, D., and McNeil Watson, F. (1983). *In* "Measurements of Suspended Particles by Quasi-elastic light Scattering" (B. E. Dahneke, ed.), Chapter 4, Wiley, New York.

Provencher, S. W. (1976a). *Biophys. J.* **16**, 27.

Provencher, S. W. (1976b). *J. Chem. Phys.* **64**, 2772.

Provencher, S. W., Hendrix, J., DeMaeyer, L., and Paulussen, N. (1978). *J. Chem. Phys.* **69**, 4273.

Provencher, S. W. (1979). *Makromol. Chem.* **180**, 201.

Provencher, S. W. (1982a). *Comp. Phys. Comm.* **27**, 213.

Provencher, S. W. (1982b). *Comp. Phys. Comm.* **27**, 229.

Provencher, S. W. (1982c). "CONTIN Users Manual," EBL Technical Report DA05, European Molecular Biology Laboratory, Heidelberg.

Provencher, S. W. (1983). *In* "Proceedings of the 5th International Conference on Photon Correlation Techniques in Fluid Mechanics" (E. O. Schulz-DuBois, ed.), Springer Ser. Opt. Sci., p. 322, Springer-Verlag, Berlin and New York.

Pusey, P.N., Fijnaut, N. M., and Vrij, A. (1982). *J. Chem. Phys.* **77**, 4270.

Ruf, H. (1989). *Biophys. J.* **56**, 67.

Saleh, B. E. A. and Cardoso, M. F. (1973). *J. Phys. A: Math. Nucl. Gen.* **6**, 1897.

Schalzel, K. Drewel, M., and Stimac, S. (1988). *J. Modern Optics* **35**, 711.

Schulz-DuBois, E. O., ed. (1983) Photon correlation spectroscopy of Brownian motion: Polydispersity analysis and studies of particle dynamics. *In* "Proceedings of the 5th International Conference on Photon Correlation Techniques in Fluid Mechanics," Springer Ser. Opt. Sci., pp. 286–315. Springer-Verlag, Berlin and New York.

Skilling, J., ed. (1989). "Maximum-Entropy and Bayesian Methods," Cambridge, England, Dordrecht: Klewer Academic, Hingham, MA.

Skilling, J. and Bryan, R. K. (1984). *Mon. Not. R. Astr. Soc.* **211**, 111.

Skilling, J. and Gull, S. F. (1985). *In* "Maximum-Entropy and Bayesian Methods in Inverse Problems" (C. R. Smith and W. T. Grandy, Jr., eds.), D. Reidel, Dordrecht.

Steenstrup, S. (1985). *Aust. J. Phys.* **38**, 319.

Stock, G. B. (1976). *Biophys. J.* **16**, 535.

Stock, G. B. (1978). *Biophys. J.* **18**, 79.

Tarantola, A. and Valette, B. (1982). *J. Geophys.* **50**, 159.

Toutenburg, H. (1982). "Prior Information in Linear Models," Wiley, Chichester.

Vancso, G., Tomka, I., and Vancso-Polacsek, K. (1988). *Macromolecules* **21**, 415.

Vrij, A. (1982). *J. Colloid and Interface Sci.* **90**, 110.

Wahba, G. (1977). *SIAM J. Number. Anal.* **14**, 651.

Weiner, B. B. and Tscharnuter, W. W. (1987). *In* "Particle Size Distributions" (T. Provder, ed.), ACS Symposium Series 332, pp. 48–61, Amer. Chem. Soc., Washington, D. C.

Wismer, D. A. and Chattergy, R. (1978). "Introduction to Nonlinear Optimization: A Problem Solving Approach," North-Holland, New York.

VIII

CHARACTERIZATION OF POLYMER MOLECULAR-WEIGHT DISTRIBUTION (PARTICLE SIZING)

8.1. INTRODUCTION

In the previous chapters, we have presented the topics as follows.

Chapter I. An overview of laser light scattering, including its relationship with other scattering techniques, namely small-angle x-ray scattering and small-angle neutron scattering, as well as a listing of reviews, books and proceedings covering some aspects of laser light scattering.

Chapter II. A presentation of the light scattering theory from Einstein's fluctuation viewpoint. This approach provided us not only the basic equations for the time-averaged intensity (Eqs. (2.1.6)–(2.1.9)) and the spectrum of scattered light (Eqs. (2.4.46)–(2.4.50)), but also the equations needed for coherence considerations due to the finite size of the scattering volume [Eq. (2.2.10)].

Chapter III. A detailed development on optical wavefront matching based on the phase integral [Eq. (2.2.6)], in which the finite dimensions of the scattering volume were taken into account (e.g. see Eq. (3.1.7) or Eq. (3.1.8)). With the introduction of photon statistics, fundamental equations governing optical mixing spectrometers, including self-beating (Section 3.3.2) and homodyne–heterodyne spectrometers, were derived.

Chapter IV. Basic equations for photon correlation spectroscopy were presented. In addition to the full and clipped photon correlation (Section 4.2

and 4.4), schemes such as complementary clipping (Section 4.3), random clipping (Section 4.5), scaled photon correlation (Section 4.6), add–subtract photon correlation (Section 4.7), time of arrival of photoelectrons (Section 4.9) and the photon structure function (Section 4.10) were described.

Chapter V. Principles of (Fabry–Perot) interferometry were included in this chapter in order to maintain completeness in covering a broad spectral range.

Chapter VI. The experimental methods—from the laser source, via (fiber) optics for incident and scattered light beams, as well as sample-cell geometry, to the use of photodetectors—were discussed. Experimentalists, especially uninitiated ones, should find this chapter of practical importance in the choice of lasers, optics, photodetectors, and correlators.

Chapter VII. The key to the data analysis of photon correlation functions is to recognize the ill-posed nature of the Laplace inversion problem. Several schemes were presented in order to ensure awareness of the complexity of this topic. While inversion of the Laplace transform is mathematical in nature and outside the normal expertise of most biologists and polymer and colloid scientists, reliable solutions to obtain estimates of the characteristic linewidth distribution function from the intensity–intensity time correlation functions can be achieved.

In this chapter, instead of exploring the many applications in which laser light scattering can be used effectively, a description of the characterization of polymer molecular-weight distributions is presented. This characterization is closely related to particle sizing. In fact, particle sizing represents the major analytical application for laser light scattering (or, more precisely, photon correlation spectroscopy) in industrial laboratories. For a suspension of anisotropic submicrometer-sized particles, a truncated pulsed electrooptic birefringence method has been developed (Huntley-James and Jennings, 1990).

At small enough scattering angles, i.e., for $KR_g < 1$, the intensity–intensity time correlation function measures a sum (or integral) of translational Brownian motions of (noninteracting) colloidal particles in suspension or macromolecules in solution, with each particle or macromolecule being weighted by the scattering strength of the particle or macromolecule at the scattering vector $\mathbf{K}$.

This application is certainly not the only use for laser light scattering. Interested readers may wish to consult a recent Milestone book on "Laser Light Scattering by Macromolecular, Supramolecular, and Fluid Systems," where a collection of reprints on laser light scattering, including applications to critical phenomena, polymer solutions, colloids, and biological systems, has been made available in one convenient volume (Chu, 1990a, and see Appendix 8.A).

The development of this chapter follows the experimental procedure in

1. finding a solvent for the polymer (Section 8.2),
2. dissolving the polymer and clarifying the resultant polymer solution (Section 8.3),
3. measuring the appropriate time-averaged scattered intensity and the intensity–intensity time correlation function (Section 8.4),
4. performing the Laplace inversion (Section 8.5),
5. determining the scaling relation between the translational diffusion coefficient at infinite dilution (D^0) and the molecular weight (Section 8.6),
6. estimating the polymer molecular weight distribution from the characteristic linewidth distribution (Section 8.6),
7. summarizing the polymers characterized by laser light scattering (Section 8.6)* and
8. sketching the use of dynamic light scattering to study nonergodic media (Section 8.7).

8.2. FINDING A SOLVENT

The requirements for finding a solvent for a polymer for preparation in light scattering experiments can be summarized as follows:

(1) The solvent should have a boiling point above the operating temperature of the intended experiment. This specification for the solvent may appear to be either trivial or superfluous for room-temperature experiments. However, many modern speciality polymers are difficult to dissolve under normal experimental conditions. Therefore, the use of temperature (and/or pressure) as an additional variable can be advantageous.

(2) The solvent should have a refractive index different from that of the polymer. If the solvent and the polymer were isorefractive, there would be no light scattering due to local concentration fluctuations.

(3) The solvent, though it need not be composed of small molecules, should not exhibit a time correlation function in the delay-time range of the polymer under investigation.

In laser light scattering characterization of poly(tetrafluoroethylene) (PTFE) in solution (Chu et al., 1987, 1988a, 1989b), oligomers of poly(chlorotrifluoroethylene) and of PTFE, namely perfluorotetracosane (n-$C_{24}F_{50}$), were used as solvents for PTFE. The oligomers, which satisfied the above three requirements, were used successfully to dissolve different molecular-weight fractions of PTFE. It was interesting to note that mixed

---

* Similar schemes have been reviewed previously (Chu, 1989a,b; 1990b).

oligomers of different molecular-weight fractions could be used as a pseudo-one-component solvent because preferential adsorption by solvents of very similar chemical properties, except for the molecular weight, would be expected to be negligible. Thus, the use of oligomers as solvents may provide an additional avenue for the dissolution of intractable polymers.

8.3. DISSOLVING THE POLYMER AND CLARIFYING THE RESULTANT POLYMER SOLUTION

In polymer-solution preparation that is designed specifically for light scattering experiments, it is only logical to expect that the excess scattered intensity due to the solute (polymer) comes only from the solute. Scattering from other foreign macromolecules or particles (as a form of contamination in the initial polymer sample), especially those comparable in size with the polymers of interest, will be difficult to exclude by filtration. Large dust particles or foreign macromolecules with sizes substantially greater than those of the polymer of interest can be separated by filtration. If the foreign particles have similar sizes to those of the polymer (or particle) of interest but very different densities from that of the polymer, they can be removed by centrifugation. Occasionally, we can use dust discrimination by electronic means. Large dust particles passing through the scattering volume will produce a sudden increase in the forward-scattered intensity. Thus, using estimates of dust-particle sizes and the scattering-volume dimensions, one can exclude from the time-averaged scattered intensity those scattered-intensity jumps that fall within a preset magnitude and duration of the signal, as attributed to the large dust particles. Nevertheless, discrimination of dust particles by electronic means introduces a subjective bias and should be used only when the polymer solution cannot be clarified following established procedures. The best way to get reliable light scattering results is to be able to clarify the polymer solution properly.

In the clarification procedure, the reader should recall the origin of dust (or impurity) particles. They can come from the polymer sample, the solvent, the light-scattering cell, and the air above the polymer solution in the light-scattering cell. In principle, once the solution passes through the filter, it should never be exposed to unfiltered air (or other gas).

Figure 8.3.1 shows a glass dissolution–filtration apparatus that permits dissolving the polymer in a solvent in solution vessel (B). After dissolution, the polymer solution is filtered by introducing additional gas through a coarse filter (F_2). The polymer solution in B is forced through the gap between A and B, overflowed to the container (C), and then filtered through a fine-grade sintered-glass filter (F_1) directly to the dust-free light-scattering cell (D). The

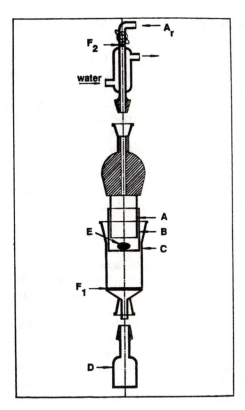

FIG. 8.3.1.   Separate components of a high-temperature dissolution filtration apparatus. The assembled apparatus can be placed in a constant-temperature oven. A, cylindrical insert (without bottom) with a diameter ≈ 2 mm less than that of the solution vessel B; C, filtration section with fine-grade sintered-glass filter (F_1) and ground-glass joints to light-scattering cell (D), and ground-glass joint adapter for the water-cooled condenser, which is located outside of the temperature-controlled oven; E, magnetic stirring bar. Hatched areas denote volume reduction, so that the volume accessible to the vapor phase is no more than a few times that of the fluid phase. The miniature water-cooled condenser has a coarse-grade sintered-glass filter (F_2), so that the entire system is always isolated from the external dusty environment. The greaseless stopcock above F_2 is for closing off apparatus, for connecting to low vacuum in order to degas the solvent before dissolution, for filling the apparatus with inert gas, such as argon, in order to alleviate chemical decomposition, and for releasing possible pressure buildup if chemical decomposition does take place. The entire apparatus is portable and can be inserted into the light-scattering spectrometer with the light-scattering cell (D) and part of the filtration component (C) controlled at predetermined temperatures (up to $\approx 350°$C). On occasion, a ground-glass joint stopper with a pressure-relief stopcock is used to replace A and B. (Reprinted with permission from Chu *et al.*, © 1988a, American Chemical Society).

interesting aspects of this dissolution–filtration apparatus can be summarized as follows.

(1) The apparatus, including the filter (F_1), is all glass and contains no O-ring seals; no sealants are used. Thus, it can be heated to high temperatures ($\approx 350°C$) without worrying about thermal expansion, and can be used with most corrosive solvents. The apparatus can also be cleaned by pyrolysis. For example, in the light scattering characterization of PTFE, there were no solvents for the oligomers of PTFE or poly(chlorotrifluoroethylene). Without pyrolysis, the apparatus could have been used only once.

(2) The apparatus could provide dissolution and filtration in one system, so that filtration of the resultant polymer solution could be accomplished under an inert atmosphere without exposure to the surroundings. This arrangement greatly reduced dust contamination of the filtered polymer solution at elevated temperatures.

(3) The entire apparatus had no moving parts (except for the stopcock above F_2 and the decoupled magnetic stirrer). Transfer and filtration of the polymer solution was accomplished by applying a gentle pressure through F_2. Moving parts would require seals, which would have to be made of elastic materials. For the PTFE experiment, the filtration–dissolution apparatus was heated above 350°C, which would decompose or melt all known elastomers. The use of metal O-rings would also be inconvenient. However, for other, lower-temperature applications, this requirement could be relaxed.

(4) The reflux condenser was introduced in order to permit operation of the dissolution–filtration apparatus as an open system. Pressure buildup because of possible solvent and/or polymer decomposition, resulting in a lowering of the vapor pressure of the polymer solution, could then be accommodated.

The above dissolution–filtration apparatus was used successfully for preparation of PTFE solutions. Most polymer solution preparation should involve much less stringent requirements, as has been discussed. Therefore, one might modify of the dissolution–filtration apparatus by changing the sintered glass filter (F_1) to other types of filters. In any case, the key to successful preparation of a polymer solution for light scattering studies is to avoid (or at least to minimize) exposure of a clarified polymer solution to the dusty external atmosphere.

If a one-step filtration process is insufficient to clarify a polymer solution, as is sometimes the case for polymers dissolved in a polar solvent, a closed-circulation filtration apparatus (Naoki et al., 1985), as shown in Fig. 8.3.2, can be useful.

Figure 8.3.2 shows the essential features of a closed-circulation filtration apparatus: a silicone-tubing metering pump (A), a filter (B), an air-trapping–reservoir chamber (C), and the light-scattering cell in which the light-

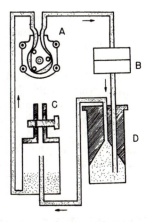

FIG. 8.3.2.   Schematic diagram of the closed-circuit filtration system: *A*, silicone-tubing metering pump; *B*, filter, *C*, air-trapping chamber; *D*, light-scattering cell. (Reprinted by permission of John Wiley & Sons, Inc., M. Naoki, I.-H. Park, S. L. Wunder and B. Chu, *J. Polym. Sci.: Polym. Phys. Ed.* **23**, 2567, Copyright ©1985).

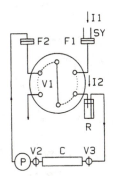

FIG. 8.3.3.   Closed filtration circuit used for sample insertion and dust removal. F1 and F2, filter holder (various Millipore and Nuclepore filters can be adapted); V1, three-way Teflon rotary valve (Rheodyne Inc. model 50); V2 and V3, single Teflon valves; C, prism cell; P, piston lab pump (Fluid Metering, Inc. model RH0CYTC); R, small reservoir of ~1-ml volume with septum cap (sample solution can be introduced directly through the self-sealing Teflon-lined septum); SY, syringe; I1, inlet position 1 for sample insertion; I2, inlet position 2 for sample insertion without filtration through F1. Dotted arcs indicate valve position 1, in which ports 1 and 2, ports 3 and 6, and ports 4 and 5 are connected, for flushing the system and using the cell as part of the flow system; dashed arcs indicate valve position 2, in which ports 2, 3, 6, and 5 are connected, for filtration of solution through F2 in a closed circulating circuit. (After Chu *et al.*, 1988b.)

scattering experiment is to be performed. The flexible silicon-tubing pump can be replaced by a piston pump, as shown schematically in Fig. 8.3.3, in which more complex maneuvers in the flow path could be achieved by means of a three-way Teflon rotary valve. The light-scattering cell (D in Fig. 8.3.2) was filled with polymer solution. Thermal expansion was taken into account by the air-trapping–reservoir chamber, which could also be used as a dilution chamber. It should be noted that the closed-circuit filtration systems did not include a constant-temperature capability, but could be modified to accommodate such a change for temperatures up to $\approx 150°C$ using a Teflon piston lab pump. The volume of the air-trapping–reservoir chamber (C in Fig. 8.3.2 or R in Fig. 8.3.3) could be increased to accommodate dilution of the filtered polymer solution by further introduction of solvent into the filtration system.

8.4. MEASURING THE APPROPRIATE TIME-AVERAGED SCATTERED INTENSITY $\mathscr{R}_{VV}(K)$ AND THE INTENSITY–INTENSITY TIME CORRELATION FUNCTION $G^{(2)}(K, \tau)$

8.4.1. TIME-AVERAGED SCATTERED INTENSITY $\mathscr{R}_{VV}(K)$

According to Eqs. (2.1.6)–(2.1.9), measurements of the angular distribution of scattered intensity over a range of concentrations in the dilute solution regime permit determination of M, A_2, and R_g. For polydisperse particles, each with the same optical constant H ($\equiv k^4 n_0^2 (dn/dC)^2/(4\pi^2 N_a)$), Eq. (2.1.6) becomes

$$\lim_{\substack{C \to 0 \\ K \to 0}} \mathscr{R}_{ex} = \sum_i \mathscr{R}_{ex,i} = H \sum_i C_i M_i, \tag{8.4.1}$$

where the subscript i denotes species i. With

$$M_w = \frac{\sum_i C_i M_i}{\sum_i C_i}, \tag{8.4.2}$$

we have in the limits of $C \to 0$ and $K \to 0$

$$\mathscr{R}_{ex}^0 = HCM_w, \tag{8.4.3}$$

where $C = \sum_i C_i$. Similarly, according to Eq. (2.1.8),

$$\lim_{C \to 0} \mathscr{R}_{ex} = H \sum_i M_i C_i P_i(K) \tag{8.4.4}$$

where $P_i(K)$ is the particle scattering factor at scattering vector K for species i.

For $KR_{g,i} \lesssim 1$,

$$P_i(K) \simeq 1 - \frac{K^2 R_{g,i}^2}{3} \qquad (8.4.5)$$

Thus, Eq. (8.4.4) becomes

$$\lim_{C \to 0} \frac{HC}{\mathscr{R}_{ex}} = \frac{1}{M_w}\left(1 + \frac{K^2\langle R_g^2\rangle_z}{3}\right), \qquad (8.4.6)$$

where $\langle R_g^2\rangle_z$ $(\equiv \sum_i M_i C_i R_{g,i}^2 / \sum_i M_i C_i)$ is the z-average square of the radius of gyration.

8.4.1.A. ANISOTROPIC POLYMERS

For anisotropic polymers in dilute solution, the Rayleigh ratio using vertically polarized incident and scattered light at $KR_g \lesssim 1$ has the form

$$\frac{HC}{\mathscr{R}_{VV}} = M_{app}^{-1}\left(1 + \frac{K^2 R_{g,app}^2}{3}\right) + 2A_2 C \qquad (8.4.7)$$

with the subscript app denoting apparent quantities,

$$M_{app} = M_w(1 + \tfrac{4}{5}\delta^2), \qquad (8.4.8)$$

$$\langle R_{g,app}^2 \rangle = \frac{(1 - \tfrac{4}{5}\delta + \tfrac{4}{7}\delta^2)\langle R_g^2\rangle}{1 + \tfrac{4}{5}\delta^2}, \qquad (8.4.9)$$

where the molecular anisotropy δ is related to the depolarized light scattering:

$$\lim_{\substack{C \to 0 \\ K \to 0}} \frac{\mathscr{R}_{HV}}{HC} = \tfrac{3}{5}\delta^2 M_w \qquad (8.4.10)$$

with $\mathscr{R}_{HV}$ being the depolarized Rayleigh ratio having *vertically* polarized incident light and *horizontally* polarized scattered light. The determination of δ is not as easy as it appears in Eq. (8.4.10), because of the limits $C \to 0$ and $K \to 0$. Furthermore, the value of $\mathscr{R}_{HV}$ is much smaller than that of $\mathscr{R}_{VV}$, requiring the exclusion of stray depolarized light, which could come from leakage of polarizer–analyzer, reflection from imperfect mirrors, and scattering from imperfect surfaces. The allowance for polymer molecular anisotropy expands the applicable range of polymer characterization to wormlike chains (for examples, see Ying et al., 1985; Chu et al., 1985b).

8.4.1.B. BLOCK COPOLYMERS

For block copolymers,

$$v = \left(\frac{dn}{dC}\right)_{polymer} = \sum_{j=1} W_j v_j \qquad (8.4.11)$$

provided that the refractive-index increment v of the components of a multi-component polymer in a single solvent is additive. W_j is the weight fraction of monomer type j in the polymer with $\sum_j W_j = 1$. Thus, Eq. (8.4.1) for a block copolymer becomes

$$\lim_{\substack{C \to 0 \\ K \to 0}} \mathscr{R}_{ex} = H^* \sum_i C_i M_i v_i^2 = HCM_{app}, \qquad (8.4.12)$$

where $H^* = 4\pi^2 n_0^2/(N_a \lambda_0^4)$, $v_i = \sum_j W_{ij} v_j$, $H = H^* v^2$, and

$$M_{app} = \frac{1}{v^2} \sum_i W_i M_i v_i^2 \qquad (8.4.13)$$

with $W_i = C_i/C$, and v being the refractive-index increment of the copolymer solution. The subscripts i and j refer to the chain length and monomer types, respectively. Thus, for each molecular weight M_i, there can be a heterogeneity in composition. For example, for an ABC triblock copolymer,

$$v_i = W_{iA} v_A + W_{iB} v_B + W_{iC} v_C, \qquad (8.4.14)$$

where j ($= A, B, C$) denotes the monomer types A, B, and C. However, Eq. (8.4.14) is not unique, because different chain compositions (and/or sequences) can yield the same M_i and W_i. Similarly, polymers of different chain compositions and molecular weights can have the same radius of gyration or hydrodynamic size. Thus, the complications imposed by different monomer types with different refractive indices in a (block) terpolymer are not trivial, even for molecular-weight determinations.

If we take W_{ij} to be a representative average chain composition with $W_{iA} + W_{iB} + W_{iC} = 1$, then v_i in Eq. (8.4.14) represents an average refractive-index increment for the terpolymer with molecular weight M_i and weight fraction W_i. Similarly, for the whole polymer, we have $W_A + W_B + W_C = 1$, which implies

$$\delta W_{iA} + \delta W_{iB} + \delta W_{iC} = 0. \qquad (8.4.15)$$

With the above equations, we can rewrite Eq. (8.4.13) and get for the terpolymer

$$\begin{aligned}
M_{app} &= \sum_i W_i M_i \left(\frac{W_{iA} v_A + W_{iB} v_B + W_{iC} v_C}{v} \right)^2 \\
&= M_w + 2P_A \left(\frac{v_A - v_C}{v} \right) + 2P_B \left(\frac{v_B - v_C}{v} \right) + Q_A \left(\frac{v_A - v_C}{v} \right)^2 \qquad (8.4.16) \\
&\quad + Q_B \left(\frac{v_B - v_C}{v} \right)^2 + 2 \left(\frac{v_A - v_C}{v} \right) \left(\frac{v_B - v_C}{v} \right)
\end{aligned}$$

with

$$P_A = \sum_i W_i M_i \, \delta W_{iA},$$

$$P_B = \sum_i W_i M_i \, \delta W_{iB},$$

$$Q_A = \sum_i W_i M_i (\delta W_{iA})^2, \qquad (8.4.17)$$

$$Q_B = \sum_i W_i M_i (\delta W_{iB})^2,$$

$$R_{AB} = \sum_i W_i M_i \, \delta W_{iA} \, \delta W_{iB}.$$

According to Eq. (8.4.16), we need a minimum of six solvents of varying refractive index in order to determine the six unknowns: $P_A, P_B, Q_A, Q_B, R_{AB}$, and M_w (Kambe *et al.*, 1973). Such a characterization would be quite time-consuming and could involve large error limits. Thus, one usually tries to avoid this approach. For a diblock AC copolymer, the complication is reduced (though not eliminated). Then, we have (Bushuk and Benoit, 1958)

$$M_{app} = M_w + 2P_A \left(\frac{\nu_A - \nu_C}{\nu} \right) + Q_A \left(\frac{\nu_A - \nu_C}{\nu} \right)^2, \qquad (8.4.18)$$

where a minimum of three solvents with three different refractive indices are needed to estimate M_w, P_A, and Q_A. In general, more solvents of different refractive index would be needed to solve Eq. (8.4.18). Therefore, light scattering should be used for an estimate of the *absolute* determination of molecular weight for copolymers only as a last resort. However, it should be noted that light scattering remains the only physical technique capable of an absolute determination of the molecular weight of large macromolecules in solution.

8.4.1.C. CALIBRATION STANDARD

Precision measurements of the Rayleigh ratio (see Section 2.1.2 and Table 2.1.1) are usually achieved by comparison of the scattered intensity of the polymer solution of interest with a calibration standard, which can be a pure liquid, such as benzene; a polymer solution of known molecular weight and refractive-index increment, such as a NBS polystyrene standard dissolved in a prespecified solvent (e.g. benzene); or a colloidal suspension, such as an aqueous suspension of polystyrene latex spheres. The colloidal suspension is used mainly as a secondary standard for periods of weeks because of its strong scattered intensity and ease of preparation. However, the refractive index of the polymer solution and that of the calibration standard may be different, and the scattering volume as viewed by the light-scattering spectrometer will then be different. Furthermore, the difference depends on the optics of the spectrometer. Thus, we need to normalize the scattering volume. As the light passing through the scattering cell may be attenuated, the scattered intensity should be normalized to the center of a cylindrical light-scattering cell of radius r_c (cm) by a factor $\exp(2r_c\gamma C)$ with γ and C being the absorption coefficient (cm^2/g solute) and the solute concentration (g solute/cm^3), respectively. At a fixed wavelength, the Rayleigh ratio of a scattering medium measured at temperature t and scattering angle θ, $\mathscr{R}_\theta^t$, can be related to the calibration standard measured at 25°C and θ by

$$\mathscr{R}_\theta^t = I_\theta^t \frac{\mathscr{R}_\theta^{25}(\text{standard})}{I_\theta^{25}(\text{standard})} \exp(2\gamma r_c C)\left[\frac{n_s^t}{n^{25}(\text{standard})}\right]^x, \qquad (8.4.19)$$

where I_θ^t, $\mathscr{R}_\theta^{25}(\text{standard})$, n_s^t, $n^{25}(\text{standard})$, and x are, respectively, the scattered intensity of the scattering medium measured at temperature t and scattering angle θ, the known Rayleigh ratio of the calibration standard at θ and 25°C, the refractive index of the scattering medium measured at temperature t, the refractive index of the calibration standard measured at 25°C, and a parameter that normally varies from 1 to 2 depending on the detection optics geometry. The refraction correction term $[n_s^t/n^{25}(\text{standard})]^x$ has $x = 1$ if the detection optics utilizes a slit that cuts a section of the incident laser beam. Thus, the scattering volume becomes the laser-beam cross section times a (fixed) slit width, which depends on the refractive index of the scattering medium. However, if the detection optics is a set of pinholes that cut the incident laser beam not only along the (horizontal) propagation direction but also in the vertical direction, then the field observed by the detection optics in the vertical direction depends on the laser-beam intensity profile, making $1 \lesssim x \lesssim 2$. If the pinhole of the detection optics is so small that the incident laser-beam intensity profile in the vertical direction (i.e., perpendicular to the scattering plane) is constant over the field of view accepted by the pinhole, then $x = 2$. For this reason, it is easier to use a vertical slit (with $x = 1$ in Eq. (8.4.19)) as the field stop in the detection optics.

8.4.2. INTENSITY-INTENSITY TIME CORRELATION FUNCTION

In Chapter VII, the methods of data analysis have been discussed in some detail. However, before one is ready to perform an inversion of the Laplace transform

$$g^{(1)}(\tau_i) = \int_0^\infty G(\Gamma)e^{-\Gamma\tau_i}\,d\Gamma, \qquad (7.1.1)$$

it is important to recognize that we must measure $g^{(1)}(\tau_i)$ with sufficient precision and over a correct range of delay-time increment. In fact, we measure the intensity–intensity time correlation function $G^{(2)}(\mathbf{K}, \tau)$, while the quantity of interest is the first-order normalized electric field correlation function $g^{(1)}(\mathbf{K}, \tau_i)$, with the relation

$$G^{(2)}(\mathbf{K}, \tau_i) = \langle n(t)n(t + \tau_i)\rangle = N_s\langle n(t)\rangle^2[1 + \beta|g^{(1)}(\mathbf{K}, \tau_i)|^2], \quad (7.1.2)$$

where $N_s\langle n(t)\rangle^2$ is a background, which can be computed from measurements of the total number of samples, N_s, and the mean number of counts per sample time, $\langle n(t)\rangle$, or from measurements of $G^{(2)}(\tau_i)$ at large values of τ_i such that $\lim_{\tau_i \to \infty} G^{(2)}(\tau_i)$ agrees with $N_s\langle n(t)\rangle^2$ to within $\sim 0.1\%$. The 0.1% is a practical limit that takes account of the statistical counting uncertainties. Agreement of the measured baseline (the limit of $G^{(2)}(\tau)$ for large τ) with the computed baseline ($N_s\langle n(t)\rangle^2$) demonstrates that we have covered the slow-decay characteristic times ($1/\Gamma_{\text{slowest}}$) to within experimental error limits. This requirement is illustrated in Fig. 8.4.1 (point (a)).

If we obtain a $g^{(1)}(\mathbf{K}, \tau)$ curve from $g^{(1)}(\mathbf{K}, \tau \to 0) \to 1$ to $g^{(1)}(\mathbf{K}, \tau_{\text{max}}) \simeq 0$, we know that we have covered the range of dynamical motions of the polymer system of interest. Unfortunately, for a correlator (e.g. with equidistant delay-time increments) the measured time correlation function covers a delay-time range from $\tau_{\text{min}} = \Delta\tau$ to $\tau_{\text{max}} = N \Delta\tau$ with N being the maximum channel

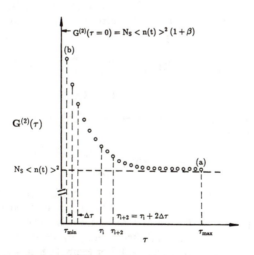

FIG. 8.4.1. Schematic plot of $G^{(2)}(\tau)$ versus τ for a 24-channel correlator with equidistant $\Delta\tau$. Requirements for an acceptable measurement of $G^{(2)}(\tau)$, covering an appropriate bandwidth so that the characteristic linewidths of the polymer system of interest are measured to within typical experimental uncertainties, are: (a) $|G^{(2)}(\tau_{\text{max}}) - N_s\langle n(t)\rangle^2|/N_s\langle n(t)\rangle^2 \simeq 0.1\%$ in order to cover slow decay times; (b) $[G^2(\mathbf{K}, \tau_{\text{min}})/N_s\langle n(t)\rangle^2 - 1]/b \simeq 1\%$ in order to cover fast decay times.

number. In any case, the correlator has a bandwidth limit, and the first channel $G^{(2)}(\mathbf{K}, \Delta\tau)$ is not the same as $G^{(2)}(\mathbf{K}, \Delta\tau = 0)$. Since $g^{(1)}(\mathbf{K}, \tau) \to 1$ only when $\tau \to 0$, extrapolation of $g^{(1)}(\mathbf{K}, \tau)$ to $\tau \to 0$ is not trivial. We have to estimate $g^{(1)}(\mathbf{K}, \tau = 0)$ from a least-squares fit of $G^{(2)}(\mathbf{K}, \tau)$ over the delay-time range $(\tau_{\min} = \Delta\tau$ to $\tau_{\max} = N \Delta\tau)$ and compare the estimated $\hat{g}^{(1)}(\mathbf{K}, \tau = 0)$ with one in order to see whether we have missed any high-frequency components in $G(\Gamma)$. Thus, in an experimental estimate of $G^{(2)}(\mathbf{K}, \tau = 0)$, we need to fit the measured intensity–intensity time correlation function for extrapolation to $\tau = 0$ and to know the magnitude of the spatial coherence factor β.

If the scattered intensity of a polymer solution, $I(\mathbf{K})$, comes from that of the solvent (I_2) and of the solute (I_3), then $\langle I(t)\rangle^2 = (I_2 + I_3)^2$ and Eq. (7.1.2) becomes

$$G^{(2)}(\mathbf{K}, \tau_i) = [I_2 + I_3(\mathbf{K})]^2 \left\{ 1 + \beta\left[\left(\frac{I_2}{I_2 + I_3(K)}\right)|g_2^{(1)}(\mathbf{K}, \tau_i)| \right.\right.$$
$$\left.\left. + \left(\frac{I_3(\mathbf{K})}{I_2 + I_3(\mathbf{K})}\right)|g_3^{(1)}(\mathbf{K}, \tau_i)| \right]^2 \right\},$$
(8.4.20)

where we have assumed the solvent scattered intensity to be independent of $\mathbf{K}$ for vertically polarized incident and scattered light. If the solvent molecules are much smaller than the polymer molecules, the translational motion of the solvent molecules is much faster than that of the polymer molecules, and then even the first channel of the correlator can measure only the decay of the polymer molecules. So

$$G^{(2)}(\mathbf{K}, \tau_i) \simeq [I_2 + I_3(\mathbf{K})]^2 \left[1 + \beta\left(\frac{I_3(\mathbf{K})}{I_2 + I_3(\mathbf{K})}\right)^2 |g_3^{(1)}(\mathbf{K}, \tau_i)|^2 \right]$$
$$\simeq [I_2 + I_3(\mathbf{K})]^2 [1 + b|g_3^{(1)}(\mathbf{K}, \tau_i)|^2],$$
(8.4.21)

which is different from Eq. (4.1.6) with $g^{(2)}(\mathbf{K}, \tau_i)$ defined by Eq. (3.3.6). In Eq. (8.4.21), we note that

$$b(\mathbf{K}) = \beta(\mathbf{K})\left(\frac{I_3(\mathbf{K})}{I_3(\mathbf{K}) + I_2}\right)^2.$$

$b(\mathbf{K}) \simeq \beta(\mathbf{K})$ only when $I_3(\mathbf{K}) \gg I_2$. Thus, $G^{(2)}(\mathbf{K}, \tau_{\min})/N_s\langle n(t)\rangle^2 - 1 = b$, which requires that we know both I_2 and I_3, i.e., the scattered intensity due to polymer solution ($I_2 + I_3$) and that due to the solvent, as well as the spatial coherence factor β. For a weakly scattering polymer solution, an apparent spatial coherence factor b is observed. For example, if $\beta = 0.80$ and $[I_3/(I_3 + I_2)]^2 = 0.40$, then $b = 0.80 \times 0.40 = 0.32$, i.e., extrapolation of $G^{(2)}(\tau = 0)$ should yield $G^{(2)}(\tau = 0)/N_s\langle n(t)\rangle^2 - 1 = b = 0.32$ instead of 0.80. If we get a value close to 0.32 (and not 0.80), we know that we have covered the high-frequency components of the characteristic linewidth to within experimental error limits. The value of β depends on the detection optics, the scattering volume, and the scattering angle. Thus, we need to measure $\beta(\mathbf{K})$ in a separate experiment using strongly scattering systems, e.g., an aqueous suspension of monodisperse latex spheres. One would expect a precision of $\sim 1\%$. Thus, the estimated precision on b should be less than 2%, implying uncertainties in the high-frequency components of the characteristic linewidth distribution of at least a few percent for weakly scattering polymer solutions.

In a dilute polymer solution, the presence of dust may also contribute to the intensity–intensity time correlation function:

$$G^{(2)}(\mathbf{K}, \tau_i) = [I_2 + I_3(\mathbf{K}) + I_d(\mathbf{K})]^2 \left[1 + \beta \left(\frac{I_2 |g_2^{(1)}(\mathbf{K}, \tau_i)|}{I_2 + I_3(\mathbf{K}) + I_d(\mathbf{K})} \right. \right.$$
$$\left. \left. + \frac{I_3(\mathbf{K})|g_3^{(1)}(\mathbf{K}, \tau_i)|}{I_2 + I_3(\mathbf{K}) + I_d(\mathbf{K})} + \frac{I_d(\mathbf{K})|g_d^{(1)}(\mathbf{K}, \tau_i)|}{I_2 + I_3(\mathbf{K}) + I_d(\mathbf{K})} \right)^2 \right] \quad (8.4.22)$$

with the subscript d denoting the dust particles. If $I_d(\mathbf{K}) \sim I_3(\mathbf{K}) \gg I_2$ and the dust particles are much larger than the polymer molecules, Eq. (8.4.22) is reduced to

$$G^{(2)}(\mathbf{K}, \tau_i) \simeq [I_3(\mathbf{K}) + I_d]^2 \left\{ 1 + \beta \left[\frac{I_3(\mathbf{K})}{I_3(\mathbf{K}) + I_d(\mathbf{K})} |g_3^{(1)}(\mathbf{K}, \tau_i)| \right. \right.$$
$$\left. \left. + \frac{I_d(\mathbf{K})}{I_3(\mathbf{K}) + I_d(\mathbf{K})} |g_d^{(1)}(\mathbf{K}, \tau_i)| \right]^2 \right\}, \quad (8.4.23)$$

which is equivalent in form to Eq. (8.4.20) except that the subscript 2 is now replaced by d. Thus $g_2^{(1)}(\mathbf{K}, \tau_i)$ decays quickly because the solvent molecules are small, i.e., $g_2^{(1)}(\mathbf{K}, \tau_{min} = \Delta\tau) \approx 0$, but $g_d^{(1)}(\mathbf{K}, \tau_i)$ decays very slowly because the dust particles are often very large compared with the polymer size, i.e., $g_d^{(1)}(\mathbf{K}, \tau_{max}) \simeq g_d^{(1)}(\mathbf{K}, \tau_{min})$, so that the dust particles contribute an almost constant background.

Finally, in order to utilize the scaling relation between the translational diffusion coefficient at infinite dilution (D^0) and the molecular weight, $D^0 = k_D M^{-\alpha_D}$ (with k_D and α_D being constants), we need to exclude the presence of internal motions in $G(\Gamma)$. Thus, the $G^{(2)}(\mathbf{K}, \tau_i)$ measurements should be performed at $KR_g(\text{largest size}) \lesssim 1$. This requirement may be difficult to fulfill and may involve several iterations, since we do not know in advance what $R_g(\text{largest size})$ is and can only estimate a range for it in any case. Nevertheless, in practice, if we can achieve similar molecular-weight distributions using two different scattering angles, the results can at least be considered self-consistent.

In the presence of intermolecular interactions and internal motions, we may write

$$D_i \simeq D_i^0 (1 + \bar{k}_d C) \quad (8.4.24)$$

and

$$\Gamma_i \simeq D_i K^2 (1 + f R_{g,i}^2 K^2) \tag{8.4.25}$$

instead of $D_i = D_i^0$ and $\Gamma_i = D_i K^2$, respectively. The term $\bar{k}_d C$ represents the overall intermolecular interactions of the polydisperse solution at concentration C, and the average diffusion second virial coefficient $\bar{k}_d$. The width Γ_i represents the characteristic linewidth of species i and no longer has a simple K^2 dependence. However, at $KR_{g,i} > 1$, species i may have characteristic linewidths related to internal motions in addition to the translational motions. Then, the mixing of characteristic times due to translation and internal motions will render the determination of size distributions ambiguous at best.

We have already discussed the light-scattering spectrometers in Chapter 6. For examples, see Figures 6.4.3–6.4.6 and 6.9.5.

8.5. PERFORMING THE LAPLACE INVERSION AND USING THE SCALING RELATION FOR TRANSFORMATION OF $G(\Gamma)$ TO THE MOLECULAR-WEIGHT DISTRIBUTION

Inversion of the Laplace transform generally involves a very difficult and ill-conditioned problem. Avoidance of the ill-posed nature of Laplace inversion is not trivial, as has been discussed in some detail in Chapter VII. However, under certain conditions, we can reliably retrieve useful information by performing the Laplace inversion on precisely measured experimental data. Based on the discussions presented in Chapter VII, we can summarize the conditions as follows:

(1) The measured time correlation function should be known over a range of delay times covering the expected characteristic linewidths. As the exact range of characteristic linewidths is not known a priori, we make $G^{(2)}(\Delta \tau)/A$-1 exceed 0.998β, and $G^{(2)}$(last channel)/A-1 agree to within a few tenths of a percent with the background A. (For details, see Section 7.3.)

(2) The number of measured channels in the intensity–intensity time correlation function is not crucial. However, there should be sufficient channels to depict the shape of $G^{(2)}(\tau)$.

(3) The experimental data should be used mainly for resolving unimodal characteristic linewidth distributions. Generally, the average linewidth $\bar{\Gamma}$ can be determined to within 1–2%, the width of the distribution $(\mu_2/\bar{\Gamma}^2)$ to 5–10%, and the skewness to 30–50%. One usually cannot reliably extract more than three parameters. For bimodal distributions, the two characteristic linewidths have to be separated by a factor of at least 2–3. The reliability of the values of the two characteristic linewidths, as well as their amplitudes,

becomes more acceptable as the separation distance of the two characteristic linewidths increases from 3 to 5. Needless to say, the precision depends on many factors. The experimentalist should use simulated data with appropriate statistical noise to examine each individual situation.

The resolution of photon correlation spectroscopy (PCS) is coarse. The upper or/and the lower bound for the characteristic linewidth distribution is difficult to establish. However, PCS can cover a very broad size range from tens of angstroms to micrometers in diameter, with particular emphasis on the larger-size particles because the scattered intensity of species i is proportional to the square of the polarizability of species i, i.e., $\alpha_i^2 \propto (\text{size})_i^6$. In a recent simulation, we showed that the variation of $G(\Gamma)$ with and without the 1% hatched section shown in Fig. 8.5.1(b) corresponds to a baseline adjustment of only $\approx 0.05\%$. Thus, we cannot use PCS to determine the upper and lower limits of particle sizes precisely. Nevertheless, we can say that $\gtrsim 99\%$ of the polymer (or particle-size) fractions are contained within a certain molecular weight (or size) range, as demonstrated by Figure 8.5.1(c) and (d).

In Fig. 8.5.1, we have converted $G(\tau)$ to $F_w(M)$ by using the scaling relation $D_T = k_D M^{-\alpha_D}$ with k_D and α_D being the two scaling constants and D_T the translational diffusion coefficient. In estimating the molecular-weight distribution (MWD) from $G(\Gamma)$, the transformation depends on whether $G(\Gamma)$ is continuous or discrete. We shall outline a conversion procedure as follows.

For a *discrete* $G(\Gamma)$, defined as $F(\Gamma)$, we have

$$\sum_i F(\Gamma_i) = \sum_i \left(\frac{N_i M_i^2}{\sum_j N_j M_j^2} \right) = 1, \tag{8.5.1}$$

where N_i, M_i, and $\Gamma_i (\equiv K^2 D_{T,i})$ are the number of molecules, the molecular weight, and the characteristic linewidth of species i, respectively. By definition, $M_w = \sum_i F(M_i)/\sum_j [F(M_j)/M_j]$. Then, we can write

$$M_w = \frac{\sum N_i M_i^2}{\sum N_i M_i} = \frac{\sum N_i M_i^2}{\sum \left[F(\Gamma_i)\left(\sum_j N_j M_j^2 \right) \middle/ M_i \right]}$$

$$= \left(\sum \frac{F(\Gamma_i)}{M_i} \right)^{-1} = k_D^{1/\alpha_D} [\sum F(\Gamma_i) D_i^{1/\alpha_D}]^{-1}. \tag{8.5.2}$$

A *logarithmically spaced* discrete distribution $F(\Gamma_i)$ is related to $G(\Gamma_i)$ by

$$F(\Gamma_i) = \Gamma_i G(\Gamma_i). \tag{8.5.3}$$

Thus, Eq. (8.5.2) becomes

$$M_w = \frac{k_D^{1/\alpha_D}}{\sum G(\Gamma_i)\Gamma_i D_i^{1/\alpha_D}} = \frac{(K^2 k_D)^{1/\alpha_D}}{\sum G(\Gamma_i)\Gamma_i^{1+1/\alpha_D}}. \tag{8.5.4}$$

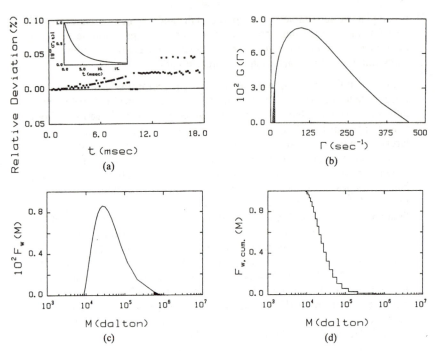

Fig. 8.5.1. (a) Plot of relative deviation in $|g^{(1)}(t)|$ for two characteristic linewidth distributions: $G(\Gamma)$ with and without the 1% hatched section. The shaded contribution of $G(\Gamma)$ corresponds to a baseline adjustment of $\approx 0.05\%$. A plot of $|g^{(1)}(t)|$ vs t is inserted to demonstrate the precision required in the time-correlation-function measurements, i.e., fittings can always be achieved such that the measured and the computed $|g^{(1)}(t)|$ are indistinguishable within the experimental error limits. Measured and computed baseline must agree within $\sim 0.1\%$. Here $t \equiv \tau$. (b) Plot of $G(\Gamma)$ versus Γ. The hatched area represents 1% of the area $\int G(\Gamma)\,d\Gamma$. The results were obtained by means of Eq. (7.1.1) using data of (a). (c) Molecular-weight distribution of poly(1,4-phenyleneterephthalamide) (PPTA) based on $G(\Gamma)$ from (b). The hatched area represents the baseline uncertainty in the time-correlation-function measurement, which can be translated into uncertainties in the high-molecular-weight tail. $F_w(M)$ is the weight distribution. (d) Cumulative molecular-weight distribution of PPTA from (c). Molecular-weight fractions vary from 10^4 to $\approx 5 \times 10^5$. The high-molecular-weight tail corresponding to polymers with $M > 5 \times 10^5$ is seen to be within the experimental error limits. $F_{w,\,cum} = \int_\infty^M F_w(M)M\,dM / \int_\infty^0 F_w(M)\,dM$. At $M > 5 \times 10^5$ g mol^{-1}, the amount of polymer in $F_{w,\,cum}$ has become negligibly small. (Reprinted with permission from Benjamin Chu, Photon Correlation Spectroscopy of Polymer Solutions *in* Comprehensive Polymer Science, Polymer Characterization, Vol. 1, eds. C. Price and C. Booth ©1989, Pergamon Press plc).

As M_w can be determined from absolute light scattering intensity measurements, k_D can be cancelled by taking the ratio of two polymer (or particle-size) fractions:

$$\frac{M_{w1}}{M_{w2}} = \left(\frac{\sum G(\Gamma_i)\Gamma_i}{\sum G(\Gamma_i)\Gamma_i^{1+1/\alpha_D}} \right)_1 \left(\frac{\sum G(\Gamma_i)\Gamma_i^{1+1/\alpha_D}}{\sum G(\Gamma_i)\Gamma_i} \right)_2 \left(\frac{K_1}{K_2} \right)^{2/\alpha_D}, \quad (8.5.5)$$

yielding α_D. Then, k_D can be found from Eq. (8.5.4). Similarly, we can calculate M_n:

$$
\begin{aligned}
M_n &= \sum N_i M_i / \sum N_i \\
&= \frac{\sum F(\Gamma_i)/M_i}{\sum F(\Gamma_i)/M_i^2} \\
&= (k_D K^2)^{1/\alpha_D} \frac{\sum G(\Gamma_i)\Gamma_i^{1+1/\alpha_D}}{\sum G(\Gamma_i)\Gamma_i^{1+2/\alpha_D}}.
\end{aligned}
\quad (8.5.6)
$$

The continuous molecular-weight distribution is related to $G(\Gamma)$ by

$$MG(M) \propto \Gamma G(\Gamma). \quad (8.5.7)$$

8.6. SUMMARY OF POLYMERS CHARACTERIZED (AFTER CHU, 1990b)

Laser light scattering (LLS) offers several advantages as an analytical tool for polymer-solution studies. It is a nonintrusive technique; the molecular-weight determination is absolute; the amount of solution required can be quite small (often $\lesssim 1$ ml); the glass solution cell can be flame-sealed; and the experiments can be performed at high temperatures.

LLS has been used successfully to characterize many intractable polymers in dilute solution, such as poly(ethyleneterephthalate) in hexafluoro-2-propanol (Naoki et al. 1985), polyethylene in 1,2,4-trichlorobenzene (Chu et al., 1984a; Pope and Chu, 1984), poly(1,4-phenylene terephthalamide) (PPTA) in concentrated sulfuric acid (Chu et al., 1984a,b; Ying and Chu, 1984; Chu et al., 1985a,b; Ying et al., 1985), and an alternating copolymer of ethylene and tetrafluoroethylene (PETFE) in diisobutyl adipate (Chu and Wu, 1986, 1987; Wu et al., 1987; Chu et al., 1989a). The polymers are difficult to characterize for a variety of reasons. For examples, the solvent hexafluoro-2-propanol has a low vapor pressure at room temperature and a low refractive index; 1,2,4-trichlorobenzene is carcinogenic; and concentrated sulfuric acid is corrosive. A proper seal (flame seal, if necessary) of the sample solution in a closed glass LLS cell can alleviate most such problems. PPTA in concentrated sulfuric acid also exhibits solution behavior involving electrostatic interactions. In the light scattering characterization of PPTA, it was essential to consider its aggregation behavior. PETFE in diisobutyl adipate has a theta temperature of 231°C (Chu *et al.*, 1989a). The high-temperature capability of LLS makes PETFE characterization in dilute solution feasible.

The LLS characterization of poly(tetrafluoroethylene) (PTFE) in dilute solution was accomplished according to the following experimental procedure (Chu *et al.*, 1987, 1988a, 1989b), as has been outlined in this chapter:

1. finding a solvent for PTFE,
2. developing a new all-glass dissolution–filtration apparatus for solution preparation and clarification,
3. constructing a high-temperature light-scattering spectrometer,
4. improving the reliability of Laplace-inversion algorithms, and
5. determining the scaling relation between the translational diffusion coefficient at infinite dilution (D^0) and the molecular weight (M).

For the two PTFE polymer samples, we were able to determine the weight-average molecular weight (M_w), the z-average radius of gyration (R_g), the second virial coefficient (A_2), the z-average translational diffusion coefficient at infinite dilution (D^0), and its variance (Var) in perfluorotetracosane (n-$C_{24}F_{50}$), as well as the molecular-weight distribution (MWD). The light scattering results for the low-molecular-weight PTFE sample were checked using oligomers of poly(chlorotrifluoroethylene) with two boiling fractions (and therefore two different refractive indices) and perfluorotetracosane, a one-component solvent.

The PTFE characterization is presented to demonstrate the unique capabilites of LLS. Figure 8.6.1 shows Zimm plots of DLX-6000 PTFE in two oligomers of poly(chlorotrifluoro-ethylene) with slightly different boiling

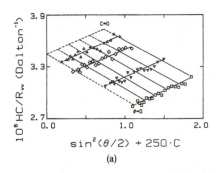 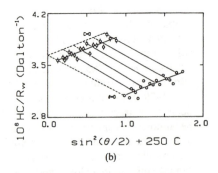

(a)   (b)

FIG. 8.6.1.   (a) Zimm plot of DLX-6000 PTFE in oligomer with $n_s = 1.320$ and $(\partial n/\partial C)_T = 5.3 \times 10^{-2}$ ml g^{-1}. (After Chu *et al.*, 1987.) (b) Zimm plot of DLX-6000 PTFE in oligomer with $n_s = 1.325$ and $(\partial n/\partial C)_T = 4.7 \times 10^{-2}$ ml g^{-1}. (After Chu *et al.*, 1988a.)

Table 8.6.1

Static Properties (M_w, A_2, and R_g) of PTFE in Various Solvents

solv[a]	temp (°C)	$M_w(10^5$ g/mol)	$A_2(10^{-5}$ mol ml/g^2	R_g(nm)
		Low Molecular Weight Sample (DLX-6000)		
MO-I	340	2.84 ± 10%	−6.7 ± 20%	17.8 ± 10%
MO-II	340	2.78 ± 10%	−6.2 ± 20%	18.3 ± 10%
n-C$_{24}$F$_{50}$	325	2.6 ± 15%	1.9 ± 21%	20 ± 25%
		High Molecular Weight Sample (PTFE-5)[b]		
n-C$_{24}$F$_{50}$	325	21 ± 15%	1.0 ± 20%	59 ± 14%

(a) The symbols I and II denote the two boiling fractions of MO used to characterize the same DLX-6000 PTFE in Chu, et al. (1988a). MO denotes mixed oligomers of poly(chlorotrifluoroethylene)

(b) PTFE-5 is identical to the sample of same designation described by Tuminello, et al. (1988).

points, refractive indices ($n_0 = 1.320$ (a) and 1.325 (b)), and refractive-index increments ($(\partial n/\partial C)_T = 5.3 \times 10^{-2}$ ml/g (a) and 4.7×10^{-2} ml/g (b)). The results are listed in Table 8.6.1, in which we have also established an absolute determination of the weight-average molecular weight (M_w) of DLX-6000 PTFE in perfluorotetracosane. We have considered the oligomers of poly(chlorotrifluoroethylene) as pseudomixed solvents. The slight changes in the solvent composition are depicted by changes in boiling point and refractive index. The fact that the molecular weight remained the same in the two oligomer solvents suggested negligible preferential absorption of solvent of different molecular weights by PTFE. Preferential absorption for such pseudomixed solvents was expected to be very small, if any. What we did was to show that our supposition, i.e., a lack of preferential absorption, was reasonable, and to recognize that low-molecular-weight oligomers could be used as solvents. Finally, the molecular weight of PTFE was determined in perfluorotetracosane, as listed in Table 8.6.1, confirming the first absolute molecular-weight determination of PTFE since its discovery about 50 years ago.

Figure 8.6.2(a) shows a typical plot of the unnormalized net intensity—intensity time correlation function $b|g^{(1)}(\mathbf{K}, t)|^2$ vs $t(\equiv \tau)$, the delay time. The experimentally measured $b|g^{(1)}(t)|^2$ can be fitted within the noise limit using MSVD, RILIE, or CONTIN, as shown in Figure 8.6.2(b) with the $G(\Gamma)$-values shown in Figure 8.6.3. Unfortunately, with one PTFE sample, $G(\Gamma)$ could not be transformed to the molecular-weight distribution, since the D^0-vs-M scaling relation could not be established. Consequently, it was necessary to characterize another molecular-weight fraction of PTFE (Chu et al., 1989b).

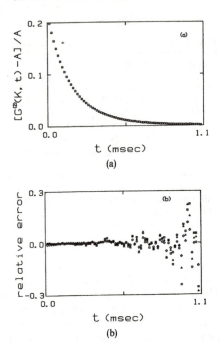

FIG. 8.6.2. (a) Plot of $b|g^{(1)}(\mathbf{K},t)|^2$ vs t for DLX-6000 PTFE in oligomers of poly(chlorotrifluoroethylene) at $C = 7.51 \times 10^{-4}$ g/ml, $\theta = 30°$, and $340°C$. t is the delay time. (b) Relative errors in fitting the measured net intensity–intensity time correlation function of (a): squares, CONTIN; diamonds, RILIE; triangles, MSVD. (After Chu et al., 1988a.)

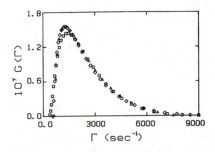

FIG. 8.6.3. Plots of $G(\Gamma)$ vs Γ for 7.5×10^{-4}-g/ml DLX-6000 PTFE ($M_w = 2.8 \times 10^5$ g/mol) in oligomers of poly(chlorotrifluoroethylene) at $340°C$ using three different methods of Laplace inversion: squares, CONTIN; diamonds, RILIE; triangles, MSVD. (After Chu et al., 1988a.)

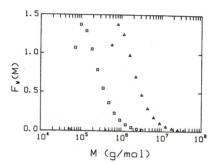

Fig. 8.6.4.   Molecular-weight distribution of the two PTFE polymer samples in n-$C_{24}F_{50}$ at 325°C and $\theta = 30°$. We used the empirical relation $D^0 = 3.99 \times 10^{-4} \, M^{-0.55}$ with D^0 and M expressed in cm²/sec and g/mol, respectively. Squares denote PTFE with $M_w = 2.6 \times 10^5$ g/mol and $M_z : M_w : M_n = 3.8 : 2.1 : 1.0$. Triangles denote PTFE with $M_w = 2.1 \times 10^6$ g/mol and $M_z : M_w : M_n = 4.4 : 2.3 : 10$. (After Chu et al., 1989b.)

This was accomplished by dissolving two PTFE samples (DLX-6000 and PTFE-5) in poly(fluorotetracosane) and performing the laser light scattering measurements. By applying an integration method,

$$D^0 = 3.99 \times 10^{-4} M^{-0.55} \text{ cm}^2/\text{sec}, \qquad (8.6.1)$$

with M expressed in g/mol, for PTFE in n-$C_{24}F_{50}$ at 325°C. With Eq. (8.6.1), a determination of MWD of DLX-6000 PTFE and PTFE-5, as shown in Fig. 8.6.4, was accomplished.

Finally, as has been discussed in Chu (1989b, pp. 82–84), laser light scattering permits us to estimate the MWD by making only one linewidth measurement at one scattering angle and one concentration, provided we know the magnitudes of Eqs. (8.4.24), (8.4.25) and $D_T = k_D M^{-\alpha_D}$. Thus, this noninvasive technique has the potential to become a powerful probe or monitor. Furthermore, it is permissible to use *difference* correlation functions to measure changes of molecular weight and polydispersity without any of the complications associated with the detailed analysis and complex mathematics. By using the cumulants expansion and the scaling relation, and in the absence of internal motions and intermolecular interactions, the difference term can be represented by

$$\log|g_{M_1}^{(1)}(\Gamma, K^2 t)| - \log|g_{M_2}^{(1)}(\Gamma, K^2 t)|$$

$$\cong -\frac{k_D}{2.303}(M_1^{-\alpha_D} - M_2^{-\alpha_D})K^2 t + \frac{1}{4.606}(\mu_{2,M_1} - \mu_{2,M_2})K^4 t^2. \qquad (8.6.2)$$

In Eq. (8.6.2), the first term on the right-hand side corresponds to the initial

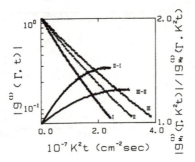

FIG. 8.6.5. Scaling of $|g^{(1)}(\Gamma, t)|$ at different molecular weights; $\theta = 30°$. Note: Concentration has negligible effects. $Y = \log|g^{(1)}(\Gamma, t)| - \log|g^{(1)}(\Gamma, t)|$.

PETFE	$\bar{M}_w$ (g/mol)		C (mg/ml)	
I	5.40×10^5			4.03×10^{-3}
II	9.00×10^5			3.50×10^{-3}
III	1.16×10^6			4.28×10^{-3}
Slope (cm²/s)		$\bar{M}_w$ (g/mol)		
II–I 1.27×10^{-8}	I = 5.62×10^5	5.40×10^5	5.3×10^5	
III–II 4.97×10^{-9}	III = 1.13×10^6	1.16×10^6	1.18×10^6	

slope in a plot of $\log|g_{M_1}^{(1)}(\Gamma, K^2 t)| - \log|g_{M_2}^{(1)}(\Gamma, K^2 t)|$ vs $K^2 t$ as shown in Figure 8.6.5 with $Y (\equiv \log|g_{M_1}^{(1)}(\Gamma, K^2 t)| - \log|g_{M_2}^{(1)}(K^2 t)|)$ being the y-coordinate. Thus, if we know that M_w (PETFE II) $= 9.00 \times 10^5$ g/mol, we can determine PETFE I and III to be 5.62×10^5 and 1.13×10^6 g/mol, using the initial slope and Eq. (8.6.2) with $D_T^0 = 3.35 \times 10^{-4} M_w^{-0.60}$ (cm²/s) and M_w expressed in grams per mole. Similarly, we can determine $\mu_2/\bar{\Gamma}^2$ (e.g., $\mu_2/\bar{\Gamma}^2$(II–I) is 0.11 from Eq. (8.6.2)), in agreements with more detailed computations.

8.7. LIGHT SCATTERING BY NONERGODIC MEDIA

In Chapter VIII, only one particle-sizing aspect of laser light scattering (LLS) was presented. Needless to say, LLS has many applications. The previous chapters have provided the basic practice and theory of LLS. Interested readers may wish to consult the selected papers for specific topics. Selected papers on laser light scattering by macromolecular, supramolecular, and fluid systems have been collected in one volume (Chu, 1990a). A table of contents together with additional pertinent references is presented in Appendix 8.A.

Dynamic light scattering (DLS) has been applied to the study of solutions and suspensions. Even polymer melts can be investigated, although the interpretation of dynamic parameters due to density, concentration, and segmental motion fluctuations is more complex and will not be covered in this introductory book. (See the review by Meier and Fytas (1989) on the use of PCS for amorphous polymers and compatible polymer blends.)

The "fluidlike" media in which the scatterers are able to diffuse throughout the medium, including entangled polymer coils in semidilute solutions, can be measured in the format discussed in Chapter VI, i.e., the time-averaged intensity correlation function $(ICF)_T$ obtained from a single DLS measurement is able to explore all the possible spatial configurations. Then, $(ICF)_T$ is equal to its ensemble average $(ICF)_E$. However, if DLS is used to study polymer gels or colloidal glasses, the scatterers are localized near some fixed average positions and are only able to execute limited Brownian motions about these positions. The more restricted motions of the scatterers prevent the usual DLS measurement from sampling the full ensemble of configurations in a single scattering-volume element. Only a restricted region of phase space, or *subensemble* (p) will be detected in a single scattering volume, i.e., different spatial regions in the same sample will be described by different subensembles. Thus, $(ICF)_{T,p} \neq (ICF)_E$. For semidilute or concentrated polymer solutions in very viscous media, the scattering elements may only be able to make limited Brownian excursions over the longest delay-time range of a correlator. $(ICF)_T$ may again not be representative of its true full ensemble average because of localized inhomogeneities that take a very long time for equilibration. The work by Pusey and van Megen (1989) deals extensively with this problem.

In a light scattering experiment, the scattering volume ($\sim (100\ \mu m)^3$) is usually much smaller than the sample size ($\sim 1\ cm^3$). Then an ensemble average is represented by viewing the sample over different spatial regimes, i.e., a series of illuminated positions, $p = 1$ to P, within the sample needs to be viewed by the detector. At each viewing position, one obtains an $(ICF)_{T,p}$. An average of $(ICF)_{T,p}$ over the full ensemble yields $(ICF)_E$.

Pusey and van Megen introduced the variable $X_p(t)$, which describes the time evolution of a property of the sample. The label p denotes a particular subensemble. The time average of $X_p(t)$,

$$\langle X_p \rangle_T \equiv \lim_{T \to \infty} \frac{1}{T} \int_0^T dt\, X_p(t), \qquad (8.7.1)$$

is an average over the subensemble, or $\langle X_p \rangle_T = \langle X \rangle_{T,p}$. The value of X over the full ensemble is

$$\langle X \rangle_E = \lim_{P \to \infty} \frac{1}{P} \sum_{p=1}^{P} X_p(t). \qquad (8.7.2)$$

The ensemble average of $\langle X_p \rangle_T (\equiv \langle X \rangle_{T,p})$ is the ensemble average of X:

$$
\begin{aligned}
\langle\langle X_p \rangle_T \rangle_E &= \lim_{P \to \infty} \frac{1}{P} \sum_{p=1}^{P} \langle X_p \rangle_T \\
&= \lim_{P \to \infty} \frac{1}{P} \sum_{p=1}^{P} \lim_{T \to \infty} \frac{1}{T} \int_0^T dt\, X_p(t) \qquad (8.7.3) \\
&= \lim_{T \to \infty} \frac{1}{T} \int_0^T dt\, \langle X \rangle_E = \langle\langle X \rangle_E \rangle_T = \langle X \rangle_E.
\end{aligned}
$$

Figure 8.7.1 (after Pusey and van Megen, 1989) shows the normalized time- and ensemble-average ICFs for (a) and (b), a rigid scatterer giving rise to a nonfluctuating speckle pattern and a constant intensity at the detector; (c) and (d), an ergodic medium producing a fully fluctuating Gaussian speckle pattern; and (e) and (f), a nonergodic medium giving rise to a partially fluctuating speckle pattern. In Fig. 8.7.1(a),

$$
\frac{\langle I_p(\mathbf{K},0) I_p(\mathbf{K},\tau)\rangle_T}{\langle I_p(\mathbf{K})\rangle_T^2} = \frac{I_p^2(\mathbf{K})}{I_p^2(\mathbf{K})} = 1. \qquad (8.7.4)
$$

In Fig. 8.7.1(b),

$$
\frac{\langle\langle I_p(\mathbf{K},0) I_p(\mathbf{K},\tau)\rangle_T \rangle_E}{\langle\langle I_p(\mathbf{K})\rangle_T \rangle_E^2} = \frac{\dfrac{1}{P}\sum_p I_p^2(\mathbf{K})}{\left[\dfrac{1}{P}\sum_p I_p(\mathbf{K})\right]^2} = \frac{\langle I^2(\mathbf{K})\rangle_E}{\langle I(\mathbf{K})\rangle_E^2} = 2, \qquad (8.7.5)
$$

which illustrates the difference between the normalized $(ICF)_T$ and $(ICF)_E$. In a comparison of (c) and (d) with (e) and (f), one should pay particular attention to the difference between (c) and (f) and the difference between (c) and (e). A more quantitative discussion follows.

For N particles in a scattering volume v, the electric field has the form

$$
E(\mathbf{K},\tau) = \sum_{j=1}^{N} b_j \exp[i\mathbf{K} \cdot \mathbf{r}_j(\tau)], \qquad (8.7.6)
$$

where b_j is the field amplitude of the light scattered by the jth particle at scattering vector $\mathbf{K}$, and $\mathbf{r}_j(\tau)$ is the corresponding position vector of the center of mass of the jth particle at time τ. The instantaneous intensity $I(\mathbf{K},t) \equiv |E(\mathbf{K},t)|^2$, and its ensemble average is given by

$$
\langle I(\mathbf{K})\rangle_E = \sum_{j=1}^{N} \sum_{k=1}^{N} \langle b_j b_k \exp[i\mathbf{K} \cdot (\mathbf{r}_j - \mathbf{r}_k)]\rangle_E. \qquad (8.7.7)
$$

The ensemble-averaged intensity correlation function (Pusey, 1977)

$$
\langle I(\mathbf{K},0) I(\mathbf{K},\tau)\rangle_E = \langle I(\mathbf{K})\rangle_E^2 + [N\overline{b^2} F(\mathbf{K},\tau)]^2 \qquad (8.7.8)
$$

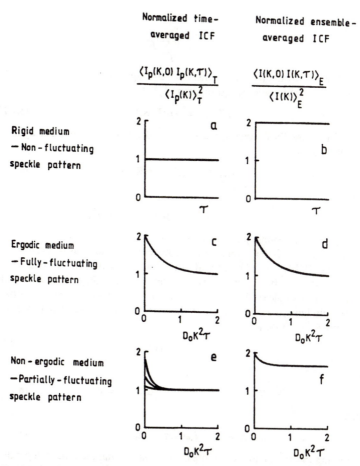

FIG. 8.7.1. Normalized time- and ensemble-averaged intensity correlation functions for three cases: (a), (b) a rigid scatterer giving rise to a nonfluctuating speckle pattern and a constant intensity at the detector; (c), (d) an ergodic medium producing a fully fluctuating Gaussian speckle pattern; (e), (f) a nonergodic medium giving rise to a partially fluctuating speckle pattern. (After Pusey and van Megen, 1989.)

has the intermediate scattering function (or dynamic structure factor) defined by

$$F(\mathbf{K}, \tau) \equiv (N\overline{b^2})^{-1} \sum_{j=1}^{N} \sum_{k=1}^{N} \langle b_j b_k \exp\{i\mathbf{K} \cdot [\mathbf{r}_j(0) - \mathbf{r}_k(\tau)]\}\rangle_{\mathrm{E}} \qquad (8.7.9)$$

with

$$\overline{b^2} \equiv N^{-1} \sum_{j=1}^{N} b_j^2. \tag{8.7.10}$$

From Eqs. (8.7.8) and (8.7.9), the ensemble-averaged scattered intensity can be identified with the static structure factor $F(\mathbf{K}, 0)$:

$$\langle I(\mathbf{K}) \rangle_{\mathrm{E}} = N\overline{b^2}F(\mathbf{K}, 0). \tag{8.7.11}$$

For future use, the following quantities are defined: the normalized dynamic structure factor

$$f_{\mathrm{E}}(\mathbf{K}, \tau) = F(\mathbf{K}, \tau)/F(\mathbf{K}, 0), \tag{8.7.12}$$

and

$$g_{\mathrm{E}}^{(2)}(\mathbf{K}, \tau) \equiv \langle I(\mathbf{K}, \tau)I(\mathbf{K}, 0) \rangle_{\mathrm{E}}/\langle I(\mathbf{K}) \rangle_{\mathrm{E}}^2, \tag{8.7.13}$$

where $\beta = 1$ for a single point ideal detector.

To construct $(\mathrm{ICF})_{\mathrm{E}}$, one can follow either a tedious procedure by measuring $(\mathrm{ICF})_{T,p}$ at different scattering volumes within the sample, or the procedure devised by Pusey and van Megen, whereby they can calculate the $f_{\mathrm{E}}(\mathbf{K}, \tau)$ from a single $(\mathrm{ICF})_T$ measurement. The Pusey–van Megen approach allows only limited excursions Δ_j of the particles about their fixed averaged positions R_j, such that

$$r_j(\tau) = R_j + \Delta_j(\tau) \tag{8.7.14}$$

with

$$R_j = \langle r_j(\tau) \rangle_T \quad \text{and} \quad \langle \Delta_j(\tau) \rangle_T = 0. \tag{8.7.15}$$

Then,

$$E(\mathbf{K}, \tau) = E_{\mathrm{F}}(\mathbf{K}, \tau) + E_{\mathrm{C}}(\mathbf{K}), \tag{8.7.16}$$

where the subscripts F and C denote the fluctuating and the time-independent component, respectively. With $\langle E(\mathbf{K}, \tau) \rangle_{\mathrm{E}} = 0$ and $\langle E(\mathbf{K}) \rangle_T = E_{\mathrm{C}}(\mathbf{K})$,

$$\begin{aligned} \langle I(\mathbf{K}) \rangle_T &= \langle |E(\mathbf{K}, \tau)|^2 \rangle_T \\ &= \langle |E_R(\mathbf{K}, \tau) + E_{\mathrm{C}}(\mathbf{K})|^2 \rangle_T \\ &= \langle I_F(\mathbf{K}) \rangle_T + I_{\mathrm{C}}(\mathbf{K}). \end{aligned} \tag{8.7.17}$$

Without writing down the details of the derivation, for which the reader should consult Pusey and van Megen (1989), we reproduce the result:

$$f_{\mathrm{E}}(\mathbf{K}, \tau) = 1 + \frac{\langle I(\mathbf{K}) \rangle_T}{\langle I(\mathbf{K}) \rangle_{\mathrm{E}}} \{ \sqrt{g_T^{(2)}(\mathbf{K}, \tau) - g_T^{(2)}(\mathbf{K}, 0) + 1} - 1 \}, \tag{8.7.18}$$

so that $f_E(\mathbf{K}, \tau)$ can be computed from measurements of $g_T^{(2)}(\mathbf{K}, \tau)$ for arbitrary $\langle I(\mathbf{K}) \rangle_T$ provided that $\langle I(\mathbf{K}) \rangle_E$ is measured. A measurement of $\langle I(\mathbf{K}) \rangle_E$ should be much easier than one of $(\mathrm{ICF})_E$ ($\equiv \langle I(\mathbf{K}, 0) I(\mathbf{K}, \tau) \rangle_E$).

The reader may feel somewhat at a loss, as these discussions on light scattering by nonergodic media are incomplete. This section can only be approached at this level because experiments on DLS analysis of Brownian particles trapped in gels are still underway (Joosten *et al.*, 1990). The details are best obtained by reading the well-written papers by Pusey and his coworkers. In this section, the reader can only be introduced the problem of studying gels with dynamic light scattering.

8.8. OTHER NEW DEVELOPMENTS

At the opposite extreme from single-particle scattering, recent developments on multiple scattering by Maret and coworkers (Maret and Wolf, 1989; Wolf *et al.*, 1988; Akkermans *et al.*, 1988, 1986) and by the Exxon group (Pine *et al.*, 1988; Weitz *et al.*, 1989; MacKintosh *et al.*, 1989; Wu *et al.*, 1990), known as diffuse-wave spectroscopy, offer the use of dynamic light scattering to study strong scattering media, such as foams (Durian *et al.*, 1990) and milk (Horne, 1989, 1990), which cannot be measured otherwise. It appears that some useful parameters can be deduced from diffuse-wave spectroscopy. Further development is underway (Pine *et al.*, 1990). The interested reader should be aware of the potential capability of this new approach to dynamic light scattering by systems exhibiting multiple scattering.

APPENDIX 8.A. BIBLIOGRAPHY

This appendix lists some useful references from Chu (1990a) and some additional recommended articles.

SECTION ONE. INTRODUCTION

Optical mixing spectroscopy, with applications to problems in physics, chemistry, biology and engineering. G. B. Benedek (*Polarization, Matter and Radiation*, Presses Universitaire de France, 1969, pp. 49–84).

Photon statistics. E. R. Pike (*Rivista del Nuovo Cimento, Serie I* **1**, 277–314, 1969).

SECTION TWO. CORRELATION TECHNIQUES AND INSTRUMENTATION

Correlation techniques in dynamic light scattering. K. Schatzel (*Applied Physics B*, **42**, 193, 1987).

Photometer for quasielastic and classical light scattering. H. R. Haller, C. Destor, D. S. Cannell (*Review of Scientific Instruments*, **54**, 973, 1983).

Prism laser light-scattering spectrometer. B. Chu, R. Xu, T. Maeda, H. S. Dhadwal (*Review of Scientific Instruments*, **59**, 716, 1988).

Dynamic light scattering using monomode optical fibers. R. G. W. Brown (*Applied Optics*, **26**, 4846, 1987).

A fiber-optic light-scattering spectrometer. H. S. Dhadwal, B. Chu (*Review of Scientific Instruments*, **60**, 845, 1989).

SECTION THREE. METHODS OF (DATA) ANALYSIS

A. ACCURACY OF MEASUREMENTS

Statistical accuracy in the digital autocorrelation of photon counting fluctuations. E. Jakeman, E. R. Pike, S. Swain (*Journal of Physics A: General Physics*, **4**, 517, 1971).

Photon-correlation spectroscopy: dependence of linewidth error on normalization, clip level, detector area, sample time and count rate. A. J. Hughes, E. Jakeman, C. J. Oliver, and E. R. Pike (*Journal of Physics A: Math. Nucl. Gen.*, **6**, 1327, 1973).

Intensity-correlation spectroscopy. V. Degiorgio, J. B. Lastovka (*Physical Review A*, **4**, 2033, 1971).

Noise in photon correlation and photon structure functions. K. Schatzel (*Optica Acta*, **30**, 155, 1983).

Photon correlation measurements at large lag times: improving statistical accuracy. K. Schatzel, M. Drewel, S. Stimac (*Journal of Modern Optics*, **35**, 711, 1988).

B. TIME-CORRELATION FUNCTION PROFILE ANALYSIS

Analysis of macromolecular polydispersity in intensity correlation spectroscopy: the method of cumulants. D. E. Koppel (*Journal of Chemical Physics*, **57**, 4814, 1972).

On the recovery and resolution of exponential relaxation rates from experimental data: a singular-value analysis of the Laplace transform inversion in the presence of noise. M. Bertero, P. Boccacci, E. R. Pike (*Proceedings of Royal Society of London A*, **383**, 15, 1982).

On the recovery and resolution of exponential relaxation rates from the experimental data. II. The optimum choice of experimental sampling points for Laplace transform inversion. M. Bertero, P. Boccacci, E. R. Pike (*Proceedings of Royal Society of London A*, **393**, 51, 1984).

On the recovery and resolution of exponential relaxation rates from the experimental data. III. The effect of sampling and truncation of data on the

Laplace transform inversion. M. Bertero, P. Brianzi, E. R. Pike (*Proceedings of Royal Society of London A*, **398**, 23, 1985).

Light scattering polydispersity analysis of molecular diffusion by Laplace transform inversion in weighted spaces. M. Bertero, P. Brianzi, E. R. Pike, G. de Villiers, K. H. Lan, N. Ostrowsky (*Journal of Chemical Physics*, **82**, 1551, 1985).

Inverse problems in polymer characterization: direct analysis of polydispersity with photon correlation spectroscopy. S. W. Provencher (*Makromolekulare Chemie*, **180**, 201, 1979).

A constrained regularization method for inverting data represented by linear algebraic or integral equations. S. W. Provencher (*Computer Physics Communications*, **27**, 213, 1982).

Correlation function profile analysis of polydisperse macromolecular solutions and colloidal suspensions. B. Chu, J. R. Ford, H. S. Dhadwal (*Methods of Enzymology, Vol. 117*, Academic Press, Florida, 1985, pp. 256–297).

Maximum entropy analysis of quasielastic light scattering from colloidal dispersions. A. K. Livesey, P. Licino, M. Delaye (*Journal of Chemical Physics*, **84**, 5102, 1986).

Maximum-entropy analysis of photon correlation spectroscopy data. S. L. Nyeo, B. Chu (*Macromolecules*, **22**, 3998, 1989).

SECTION FOUR. CRITICAL OPALESCENCE

Observation of time-dependent concentration fluctuations in a binary mixture near the critical temperature using a He–Ne laser. S. S. Alpert, Y. Yeh, E. Lipworth (*Physical Review Letters*, **14**, 486, 1965).

Observation of the spectrum of light scattered from a pure fluid near its critical point. N. C. Ford, Jr., G. B. Benedek (*Physical Review Letters*, **15**, 649, 1965).

Spatial and time-dependent concentration fluctuations of the isobutyric acid–water system in the neighborhood of its critical mixing point. B. Chu, F. J. Schoenes, W. P. Kao (*Journal of the American Chemical Society*, **90**, 3042, 1968).

Viscous damping of thermal excitations on the interface of critical fluid mixtures. J. S. Huang, W. W. Webb (*Physical Review Letters*, **23**, 160, 1969).

Dynamics of concentration fluctuations in a binary mixture in the hydrodynamical and nonhydrodynamical regimes. P. Berge, P. Calmetter, C. Laj, M. Tournarie, B. Volochine (*Physical Review Letters*, **24**, 1223, 1970).

Dynamics of fluids near the critical point: decay rate of order-parameter fluctuations. H. L. Swinney, D. L. Henry (*Physical Review A*, **8**, 2586, 1973).

Time dependence of critical concentration fluctuations in a binary liquid. H. C. Burstyn and J. V. Sengers (*Physical Review A*, **27**, 1071, 1983).

Static and dynamic light scattering studies of water-in-oil microemulsions in the critical region. Evidence of a crossover effect. J. Rouch, A. Safouane, P. Tartaglia, S. H. Chen (*Journal of Chemical Physics*, **90**, 3756, 1989).

SECTION FIVE. POLYMER SOLUTIONS

Doppler shifts in light scattering from pure liquids and polymer solutions. R. Pecora (*Journal of Chemical Physics*, **40**, 1604, 1964).

Spectral distribution of light scattered from flexible-coil macromolecules. R. Pecora (*Journal of Chemical Physics*, **49**, 1032, 1968).

Spectral distribution of light scattered by monodisperse rigid rods. R. Pecora (*Journal of Chemical Physics*, **48**, 4126, 1968).

Dynamical properties of polymer solutions in good solvent by Rayleigh scattering experiments. M. Adam, M. Delsanti (*Macromolecules*, **10**, 1229, 1977).

Static and dynamical properties of polystyrene in *trans*-decalin. 1. NBS 705 standard near Θ conditions. T. Nose, B. Chu (*Macromolecules*, **12**, 590, 1979).

Interpretation of dynamic scattering from polymer solutions. A. Z. Akcasu, M. Benmouna, C. C. Han (*Polymer*, **21**, 866, 1980).

Rotational and translational diffusion in semidilute solutions of rigid-rod macromolecules. K. M. Zero, R. Pecora (*Macromolecules*, **15**, 87, 1982).

Probe diffusion of polystyrene latex spheres in poly(ethylene oxide)–water. G. S. Ullmann, K. Ullmann, R. M. Lindner, G. D. J. Phillies (*Journal of Physical Chemistry*, **89**, 692, 1985).

Wave-vector dependence of the initial decay rate of fluctuations in polymer solutions. P. Wiltzius, D. S. Cannell (*Physical Review Letters*, **56**, 61, 1986).

Translational diffusion of linear polystyrenes in dilute and semidilute solutions of poly(vinyl methyl ether). L. M. Wheeler, T. P. Lodge, B. Hanley, M. Tirrell (*Macromolecules*, **20**, 1120, 1987).

Light-scattering characterization of poly(tetrafluoroethylene). B. Chu, C. Wu, W. Buck (*Macromolecules*, **21**, 397, 1988).

Structure and dynamics of epoxy polymers. B. Chu, C. Wu (*Macromolecules*, **21**, 1729, 1988).

Critical phenomena and polymer coil-to-globule transition. B. Chu, R. Xu, Z. Wang, J. Zuo (*Journal of Applied Crystallography*, **21**, 707, 1988).

SECTION SIX. COLLOIDAL SUSPENSIONS

Observation of diffusion broadening of Rayleigh scattered light. H. Z. Cummins, N. Knable, Y. Yeh (*Physical Review Letters*, **12**, 150, 1964).

Light scattering study of dynamic and time-averaged correlations in dispersions of charged particles. J. C. Brown, P. N. Pusey, J. W. Goodwin, R. H. Ottewill (*Journal of Physics A: Math. Gen.*, **8**, 664, 1975).

The dynamics of interacting Brownian particles. P. N. Pusey (*Journal of Physics A: Math. Gen.*, **8**, 1433, 1975).

Thermodynamic analysis of the growth of sodium dodecyl sulfate micelles. P. J. Missel, N. A. Mazer, G. B. Benedek, C. Y. Young, M. C. Carey (*Journal of Physical Chemistry*, **84**, 1044, 1980).

Quasi-elastic light scattering study of intermicellar interactions in aqueous sodium dodecyl sulfate solutions. M. Corti, V. Degiorgio (*Journal of Physical Chemistry*, **85**, 711, 1981).

Concentration dependence of the self-diffusion coefficient of hard, spherical particles measured with photon correlation spectroscopy. M. M. Kops-Werkhoven, C. Pathmamnoharan, A, Vrij, H. M. Fijnaut (*Journal of Chemical Physics*, **77**, 5913, 1982).

Light scattering investigations of the behavior of semidilute aqueous micellar solutions of cetyltrimethylammonium bromide: analogy with semidilute polymer solutions. S. J. Candau, E. Hirsch, R. Zana (*Journal of Colloid and Interface Science*, **105**, 521, 1985).

Photon correlation spectroscopy of the non-Markovian Brownian motion of spherical particles. K. Ohbayashi, T. Kohno, H. Utiyama (*Physical Review A*, **27**, 2632, 1983).

Dynamics of fractal colloidal aggregates. J. E. Martin, D. W. Schaefer (*Physical Review Letters*, **53**, 2457, 1984).

Universal diffusion-limited colloid aggregation. M. Y. Lin, H. M. Lindsay, D. A. Weitz, R. Klein, R. C. Ball, P. Meakin. (*Journal of Physics: Condensed Matter*, **2**, 3093, 1990).

Universal reaction-limited colloid aggregation. M. Y. Lin, H. M. Lindsay, D. A. Weitz, R. C. Ball, R. Klein, P. Meakin. (*Physical Review*, **41**, 2005, 1990).

Multidetector scattering as a probe of local structure in disordered phases. N. A. Clark, B. J. Ackerson, A. J. Hung (*Physical Review Letters* **50**, 1459, 1983).

Dynamic slowing down and nonexponential decay of the density correlation function in dense microemulsions. S. H. Chen, J. S. Huang (*Physical Review Letters*, **55**, 1888, 1985).

Observation of a glass transition in suspensions of spherical colloidal particles. P. N. Pusey, W. van Megen (*Physical Review Letters*, **59**, 2083, 1987).

SECTION SEVEN. BIOLOGY

Rotary-diffusion broadening of Rayleigh lines scattered from optically anisotropic macromolecules in solution. A. Wada, N. Suda, T. Tsuda, K. Soda (*Journal of Chemical Physics*, **50**, 31, 1969).

Determination of diffusion coefficients of haemocyanin at low concentration by intensity fluctuation spectroscopy of scattered laser light. R. Foord, E. Jakeman, C. J. Oliver, E. R. Pike, R. J. Blagrove, E. Wood, A. R. Peacocke (*Nature*, **227**, 242, 1970).

The simultaneous measurement of the electrophoretic mobility and diffusion coefficient in bovine serum albumin solutions by light scattering. B. R. Ware, W. H. Flygare (*Chemical Physics Letters*, **12**, 81, 1971).

Observation of protein diffusivity in intact human and bovine lenses with application to cataract. T. Tanaka, G. B. Benedek (*Investigative Ophthamology, St. Louis, Vol. 14*, 1975, pp. 449–456).

Rotational relaxation of macromolecules determined by dynamic light scattering. I. Tobacco mosaic virus. J. M. Schurr, K. S. Schmitz (*Biopolymers*, **12**, 1021, 1973).

Measurement of the velocity of blood flow (in vivo) using a fiber optic catheter and optical mixing spectroscopy. T. Tanaka, G. B. Benedek (*Applied Optics*, **14**, 189, 1975).

Spermatozoa motility in human cervical mucus. M. Dubois, P. Jouannet, P. Berge, G. David (*Nature*, **252**, 711, 1974).

Immunoassay by light scattering spectroscopy. R. J. Cohen, G. B. Benedek (*Immunochemistry*, **12**, 349, 1975).

Brownian motion of highly charged poly(L-lysine). Effects of salt and polyion concentration. S.-C. Lin, W. I. Lee, J. M. Schurr (*Biopolymers*, **17**, 1041, 1978).

Photon correlation spectroscopy in strongly absorbing and concentrated samples with applications to unliganded hemoglobin. R. S. Hall, Y. S. Oh, C. S. Johnson, Jr. (*Journal of Physical Chemistry*, **84**, 756, 1980).

Kinetics of head–tail joining in bacteriophage T4D studied by quasi-elastic light scattering: Effects of temperature, pH, and ionic strength. J. Aksiyote-Benbasat, V. A. Bloomfield (*Biochemistry*, **20**, 5018, 1981).

Determination of motile behaviour of prokaryotic and eukaryotic cells by quasi-elastic light scattering. S.-H. Chen, F. R. Hallet (*Quarterly Reviews of Biophysics*, **15**, 131, 1982).

Surface charges and calcium ion binding of disk membrane vesicles. T. Kitano, T. Chang, G. B. Caflisch, D. M. Piatt, H. Yu (*Biochemistry*, **22**, 4019, 1983).

Spectrum of light quasi-elastically scattered from suspensions of tobacco mosaic virus. Experimental study of anisotropy in translational diffusion. K. Kubota, H. Urabe, Y. Tominaga, S. Fujime (*Macromolecules*, **17**, 2096, 1984).

Dynamic light-scattering study of suspensions of fd virus. Application of a theory of the light-scattering spectrum of weakly bending filaments. T. Maeda, S. Fujime (*Macromolecules*, **18**, 2430, 1985).

A QELS–SEF study on high molecular weight poly(lysine): field strength dependent apparent diffusion coefficient and the ordinary–extraordinary phase transition. K. S. Schmitz, D. J. Ramsay (*Macromolecules*, **18**, 933, 1985).

Photon correlation spectroscopy of the polarization signal from single muscle fibres. Y. Yeh, R. J. Baskin, S. Shen, M. Jones (*Journal of Muscle Research and Cell Motility*, **11**, 137, 1990).

SECTION EIGHT. GELS, FILMS, AND OTHER TOPICS

Spectrum of light scattered from a viscoelastic gel. T. Tanaka, L. O. Hocker, G. B. Benedek (*Journal of Chemical Physics*, **59**, 5151, 1973).

Size and mass determination of cluster obtained by polycondensation near the gelation threshold. M. Adam, M. Delsanti, J. P. Munch, D. Durand (*Journal de Physique*, **48**, 1809, 1987).

Critical dynamics of the sol–gel transition. J. E. Martin, J. P. Wilcoxon (*Physical Review Letters*, **61**, 373, 1988).

Viscoelastic relaxation of insoluble monomolecular films. J. C. Earnshaw, R. C. McGivern, P. J. Winch (*Journal de Physique*, **49**, 1271, 1988).

Localized fluid flow measurements with a He–Ne laser spectrometer. Y. Yeh, H. Z. Cummins (*Applied Physics Letters*, **4**, 176, 1964).

Spectrum of light scattered quasielastically from a normal liquid. J. B. Lastovka, G. B. Benedek (*Physical Review Letters*, **17**, 1039, 1966).

Measurement of the thermal diffusivity of pure fluids by Rayleigh scattering of laser light. M. Corti, V. Degiorgio (*Journal of Physics C: Solid State Physics*, **8**, 953, 1975).

Onset of turbulence in a rotating fluid. J. P. Gollub, H. L. Swinney (*Physical Review Letters*, **35**, 927, 1975).

Rapid investigation of sedimentation by vibrational force field. A. Wada, I. Nishio, K. Soda (*Review of Scientific Instruments*, **50**, 458, 1979).

Detection of small polydispersities by photon correlation spectroscopy. P. N. Pusey, W. van Megen (*Journal of Chemical Physics*, **80**, 3513, 1984).

Brownian dynamics close to a wall studied by photon correlation spectroscopy from an evanescent wave. K. H. Lan, N. Ostrowsky, D. Sornette (*Physical Review Letters*, **57**, 17, 1986).

Light-scattering observations of long-range correlations in a nonequilibrium liquid. B. M. Law, R. W. Gammon, J. V. Sengers (*Physical Review Letters*, **60**, 1554, 1988).

Characterization of the local structure of fluids by apertured cross-correlation functions. B. J. Ackerson, T. W. Taylor, N. A. Clark (*Physical Review A*, **31**, 3183, 1985).

Dynamic light scattering by non-ergodic media. P. N. Pusey, W. van Megen (*Physica A*, **157**, 705, 1989).

REFERENCES

Akkermans, E., Wolf, P. E., and Maynard, R. (1986). *Phys. Rev. Lett.* **56**, 1471.

Akkermans, E., Wolf, P. E., Maynard, R., and Maret, G. (1988). *J. Phys.* **49**, 77.

Bushuk, W. and Benoit, H. (1958). *Can. J.* Chem. 36, 1616.

Chu, B. (1989a). Photon correlation spectroscopy of polymer solutions. *In* "Comprehensive Polymer Science: Polymer Characterization and Properties, Vol. 1" (Colin Price and Colin Booth, eds.), Chapter 8, pp. 161–171, Pergamon Press, Oxford.

Chu, B. (1989b). *Laser light scattering. In* "Determination of Molecular Weight" (A. R. Cooper, ed.), Chemical Analysis Series, Vol. 103, Chapter 5, pp. 53–86, Wiley, New York.

Chu, B. (1990a). "Selected Papers on Laser Light Scattering by Macromolecular, Supramolecular and Fluid Systems," A Milestone Series–SPIE Volume.

Chu, B. (1990b). Light scattering characterization of an intractable polymer—polytetrafluoroethylene in solution. *In* "Proceedings of the ISPAC-II," Symposium Issue, *J. Appl. Polym. Sci.* **45**, 243.

Chu, B. (1985). *Polymer J. Japan* **17**, 225.

Chu, B. and Wu, C. (1986). *Macromolecules* **19**, 1285.

Chu, B. and Wu, C. (1987). *Macromolecules*, **20**, 93.

Chu, B., Onclin, M., and Ford, J. R. (1984a). *J. Phys. Chem.* **88**, 6566.

Chu, B., Ying, Q.-C., Wu, C., Ford, J. R., Dhadwal, H. S., Qian, R., Bao, J., Zhang, J., and Xu, C. (1984b). *Polym. Commun.* **25**, 211.

Chu, B., Wu C., and Ford, J. R. (1985a). *J. Colloid Interface Sci.* **105**, 473.

Chu, B., Ying, Q.-C., Wu, C., Ford, J. R., and Dhadwal, H. S. (1985b). *Polymer* **26**, 1408.

Chu, B., Wu, C., and Zuo, J. (1987). *Macromolecules* **20**, 700.

Chu, B., Wu, C., and Buck, W. (1988a). *Macromolecules*, **21**, 397.

Chu, B., Xu, R., Maeda, T., and Dhadwal, H. S. (1988b). *Rev. Sci. Instrum.* **59**, 716.

Chu, B., Wu, C., and Buck, W. (1989a). *Macromolecules* **22**, 371.

Chu, B., Wu, C., and Buck, W. (1989b). *Macromolecules* **22**, 831.

Durian, D. J., Weitz, D. A., and Pine, D. J. (1990). *J. Phys: Condensed Matter*, in press.

Horne, D. S. (1989). *J. Phys. D* **22**, 1257.

Horne, D. S. (1990). *J. Chem. Soc. Furaday Trans.* **86**, 1149.

Huntley-James, M. and Jennings, B. R. (1990). *J. Phys. D: Appl. Phys.* **23**, 922.

Joosten, J. G. H., Gelade, E. T. F., and Pusey, P. N. (1990). *Phys. Rev. A*, **42**, 2161.

Kambe, H., Kambe, Y. and Honda, C. (1973), *Polymer* **14**, 460.

MacKintosh, F. C., Zhu, J. X., Pine, D. J., and Weitz, D. A. (1989). *Phys. Rev. B* **40**, 9342.

Maret, G. and Wolf, P. E. (1989). *Physica A* **157**, 293.

Meier, G. and Fytas, G. (1989). *In* "Methods to Characterize Polymers" (H. Bassler, ed.), Elsevier. With 110 references.

Naoki, M., Park, I.-H., Wunder, S. L., and Chu, B. (1985). *J. Polym. Sci. Polym. Phys. Ed.* **23**, 2567.

Pine, D. J., Weitz, D. A., Chaikin, P. M., and Herbolzheimer, E. (1988). *Phys. Rev. Lett.* **60**, 1134.

Pine, D. J., Weitz, D. A., Zhu, J. X., and Herbolzheimer, E. (1990). *J. Phys. (Paris)*, **51**, 2101.

Pope, J. W. and Chu, B. *Macromolecules*, **17**, 2633 (1984).

Pusey, P. N. (1977). *In* "Photon Correlation Spectroscopy and Velocimetry" (H. Z. Cummins and E. R. Pike, eds.), p. 45., Plenum, New York.

Pusey, P. N. and van Megen, W. (1989). *Physica A* **157**, 705.

Society of Polymer Science, Japan (1985). *Polymer J.* **17**, 225.

Tuminello, W. H., Treat, T. A., and English, A. D. (1988). *Macromolecules* **21**, 2606.

Weitz, D. A., Pine, D. J., Pusey, P. N., and Tough, R. J. A. (1989). *Phys. Rev. Lett.* **63**, 1747.

Wolf, P. E., Maret, G., Akkermans, E., and Maynard, R. (1988). *J. Phys.* **49**, 63.

Wu, C., Buck, W., and Chu, B. (1987). *Macromolecules* **20**, 98.

Wu, X. L., Pine, D. J., Chaikin, P. M., Huang, J. S., and Weitz, D. A. (1990). *J. Opt. Soc. Amer. B: Opt. Phys.* **7**, 15.

Ying, Q.-C. and Chu, B. (1984). *Makromol, Chem., Rapid Commun.* **5**, 785.

Ying, Q.-C., Chu, B., Qian, R., Bao, J., Zhang, J., and Xu, C. (1985). *Polymer* **26**, 1401.

CORRELATOR SPECIFICATIONS

B.1. BROOKHAVEN INSTRUMENTS (HOLTSVILLE, NEW YORK)

B.1.a. BI2030AT

1. Min sample time: 0.1 μsec
2. Max sample time: 9.9×10^5 μsec
3. Max countrate: 10 M counts/sec
4. Max number of real time channels (100% efficiency): 264
5. Multiplications: 4 bit by n (delayed terms 4 bit, n limited subject to the 10 M counts/sec limit).
6. Bits/channel: 36 integer, floating point only limited by the computer. 36 bit overflows are monitored by the software and are automatically calculated.
7. Multiple sample times: Up to 4 groups. Each subsequent group of channels may operate with a sample time that is up to 2^7 longer than the previous group. Random scaling between groups is automatic, providing for a continuous display of the correlation function in real time.
8. Baseline: 6 channels located at 1027–1032 sample times after the last channel of the last group.
9. Inputs: TTL, ECL, NIM, or any other special requirements.
10. Functions: AUTO, CROSS, TEST, MULTICHANNEL (INT and EXT), PROBABILITY DISTRIBUTION (differential).
11. Software: Cumulants up to 4th order, dual exponential, exponential sampling, non-negative least squares, regularized non-negative least square (CONTIN). MIE corrections are built in. Size distributions are available

in tabular and graphics form. Residuals are displayed for any 2 functions simultaneously.

12. Graphics: Color VGA.
13. Remote control: All operator inputs may be entered via RS232C or IEEE/488 (GPIB).
14. Built-in automatic control files allow correlator operation without the presence of an operator, including goniometer movements.
15. Static intensity measurements: ZIMM, DEBYE, BARRY plots and data reductions to the 4th order. AUTOMATIC calibration.
16. Goniometer alignment: Automatic $I*\sin(\theta)$ plots with cotangent correction.

B.1.b. BI-8000AT

Sampling and *sample times* are separated. No integration effects or overflows are possible. Each channel is independent of the other, thus enabling the measurement of correlation values at arbitrary sample times. This correlator consists of a single card and uniquely implements the time series approach, yielding the theoretically desired correlation function. At the present time, the minimum sample time is 1 μsec with a dynamic range of 40,000. Due to reduced efficiency, the .1 μsec sample time should be used with high count rates only.

SPECIFICATIONS

1. Min sample time: 0.1 μsec.
2. Max length of correlation time axis (approx.):
   a) sampling time 0.1 μsec: 4 μsec;
   b) sampling time 1 μsec: 40 μsec;
   c) sampling time 2.5 μsec: 100 μsec;
   d) sampling time 5.0 μsec: 200 μsec.
3. Baseline time: Twice the last channel.
4. Channels: Any number between 5 and 128.
5. Channel spacing modes:
   a) linear, e.g., 1, 2, 3, 4 ... μsec;
   b) constant ratio, e.g., 1, 2, 4, 8, 16 ... μsec;
   c) arbitrary from diskfile, e.g., 1, 2, 16, 18, 19 ... μsec.
6. Efficiency: At present, depends on the number of channels and the minimum sampling time: Max = 70%, min approx. 10%.
7. Functions: AUTO, CROSS, TEST, STRUCTURE, CROSS STRUCTURE.

8. Software support:
   a) process control: selection of the best sampling time is automatic subject to the limitations of the total time axis, and the number of channels selected;
   b) data reduction is the same as for the BI2030AT.

For sums of exponential decays, the number of channels need not exceed approximately 16. Fewer channels in a *constant ratio* mode rather than many channels in a *linear* mode yield better results and reduce calculation times. The BI-8000AT correlator addresses this and is being developed further to expand the dynamic range, lower the minimum sample time, and to provide 100% efficiency in all modes.

B.2. ALV (LANGEN, GERMANY)

B.2.a. ALV—3000 DIGITAL STRUCTURATOR/CORRELATOR

Hardware: Signal processing boards, central microprocessor (MC 68000 8 MHz), video display controller, communication ports and power supply in 19″ enclosure; a detached 8-bit parallel keyboard and a video monitor complete the basic system, control software in EPROM storage.

Communication: 2 serial V24 or RS-232 ports, 110–9600 baud, 1 parallel port (8-bit), IEEE-488 (option), full correlator control via any communication port available.

Processing: Correlation functions as well as structure functions available for full 4×4-bit data format (2×2-bit for 20 ns sampling time, 16×16-bit for channels greater than 87 in MULTIPLE TAU real time mode), real time operation, batch mode operation, batch length 2×32768 sampling times, auto and cross (option) correlation/structuration, 16 hardware processing channels each operating at 60 ns clock rate, channel stores with 41-bit capacity.

Preprocessing: 16-bit to 4-bit parallel subtractor/scaler, random scaling based upon 8,388, 607 cycles pseudo random sequence; maximum scale level: 4096 (scale exp: 12); maximum subtractor: 32767.

MULTIPLE TAU: Sampling time doubling by adding adjacent samples, linear sampling time within a block of 8 channels (first block 16 channels), random scaling by 2 for each block based on a 8,388, 607 cycles pseudo random number sequence.

Channel structure: Always up to 1024 linear channels for any sampling time in batch mode; real time operation is a function of the chosen sampling time (CST); for MULTIPLE TAU real time mode, the smallest sampling time is 800 ns:

Linear Tau (real time)	MULTIPLE TAU (real time)
STC 200 ns, 32 channels	STC 800 ns, 192 channels
STC 400 ns, 64 channels	
STC 1.0 μs, 196 channels	for STC = 800 ns the lag time
STC 5.1 μs, 1024 channels	range is 800 ns … 50 s!

Linear Tau (batch mode)	MULTIPLE TAU (batch mode)
STC 20 ns … 246 μs,	STC 20 ns … 246 μs,
channel number selectable	88 channels
from 16 … 1024 channels	fixed

For all modes, the sampling time is selectable in steps of 20 ns, additional delayed monitor channels for each channel greater than 87 in MULTIPLE TAU real time mode to compute symmetrically normalized correlation functions, delays of up to 8176 channels may be introduced at any channel number in linear correlation/structuration functions.

Input pulses: TTL, NIM or symmetric ECL pulses on 50 Ohm impedance with a minimum width and separation of 5 ns, pulse frequency less than 100 MHz.

Dust filter: Special sampling mode for batch operation, rejection of batches with too high count rates, with user defined parameters.

High speed real time correlation: Up to 32 correlation/structuration channels always working in real time with 60, 80, or 100 ns sampling time, since an additional correlator/structurator can be included (option).

Software: Special control software for PC/XT-AT or fully compatible computers, simultaneous measurement of static and dynamic light scattering, ALV-goniometer control with stepper-motor/angular encoder (ALV-3017 required), measurement of temperature (ALV-3017/18 required), total laser intensity and pointing stability (ALV-3017/18 required).

Optional hardware: ALV-3017, stepper-motor/angular encoder control and temperature measurement; ALV-3018, 4 independent measurements of total laser intensity and pointing stability, 1 additional temperature measurement.

B.2.b. ALV—5000 MULTIPLE TAU REAL-TIME DIGITAL CORRELATOR

Hardware: Correlator as a plug-in card in PC/XT-AT or fully compatible computer; latest of all CMOS VLSI-technique packed on a multi-layer printed circuit board.

Software: Operating system PC-DOS or MS-DOS. EGA, VGA or Hercules compatible graphic display, color graphics recommended, math processor 80287 or equivalent recommended, minimum 512 KByte of RAM storage required.

Operation modes: Single correlation function mode (200 ns initial sampling time) as well as dual correlation function mode (initial sampling time 400 ns), both modes in full real time, auto/cross correlation, positive and negative lag time cross correlation using dual cross-correlation mode.

Channel structure: MULTIPLE TAU, quasi-logarithmic sampling time scheme with 256 full real time channels (128 hardware channels + 128 software channels); fixed sampling time grid from 200 ns to 215 s (400 ns to 430 s in dual mode); blocks of eight channels with linear sampling time are used (first block has 16 channels), and the sampling time is increased by a factor of two in proportion to the successive block number.

Data format: Full 8×8-bit data format for the first 128 channels, 16×16-bit for the next 128 channels, i.e., influence of scaling is insignificant compared to shot noise in nearly all experimental situations.

Preprocessing: Autoscaling using a quick test measurement (without scaling) to determine all sample times that overflowed and set the appropriate scale-levels.

Pulse inputs: Positive TTL compatible input pulses on 50 Ohm impedance; low: 0–0.4 V; high: minimum 2.2–5 V; pulse width, high greater than 4 ns, low greater than 4 ns, pulse frequency less than 125 MHz.

Normalization: Symmetric normalization; a delayed monitor channel is calculated for each correlation channel to provide optimum statistical accuracy for large lag times using the symmetric normalization scheme.

Weighting of data: Calculation of standard errors in correlation functions with arbitrary intercepts using gamme intensity statistics and Poisson photon noise approximations.

Control software: Very flexible control software, based on interactive operation through function keys or batch mode control using simple commands, window-oriented graphic displays of correlation functions, count rate traces, histograms, particle size distributions residuals, standard deviations, etc.; interrupt driven RS-232 communication for external control (mainframe computers).

Data processing: On-line data accumulation, normalization, and data analysis with cumulant expansion; inverse Laplace transform with nonnegativity constraint and CONTIN 2DP regularization program (ALV-800 required).

ALV-800 transputer: Additional calculation board, based on transputer technology, for very high speed on-line computation of distribution functions.

Automatic measurement programs: Fully configurable measurement programs using a programming-language-oriented utility.

Goniometer control: Software support for ALV-light scattering electronics, including stepper-motor/angular encoder control, measurement of temperature, total laser intensity, and laser pointing stability, etc., for all ALV-goniometer models; optional software for simultaneous measurement of static and dynamic light scattering available, allowing Zimm-, Berry- and Guinier-plots.

B.3. MALVERN INSTRUMENTS (MALVERN, GREAT BRITAIN)

MALVERN 7032 CORRELATORS

1. *Inputs* A, B, Strobe
   All 50 ohm impedance, threshold user adjustable -5 $+5V$ and trigger edge $+$ or $-ve$ selectable. Normally set to accept "Malvern STD" pulse $0-$ $-2V$, trigger level $-1.5V$ pulse frequency <20 MHz.

2. *Operating models*
   Auto correlation on A   While strobe is valid
   Cross correlation on A $\times$ B
   Optionally:
   Probability density on A
   This is a separate card and can be operated simultaneously with a correlation experiment.

   The 7032 is a totally real time correlator at all sample times. Sample times 0.05, .1 μsec...0.99 seconds with 2 decade precision or 100 nsec (whichever is longest). Single bit operation at 50 nsec, multi bit from 100 nsec upward. Multi-bit operation is 8-bit $\times$ n where n is counts in a particular sample time.

3. *Storage scheme*
   64, 128 or 256 channels $+$ 8 pre- or post-computations or channels and monitor channels for A, B; total samples and detected events for each sample time.

   8 simultaneous sample times are available on a geometric progression of ratio 1–15.

   The storage channes can be allocated in blocks of 16 to any of these sample times.

   Normally channels are disposed in the sample times

   > 1 (Linear or serial mode)
   > 1, 2 (Dual time mode)
   > 1, 2, 3, 4 (Parallel modes)
   > 1, 2, 3, 4, 5, 6, 7, 8 (Parallel modes),

giving a range of configurations from a 256-channel classic linear correlator with 8 far-point channels, to an 8 × 32-channel array in parallel mode.

Storage capacity

> 48 bits per channel (multibit)
> 41 bits per channel at 20 nsec sample time

Each sample time section can be separately pre-scaled up to 8 bits, giving a 16-bit precision capacity per channel.

4. Multiple correlator systems can be supported in which the 256 store channels can be assigned to up to four totally separate "Frontend" correlator boards.

NOTATION

For most of the notation in Sections 6.6 and 6.7, Appendix 6.A; and Chapters 7 & 8, see text.

a: quantum efficiency × detector gain
as: add–subtract; Section 4.7
A: background; Eq. (4.9.8) or (7.1.2)
A: refractive-index gradient constant; Eq. (6.9.8)
A: CH; Eq. (7.5.8)
A_2:   second virial coefficient; Eq. (2.1.8)
A_{sca}:   cross-sectional area of scattering volume v
A_{coh}:   coherence area
C: concentration
C: capacitance; Eq. (6.6.2)
C: curvature matrix; Eq. (7.5.5)
CONTIN: a regularization program from Provencher
ck: complementary clipping
c_m: velocity of light in the scattering medium
c_0:   velocity of light in free space
C'_P: specific heat at constant pressure per unit volume
$C(\rho_r)$:   isotropic spatial correlation function; Eq. (2.D.1)
C_V: specific heat at constant volume
C'_V: specific heat at constant volume per unit volume

329

d:	optical path length; Eq. (2.3.7)
d:	$2\pi/K$, nm; Appendix I.B
d:	plate separation distance; Eq. (5.2.16)
d_A:	aperture diameter of detection system; Eq. (6.3.1)
d_i:	data set
$\hat{d}_i$:	measured estimates of d_i
dq:	heat absorbed by the system per unit volume; Eq. (2.4.23)
d_s:	spot diameter; Eq. (6.2.4)
dS:	surface element, with S being the surface encompassing volume v
dx'_α:	volume element in the scattering volume; dx'_α denotes $dx'_1 dx'_2 dx'_3 = dx' dy' dz' = d\mathbf{r}'$; Fig. 2.1.1
D:	aperture diameter; Eq. (5.2.18)
D:	laser beam diameter; Eq. (6.2.4)
D_A:	diameter of field stop
D_f:	fiber core diameter; Eq. (6.9.9)
D_I:	diameter of incident beam
D_{opt}:	optimum aperture; Eq. (6.2.6)
D_V:	discriminator level; Section 6.5.5
e:	electron charge; Eq. (2.1.14)
E_r:	total reflected electric field; Eq. (5.1.17)
E_s:	scattered electric field
E_t:	total transmitted electric field
f:	focal length of lens
F:	contrast
$F(\rho_r)$:	normalized spatial correlation function for local entropy-density fluctuations at constant pressure; Eq. (2.D.2)
FP:	Fabry–Perot (interferometry)
$\mathscr{F}$:	finesse; Eq. (5.2.14)
$\mathbf{g}$:	$U^T\mathbf{b}$; Eq. (7.5.13)
g_i:	prior probability assignment; Eq. (7.6.2)
$g^{(1)}(\tau)$:	normalized electric field time correlation function at delay time τ; Eq. (3.3.9)
$g^{(2)}(\tau)$:	normalized intensity time correlation function; Eq. (3.3.6)
G:	amplifier or photomultiplier gain;
$G(\Gamma)$:	normalized characteristic linewidth distribution function of Γ
$G(\rho_r)$:	normalized spatial correlation function for local pressure fluctuations at constant entropy; Eq. (2.D.3)
$G^{(2)}(\tau)$:	intensity time correlation function
$\Gamma\ (\equiv\Gamma_c)$:	characteristic linewidth
h:	Planck's constant
$\mathbf{h}$:	Euclidean norm of a vector
$\hbar$:	$h/2\pi$

H:	optical constant; Eq. (2.1.6)
$i(t)$:	photocurrent at time t
I_B:	intensity of one of two Brillouin components in Brillouin scattering; Eq. (2.4.50)
I_c:	intensity of central component in Brillouin scattering; Eq. (2.4.50)
I_{INC}:	intensity of incident beam
i_e:	scattered intensity of a single electron
I_e:	scattered intensity of N/V particles per unit volume, each with v electrons
I_{ex}:	excess scattered intensity; Eq. (2.1.5)
I_0:	short-time average intensity
I_s:	scattered intensity
I_t:	intensity of transmitted beam
$j(t)$:	current density at time t
J_{ij}:	Jacobian; Eq. (7.5.4)
k_B:	Boltzmann constant ($\equiv k$; this k is not $|\mathbf{k}| = 2\pi/\lambda$); Eq. (2.3.19)
$\mathbf{k}_I$:	incident wave vector; Eq. (1.1.1)
$\mathbf{k}_s$:	scattered wave vector; Eq. (1.1.1)
K:	magnitude of momentum-transfer vector $\mathbf{K}$; Eq. (1.1.1)
K_{ij}:	curvature matrix; Eq. (7.6.7)
$K_m(\Gamma)$:	mth cumulant of Γ
K_S:	adiabatic compressibility
K_T:	isothermal compressibility
$\bar{l}, \bar{m}, \bar{n}$:	indices for lattice-point location in reciprocal space; Eq. (3.1.1)
LLS:	laser light scattering
LO:	local oscillator
L_x, L_y, L_z:	dimensions of a rectangular scattering volume in the x, y, z directions; Eq. (2.2.10)
$\mathscr{L}_x$:	Lorentzian function of x
m:	electron mass; Eq. (2.1.14)
m:	magnification; Eq. (6.A.7)
$m(x)$:	prior probability distribution; Eq. (7.6.3)
M:	molecular weight
MEM:	maximum-entropy method
$M(\rho_r)$:	normalized spatial correlation function for local density fluctuations at constant temperature; Eq. (2.3.15)
$(\partial n/\partial C)_T$:	refractive-index increment
n:	number of pulses per sampling time interval T
n:	refractive index of scattering medium
$n_{ck}(t)$:	number of complementary clipped photoelectrons at time t; Eq. (4.3.1)
$n_k(t)$:	number of k-clipped photoelectron pulses at time t

n_0:	solvent refractive index
$n_s(t)$:	number of scaled photoelectron pulses at time t
N:	total number of molecules (or particles) in volume V
N:	number of photoelectrons per second; Eq. (6.6.1)
N:	number of characteristic linewidths as an approximation to $G(\Gamma)$
$(NA)_f$:	numerical aperture of fiber (in air); Eq. (6.9.8)
N_a:	Avogadro's number
N_0:	refractive index of microlens; Eq. (6.9.9)
$N(\rho_r)$:	normalized spatial correlation function for local temperature fluctuations at constant density; Eq. (2.3.16)
N_s:	number of samples
P:	power scattered into a solid angle Ω; Eq. (2.3.27)
P:	pressure
$\mathbf{P}$:	parameter vector of length n; Eq. (7.5.5)
$p(n, t, T)$:	probability distribution of registering n photoelectron pulses over a sampling time interval T at time t to $t + T$; Eq. (3.2.2)
p_i:	probability assignment; Eq. (7.6.1)
P_I:	incident power
PCS:	photon correlation spectroscopy
$P(K)$:	single-particle scattering factor as a function of K
$P(r)$:	power of a Gaussian beam passing through an aperture of radius r; Eq. (6.2.3)
$P(x)$:	univariate probability distribution; Eq. (7.6.3)
$\mathbf{q}^+, \mathbf{q}^-$:	sound wave vectors with $+$ and $-$ denoting the two directions of propagation; Eq. (2.B.10)
$Q(\lambda_1, \lambda_2)$:	generating function; Eq. (4.2.5)
$Q(x_\alpha)$:	location of the observer or point detector; Fig. 2.1.1
$Q(y)$:	posterior distribution; Eq. (7.6.4)
$Q^0(y)$:	prior distribution; Eq. (7.6.4)
r:	distance from optic axis; Eq. (6.2.1)
r_0:	beam radius where the intensity falls to $1/e^2$ of the peak on-axis value; Eq. (6.2.1)
$\mathbf{r}$:	vector from volume element dx'_α to the point detector; Fig. 2.1.1
$\mathbf{r}'$:	vector from center of scattering volume to the volume element dx'_a ($\equiv d\mathbf{r}'$) in the scattering volume; Fig. 2.1.1
$\mathscr{R}$:	Rayleigh ratio; Eq. (2.3.4)
$\mathscr{R}_{ex}$:	excess Rayleigh ratio for vertically polarized incident light; Eq. (2.1.6)
$\mathscr{R}_{ex,u}$:	excess Rayleigh ratio for unpolarized incident light; Eq. (2.1.7)

$\mathscr{R}_{HV}$:   Rayleigh ratio with horizontally polarized incident light and vertically polarized scattered light

$\mathscr{R}_{VV}$:   Rayleigh ratio with vertically polarized incident and scattered light

$\mathscr{R}^*$:   an alternate definition for the Rayleigh ratio; Eq. (2.3.21) (the reader may wish to avoid $\mathscr{R}^*$ in order to avoid confusion)

$\mathscr{R}^+$:   resolving power; Eq. (5.2.13)

R: energy reflection coefficient; Eq. (5.1.18)

R: spatial resolution ($\sim K^{-1}$); paragraph after Eq. (1.1.2)

$[R_1]$:   ray transfer matrix; Eq. (6.A.1)

$\mathrm{Re}\{g^{(1)}(\tau)\}$:   real part of $g^{(1)}(\tau)$; Eq. (4.1.7)

$R_E(\tau)$:   time correlation function of E at delay time τ

R_g: radius of gyration; Eq. (2.1.9) and Appendix 2.A

$R_x(\tau)$:   time correlation of $x(t)$ at delay time τ

$\mathbf{R}$:   vector from origin of scattering volume to the idealized point detector; Fig. 2.1.1

$\hat{\mathbf{R}}$:   unit vector of $\mathbf{R}$

s: entropy per unit volume

$\hat{s}^{(2)}(\tau_l)$:   expectation value for the normalized sample correlation function; $\hat{}$ denotes expectation value

S: entropy function; Eqs. (7.6.1) and (7.6.2)

S: Laplace operator; Eq (6.6.2)

S: diagonal matrix; Eq. (7.5.11)

$S^{(2)}(\tau)$:   structure function; Eq. (4.10.3)

$\hat{S}^{(2)}(\tau_l)$:   expectation value for the sample correlation function; Eq. (4.1.8) ($\hat{}$ does not indicate a normalized quantity here)

SV: scattering volume; same as v; Fig. 6.9.2

$S_x(\omega)$:   power spectral density; Eq. (2.4.14)

$S_{x,N}(\omega)$:   normalized power spectral density; Eqs. (2.4.16) and (2.4.17)

SANS: small-angle neutron scattering

SAXS: small-angle x-ray scattering

t: temperature in °C; Eq. (2.1.12)

t: time variable; Eq. (2.4.3)

t: time when the TOAC is ready to accept a start pulse; Eq. (4.9.1)

$\hat{t}$:   optical path length in the medium; Eq. (6.A.2)

t_0:   time at which measurement is initiated; Eq. (2.4.3)

T: absolute temperature

T: $(t + T)$ time for the first start pulse; Eq. (4.9.1)

T: energy transmission coefficient; Eq. (5.1.16)

T: sampling time

$2T$:   time period over which a quantity is averaged; Eq. (2.4.3)

TOA: time of arrival

v:	scattering volume
V:	(total) volume
V:	operating voltage; Section 6.5.5
v_F^*:	correlation volume due to local entropy-density fluctuations; Eq. (2.D.6)
v_G^*:	correlation volume due to local pressure fluctuations; Eq. (2.D.7)
v_k:	volume of shell in k-space; Eq. (3.1.4)
v_m^*:	correlation volume due to local density fluctuations; Eq. (2.3.17)
v_n^*:	correlation volume due to local temperature fluctuations; Eq. (2.3.18)
v_s:	phase velocity of sound
$\langle w \rangle$:	$\langle I \rangle T (= I_0 T)$
$w_i(x_1,\ldots,x_N)$:	joint probability distribution
$\bar{x}$:	time average of x; Eq. (2.4.3)
x_i:	random variable
$\langle x \rangle$:	ensemble average of x
$\hat{x}_{2T}(\omega)$:	Fourier frequency components of $x_{2T}(t)$
z:	length of microlens in millimeters; Eq. (6.A.3)
Z:	impedance; Eq. (6.6.2)
α:	polarizability
α_{ex}:	excess polarizability; Eq. (2.1.4)
β:	coherence factor
β:	tilt angle in radians; Eq. (5.2.17)
$\Gamma \equiv \Gamma_c$:	characteristic linewidth
Γ_P:	damping rate; Eq. (2.4.36)
$\delta\theta_{INC}$:	angular divergence subtended by the incident beam
Δ_i:	local fluctuations in a random process at point $\mathbf{r}_i$
ΔP_s:	local pressure change at constant entropy density from its equilibrium value P_0; Eq. (2.3.13)
Δs_P:	local entropy-density change at constant pressure from its equilibrium value s_0; Eq. (2.3.13)
$\Delta t = \Delta \tau = T$:	delay-time increment
ϵ:	change of phase on reflection; Eq. (5.1.19)
ϵ:	electric inductive capacity in the scattering medium
ϵ_0:	electric inductive capacity in free space
η:	shear viscosity
η':	bulk viscosity
θ:	scattering angle
ϑ:	angle between direction of polarization of incident electric field $\mathbf{E}_{INC}$ and direction of observation $\mathbf{R}$; Fig. 2.1.1

ϑ_a:	aberration angular radius
ϑ_d:	diffraction angular radius
κ_e:	dielectric constant
λ:	wavelength of light in the scattering medium; $\lambda = \lambda_0/n$
λ_0:	wavelength of light in vacuo
Λ:	thermal conductivity; Eq. (2.4.23)
Λ/C'_P:	thermal diffusivity
μ:	magnetic inductive capacity
$\mu_2/\bar{\Gamma}^2$:	variance
v:	number of electrons in a particle; Eq. (2.1.15)
$\tilde{v}$:	wave number
ζ^*:	unit-cell dimension
ρ:	density
ρ_k:	density of cell in **k**-space
$\boldsymbol{\rho}$:	ratio of reflected to incident amplitude
ρ_r:	distance between $\mathbf{r}_1$ and $\mathbf{r}_2$; $\rho_r = \mathbf{r}_1 - \mathbf{r}_2$
τ:	delay time
τ:	output time constant; Eq. (6.6.4)
$\boldsymbol{\tau}$:	ratio of transmitted to incident amplitude
τ^*:	turbidity; Eq. (2.3.5)
τ_c:	characteristic decay time; e.g., Eq. (2.4.20)
ϕ:	angle between x-axis and $\mathbf{k}_1$; Fig. 2.1.1
ϕ:	optical phase
Φ:	phase integral; Eq.(2.2.6)
Φ:	relative phase; Eq. (6.9.3)
χ_e:	electric susceptibility
$\bar{\omega}$:	angular frequency of sound waves
ω_1:	angular frequency of incident light; $\omega_1 = 2\pi v_1$
ω_s:	angular frequency of scattered light
Ω:	solid angle
Ω^*:	domain of $Q(y)$
Ω_{coh}:	coherence solid angle; Eq. (3.1.7)

INDEX

A CATALOG OF SELECTED

DOVER BOOKS
IN SCIENCE AND MATHEMATICS

Mathematics

FUNCTIONAL ANALYSIS (Second Corrected Edition), George Bachman and Lawrence Narici. Excellent treatment of subject geared toward students with background in linear algebra, advanced calculus, physics and engineering. Text covers introduction to inner-product spaces, normed, metric spaces, and topological spaces; complete orthonormal sets, the Hahn-Banach Theorem and its consequences, and many other related subjects. 1966 ed. 544pp. 6⅛ x 9¼. 0-486-40251-7

ASYMPTOTIC EXPANSIONS OF INTEGRALS, Norman Bleistein & Richard A. Handelsman. Best introduction to important field with applications in a variety of scientific disciplines. New preface. Problems. Diagrams. Tables. Bibliography. Index. 448pp. 5⅜ x 8½. 0-486-65082-0

VECTOR AND TENSOR ANALYSIS WITH APPLICATIONS, A. I. Borisenko and I. E. Tarapov. Concise introduction. Worked-out problems, solutions, exercises. 257pp. 5⅝ x 8¼. 0-486-63833-2

AN INTRODUCTION TO ORDINARY DIFFERENTIAL EQUATIONS, Earl A. Coddington. A thorough and systematic first course in elementary differential equations for undergraduates in mathematics and science, with many exercises and problems (with answers). Index. 304pp. 5⅜ x 8½. 0-486-65942-9

FOURIER SERIES AND ORTHOGONAL FUNCTIONS, Harry F. Davis. An incisive text combining theory and practical example to introduce Fourier series, orthogonal functions and applications of the Fourier method to boundary-value problems. 570 exercises. Answers and notes. 416pp. 5⅜ x 8½. 0-486-65973-9

COMPUTABILITY AND UNSOLVABILITY, Martin Davis. Classic graduate-level introduction to theory of computability, usually referred to as theory of recurrent functions. New preface and appendix. 288pp. 5⅜ x 8½. 0-486-61471-9

ASYMPTOTIC METHODS IN ANALYSIS, N. G. de Bruijn. An inexpensive, comprehensive guide to asymptotic methods–the pioneering work that teaches by explaining worked examples in detail. Index. 224pp. 5⅜ x 8½ 0-486-64221-6

APPLIED COMPLEX VARIABLES, John W. Dettman. Step-by-step coverage of fundamentals of analytic function theory–plus lucid exposition of five important applications: Potential Theory; Ordinary Differential Equations; Fourier Transforms; Laplace Transforms; Asymptotic Expansions. 66 figures. Exercises at chapter ends. 512pp. 5⅜ x 8½. 0-486-64670-X

INTRODUCTION TO LINEAR ALGEBRA AND DIFFERENTIAL EQUATIONS, John W. Dettman. Excellent text covers complex numbers, determinants, orthonormal bases, Laplace transforms, much more. Exercises with solutions. Undergraduate level. 416pp. 5⅜ x 8½. 0-486-65191-6

RIEMANN'S ZETA FUNCTION, H. M. Edwards. Superb, high-level study of landmark 1859 publication entitled "On the Number of Primes Less Than a Given Magnitude" traces developments in mathematical theory that it inspired. xiv+315pp. 5⅜ x 8½. 0-486-41740-9

CALCULUS OF VARIATIONS WITH APPLICATIONS, George M. Ewing. Applications-oriented introduction to variational theory develops insight and promotes understanding of specialized books, research papers. Suitable for advanced undergraduate/graduate students as primary, supplementary text. 352pp. 5⅜ x 8½.
0-486-64856-7

COMPLEX VARIABLES, Francis J. Flanigan. Unusual approach, delaying complex algebra till harmonic functions have been analyzed from real variable viewpoint. Includes problems with answers. 364pp. 5⅜ x 8½. 0-486-61388-7

AN INTRODUCTION TO THE CALCULUS OF VARIATIONS, Charles Fox. Graduate-level text covers variations of an integral, isoperimetrical problems, least action, special relativity, approximations, more. References. 279pp. 5⅜ x 8½.
0-486-65499-0

COUNTEREXAMPLES IN ANALYSIS, Bernard R. Gelbaum and John M. H. Olmsted. These counterexamples deal mostly with the part of analysis known as "real variables." The first half covers the real number system, and the second half encompasses higher dimensions. 1962 edition. xxiv+198pp. 5⅜ x 8½. 0-486-42875-3

CATASTROPHE THEORY FOR SCIENTISTS AND ENGINEERS, Robert Gilmore. Advanced-level treatment describes mathematics of theory grounded in the work of Poincaré, R. Thom, other mathematicians. Also important applications to problems in mathematics, physics, chemistry and engineering. 1981 edition. References. 28 tables. 397 black-and-white illustrations. xvii + 666pp. 6⅛ x 9¼.
0-486-67539-4

INTRODUCTION TO DIFFERENCE EQUATIONS, Samuel Goldberg. Exceptionally clear exposition of important discipline with applications to sociology, psychology, economics. Many illustrative examples; over 250 problems. 260pp. 5⅜ x 8½.
0-486-65084-7

NUMERICAL METHODS FOR SCIENTISTS AND ENGINEERS, Richard Hamming. Classic text stresses frequency approach in coverage of algorithms, polynomial approximation, Fourier approximation, exponential approximation, other topics. Revised and enlarged 2nd edition. 721pp. 5⅜ x 8½. 0-486-65241-6

INTRODUCTION TO NUMERICAL ANALYSIS (2nd Edition), F. B. Hildebrand. Classic, fundamental treatment covers computation, approximation, interpolation, numerical differentiation and integration, other topics. 150 new problems. 669pp. 5⅜ x 8½. 0-486-65363-3

THREE PEARLS OF NUMBER THEORY, A. Y. Khinchin. Three compelling puzzles require proof of a basic law governing the world of numbers. Challenges concern van der Waerden's theorem, the Landau-Schnirelmann hypothesis and Mann's theorem, and a solution to Waring's problem. Solutions included. 64pp. 5⅜ x 8½.
0-486-40026-3

THE PHILOSOPHY OF MATHEMATICS: AN INTRODUCTORY ESSAY, Stephan Körner. Surveys the views of Plato, Aristotle, Leibniz & Kant concerning propositions and theories of applied and pure mathematics. Introduction. Two appendices. Index. 198pp. 5⅜ x 8½. 0-486-25048-2

INTRODUCTORY REAL ANALYSIS, A.N. Kolmogorov, S. V. Fomin. Translated by Richard A. Silverman. Self-contained, evenly paced introduction to real and functional analysis. Some 350 problems. 403pp. 5⅜ x 8½. 0-486-61226-0

APPLIED ANALYSIS, Cornelius Lanczos. Classic work on analysis and design of finite processes for approximating solution of analytical problems. Algebraic equations, matrices, harmonic analysis, quadrature methods, much more. 559pp. 5⅜ x 8½. 0-486-65656-X

AN INTRODUCTION TO ALGEBRAIC STRUCTURES, Joseph Landin. Superb self-contained text covers "abstract algebra": sets and numbers, theory of groups, theory of rings, much more. Numerous well-chosen examples, exercises. 247pp. 5⅜ x 8½. 0-486-65940-2

QUALITATIVE THEORY OF DIFFERENTIAL EQUATIONS, V. V. Nemytskii and V.V. Stepanov. Classic graduate-level text by two prominent Soviet mathematicians covers classical differential equations as well as topological dynamics and ergodic theory. Bibliographies. 523pp. 5⅜ x 8½. 0-486-65954-2

THEORY OF MATRICES, Sam Perlis. Outstanding text covering rank, nonsingularity and inverses in connection with the development of canonical matrices under the relation of equivalence, and without the intervention of determinants. Includes exercises. 237pp. 5⅜ x 8½. 0-486-66810-X

INTRODUCTION TO ANALYSIS, Maxwell Rosenlicht. Unusually clear, accessible coverage of set theory, real number system, metric spaces, continuous functions, Riemann integration, multiple integrals, more. Wide range of problems. Undergraduate level. Bibliography. 254pp. 5⅜ x 8½. 0-486-65038-3

MODERN NONLINEAR EQUATIONS, Thomas L. Saaty. Emphasizes practical solution of problems; covers seven types of equations. ". . . a welcome contribution to the existing literature...."–*Math Reviews.* 490pp. 5⅜ x 8½. 0-486-64232-1

MATRICES AND LINEAR ALGEBRA, Hans Schneider and George Phillip Barker. Basic textbook covers theory of matrices and its applications to systems of linear equations and related topics such as determinants, eigenvalues and differential equations. Numerous exercises. 432pp. 5⅜ x 8½. 0-486-66014-1

LINEAR ALGEBRA, Georgi E. Shilov. Determinants, linear spaces, matrix algebras, similar topics. For advanced undergraduates, graduates. Silverman translation. 387pp. 5⅜ x 8½. 0-486-63518-X

ELEMENTS OF REAL ANALYSIS, David A. Sprecher. Classic text covers fundamental concepts, real number system, point sets, functions of a real variable, Fourier series, much more. Over 500 exercises. 352pp. 5⅜ x 8½. 0-486-65385-4

SET THEORY AND LOGIC, Robert R. Stoll. Lucid introduction to unified theory of mathematical concepts. Set theory and logic seen as tools for conceptual understanding of real number system. 496pp. 5⅜ x 8¼. 0-486-63829-4

TENSOR CALCULUS, J.L. Synge and A. Schild. Widely used introductory text covers spaces and tensors, basic operations in Riemannian space, non-Riemannian spaces, etc. 324pp. 5⅜ x 8¼. 0-486-63612-7

ORDINARY DIFFERENTIAL EQUATIONS, Morris Tenenbaum and Harry Pollard. Exhaustive survey of ordinary differential equations for undergraduates in mathematics, engineering, science. Thorough analysis of theorems. Diagrams. Bibliography. Index. 818pp. 5⅜ x 8½. 0-486-64940-7

INTEGRAL EQUATIONS, F. G. Tricomi. Authoritative, well-written treatment of extremely useful mathematical tool with wide applications. Volterra Equations, Fredholm Equations, much more. Advanced undergraduate to graduate level. Exercises. Bibliography. 238pp. 5⅜ x 8½. 0-486-64828-1

FOURIER SERIES, Georgi P. Tolstov. Translated by Richard A. Silverman. A valuable addition to the literature on the subject, moving clearly from subject to subject and theorem to theorem. 107 problems, answers. 336pp. 5⅜ x 8½. 0-486-63317-9

INTRODUCTION TO MATHEMATICAL THINKING, Friedrich Waismann. Examinations of arithmetic, geometry, and theory of integers; rational and natural numbers; complete induction; limit and point of accumulation; remarkable curves; complex and hypercomplex numbers, more. 1959 ed. 27 figures. xii+260pp. 5⅜ x 8½. 0-486-63317-9

POPULAR LECTURES ON MATHEMATICAL LOGIC, Hao Wang. Noted logician's lucid treatment of historical developments, set theory, model theory, recursion theory and constructivism, proof theory, more. 3 appendixes. Bibliography. 1981 edition. ix + 283pp. 5⅜ x 8½. 0-486-67632-3

CALCULUS OF VARIATIONS, Robert Weinstock. Basic introduction covering isoperimetric problems, theory of elasticity, quantum mechanics, electrostatics, etc. Exercises throughout. 326pp. 5⅜ x 8½. 0-486-63069-2

THE CONTINUUM: A CRITICAL EXAMINATION OF THE FOUNDATION OF ANALYSIS, Hermann Weyl. Classic of 20th-century foundational research deals with the conceptual problem posed by the continuum. 156pp. 5⅜ x 8½. 0-486-67982-9

CHALLENGING MATHEMATICAL PROBLEMS WITH ELEMENTARY SOLUTIONS, A. M. Yaglom and I. M. Yaglom. Over 170 challenging problems on probability theory, combinatorial analysis, points and lines, topology, convex polygons, many other topics. Solutions. Total of 445pp. 5⅜ x 8½. Two-vol. set. Vol. I: 0-486-65536-9 Vol. II: 0-486-65537-7

INTRODUCTION TO PARTIAL DIFFERENTIAL EQUATIONS WITH APPLICATIONS, E. C. Zachmanoglou and Dale W. Thoe. Essentials of partial differential equations applied to common problems in engineering and the physical sciences. Problems and answers. 416pp. 5⅜ x 8½. 0-486-65251-3

THE THEORY OF GROUPS, Hans J. Zassenhaus. Well-written graduate-level text acquaints reader with group-theoretic methods and demonstrates their usefulness in mathematics. Axioms, the calculus of complexes, homomorphic mapping, p-group theory, more. 276pp. 5⅜ x 8½. 0-486-40922-8

Math–Decision Theory, Statistics, Probability

ELEMENTARY DECISION THEORY, Herman Chernoff and Lincoln E. Moses. Clear introduction to statistics and statistical theory covers data processing, probability and random variables, testing hypotheses, much more. Exercises. 364pp. 5⅜ x 8½. 0-486-65218-1

STATISTICS MANUAL, Edwin L. Crow et al. Comprehensive, practical collection of classical and modern methods prepared by U.S. Naval Ordnance Test Station. Stress on use. Basics of statistics assumed. 288pp. 5⅜ x 8½. 0-486-60599-X

SOME THEORY OF SAMPLING, William Edwards Deming. Analysis of the problems, theory and design of sampling techniques for social scientists, industrial managers and others who find statistics important at work. 61 tables. 90 figures. xvii +602pp. 5⅜ x 8½. 0-486-64684-X

LINEAR PROGRAMMING AND ECONOMIC ANALYSIS, Robert Dorfman, Paul A. Samuelson and Robert M. Solow. First comprehensive treatment of linear programming in standard economic analysis. Game theory, modern welfare economics, Leontief input-output, more. 525pp. 5⅜ x 8½. 0-486-65491-5

PROBABILITY: AN INTRODUCTION, Samuel Goldberg. Excellent basic text covers set theory, probability theory for finite sample spaces, binomial theorem, much more. 360 problems. Bibliographies. 322pp. 5⅜ x 8½. 0-486-65252-1

GAMES AND DECISIONS: INTRODUCTION AND CRITICAL SURVEY, R. Duncan Luce and Howard Raiffa. Superb nontechnical introduction to game theory, primarily applied to social sciences. Utility theory, zero-sum games, n-person games, decision-making, much more. Bibliography. 509pp. 5⅜ x 8½. 0-486-65943-7

INTRODUCTION TO THE THEORY OF GAMES, J. C. C. McKinsey. This comprehensive overview of the mathematical theory of games illustrates applications to situations involving conflicts of interest, including economic, social, political, and military contexts. Appropriate for advanced undergraduate and graduate courses; advanced calculus a prerequisite. 1952 ed. x+372pp. 5⅜ x 8½. 0-486-42811-7

FIFTY CHALLENGING PROBLEMS IN PROBABILITY WITH SOLUTIONS, Frederick Mosteller. Remarkable puzzlers, graded in difficulty, illustrate elementary and advanced aspects of probability. Detailed solutions. 88pp. 5⅜ x 8½. 65355-2

PROBABILITY THEORY: A CONCISE COURSE, Y. A. Rozanov. Highly readable, self-contained introduction covers combination of events, dependent events, Bernoulli trials, etc. 148pp. 5⅜ x 8¼. 0-486-63544-9

STATISTICAL METHOD FROM THE VIEWPOINT OF QUALITY CONTROL, Walter A. Shewhart. Important text explains regulation of variables, uses of statistical control to achieve quality control in industry, agriculture, other areas. 192pp. 5⅜ x 8½. 0-486-65232-7

Physics

OPTICAL RESONANCE AND TWO-LEVEL ATOMS, L. Allen and J. H. Eberly. Clear, comprehensive introduction to basic principles behind all quantum optical resonance phenomena. 53 illustrations. Preface. Index. 256pp. 5⅜ x 8½. 0-486-65533-4

QUANTUM THEORY, David Bohm. This advanced undergraduate-level text presents the quantum theory in terms of qualitative and imaginative concepts, followed by specific applications worked out in mathematical detail. Preface. Index. 655pp. 5⅜ x 8½. 0-486-65969-0

ATOMIC PHYSICS (8th EDITION), Max Born. Nobel laureate's lucid treatment of kinetic theory of gases, elementary particles, nuclear atom, wave-corpuscles, atomic structure and spectral lines, much more. Over 40 appendices, bibliography. 495pp. 5⅜ x 8½. 0-486-65984-4

A SOPHISTICATE'S PRIMER OF RELATIVITY, P. W. Bridgman. Geared toward readers already acquainted with special relativity, this book transcends the view of theory as a working tool to answer natural questions: What is a frame of reference? What is a "law of nature"? What is the role of the "observer"? Extensive treatment, written in terms accessible to those without a scientific background. 1983 ed. xlviii+172pp. 5⅜ x 8½. 0-486-42549-5

AN INTRODUCTION TO HAMILTONIAN OPTICS, H. A. Buchdahl. Detailed account of the Hamiltonian treatment of aberration theory in geometrical optics. Many classes of optical systems defined in terms of the symmetries they possess. Problems with detailed solutions. 1970 edition. xv + 360pp. 5⅜ x 8½. 0-486-67597-1

PRIMER OF QUANTUM MECHANICS, Marvin Chester. Introductory text examines the classical quantum bead on a track: its state and representations; operator eigenvalues; harmonic oscillator and bound bead in a symmetric force field; and bead in a spherical shell. Other topics include spin, matrices, and the structure of quantum mechanics; the simplest atom; indistinguishable particles; and stationary-state perturbation theory. 1992 ed. xiv+314pp. 6⅛ x 9¼. 0-486-42878-8

LECTURES ON QUANTUM MECHANICS, Paul A. M. Dirac. Four concise, brilliant lectures on mathematical methods in quantum mechanics from Nobel Prize-winning quantum pioneer build on idea of visualizing quantum theory through the use of classical mechanics. 96pp. 5⅜ x 8½. 0-486-41713-1

THIRTY YEARS THAT SHOOK PHYSICS: THE STORY OF QUANTUM THEORY, George Gamow. Lucid, accessible introduction to influential theory of energy and matter. Careful explanations of Dirac's anti-particles, Bohr's model of the atom, much more. 12 plates. Numerous drawings. 240pp. 5⅜ x 8½. 0-486-24895-X

ELECTRONIC STRUCTURE AND THE PROPERTIES OF SOLIDS: THE PHYSICS OF THE CHEMICAL BOND, Walter A. Harrison. Innovative text offers basic understanding of the electronic structure of covalent and ionic solids, simple metals, transition metals and their compounds. Problems. 1980 edition. 582pp. 6⅛ x 9¼. 0-486-66021-4

HYDRODYNAMIC AND HYDROMAGNETIC STABILITY, S. Chandrasekhar. Lucid examination of the Rayleigh-Benard problem; clear coverage of the theory of instabilities causing convection. 704pp. 5⅜ x 8¼. 0-486-64071-X

INVESTIGATIONS ON THE THEORY OF THE BROWNIAN MOVEMENT, Albert Einstein. Five papers (1905–8) investigating dynamics of Brownian motion and evolving elementary theory. Notes by R. Fürth. 122pp. 5⅜ x 8½. 0-486-60304-0

THE PHYSICS OF WAVES, William C. Elmore and Mark A. Heald. Unique overview of classical wave theory. Acoustics, optics, electromagnetic radiation, more. Ideal as classroom text or for self-study. Problems. 477pp. 5⅜ x 8½. 0-486-64926-1

GRAVITY, George Gamow. Distinguished physicist and teacher takes reader-friendly look at three scientists whose work unlocked many of the mysteries behind the laws of physics: Galileo, Newton, and Einstein. Most of the book focuses on Newton's ideas, with a concluding chapter on post-Einsteinian speculations concerning the relationship between gravity and other physical phenomena. 160pp. 5⅜ x 8½. 0-486-42563-0

PHYSICAL PRINCIPLES OF THE QUANTUM THEORY, Werner Heisenberg. Nobel Laureate discusses quantum theory, uncertainty, wave mechanics, work of Dirac, Schroedinger, Compton, Wilson, Einstein, etc. 184pp. 5⅜ x 8½. 0-486-60113-7

ATOMIC SPECTRA AND ATOMIC STRUCTURE, Gerhard Herzberg. One of best introductions; especially for specialist in other fields. Treatment is physical rather than mathematical. 80 illustrations. 257pp. 5⅜ x 8½. 0-486-60115-3

AN INTRODUCTION TO STATISTICAL THERMODYNAMICS, Terrell L. Hill. Excellent basic text offers wide-ranging coverage of quantum statistical mechanics, systems of interacting molecules, quantum statistics, more. 523pp. 5⅜ x 8½. 0-486-65242-4

THEORETICAL PHYSICS, Georg Joos, with Ira M. Freeman. Classic overview covers essential math, mechanics, electromagnetic theory, thermodynamics, quantum mechanics, nuclear physics, other topics. First paperback edition. xxiii + 885pp. 5⅜ x 8½. 0-486-65227-0

PROBLEMS AND SOLUTIONS IN QUANTUM CHEMISTRY AND PHYSICS, Charles S. Johnson, Jr. and Lee G. Pedersen. Unusually varied problems, detailed solutions in coverage of quantum mechanics, wave mechanics, angular momentum, molecular spectroscopy, more. 280 problems plus 139 supplementary exercises. 430pp. 6½ x 9¼. 0-486-65236-X

THEORETICAL SOLID STATE PHYSICS, Vol. 1: Perfect Lattices in Equilibrium; Vol. II: Non-Equilibrium and Disorder, William Jones and Norman H. March. Monumental reference work covers fundamental theory of equilibrium properties of perfect crystalline solids, non-equilibrium properties, defects and disordered systems. Appendices. Problems. Preface. Diagrams. Index. Bibliography. Total of 1,301pp. 5⅜ x 8½. Two volumes. Vol. I: 0-486-65015-4 Vol. II: 0-486-65016-2

WHAT IS RELATIVITY? L. D. Landau and G. B. Rumer. Written by a Nobel Prize physicist and his distinguished colleague, this compelling book explains the special theory of relativity to readers with no scientific background, using such familiar objects as trains, rulers, and clocks. 1960 ed. vi+72pp. 5⅜ x 8½. 0-486-42806-0

CATALOG OF DOVER BOOKS

A TREATISE ON ELECTRICITY AND MAGNETISM, James Clerk Maxwell. Important foundation work of modern physics. Brings to final form Maxwell's theory of electromagnetism and rigorously derives his general equations of field theory. 1,084pp. 5⅜ x 8½. Two-vol. set. Vol. I: 0-486-60636-8 Vol. II: 0-486-60637-6

QUANTUM MECHANICS: PRINCIPLES AND FORMALISM, Roy McWeeny. Graduate student-oriented volume develops subject as fundamental discipline, opening with review of origins of Schrödinger's equations and vector spaces. Focusing on main principles of quantum mechanics and their immediate consequences, it concludes with final generalizations covering alternative "languages" or representations. 1972 ed. 15 figures. xi+155pp. 5⅜ x 8½. 0-486-42829-X

INTRODUCTION TO QUANTUM MECHANICS With Applications to Chemistry, Linus Pauling & E. Bright Wilson, Jr. Classic undergraduate text by Nobel Prize winner applies quantum mechanics to chemical and physical problems. Numerous tables and figures enhance the text. Chapter bibliographies. Appendices. Index. 468pp. 5⅜ x 8½. 0-486-64871-0

METHODS OF THERMODYNAMICS, Howard Reiss. Outstanding text focuses on physical technique of thermodynamics, typical problem areas of understanding, and significance and use of thermodynamic potential. 1965 edition. 238pp. 5⅜ x 8½. 0-486-69445-3

THE ELECTROMAGNETIC FIELD, Albert Shadowitz. Comprehensive undergraduate text covers basics of electric and magnetic fields, builds up to electromagnetic theory. Also related topics, including relativity. Over 900 problems. 768pp. 5⅝ x 8¼. 0-486-65660-8

GREAT EXPERIMENTS IN PHYSICS: FIRSTHAND ACCOUNTS FROM GALILEO TO EINSTEIN, Morris H. Shamos (ed.). 25 crucial discoveries: Newton's laws of motion, Chadwick's study of the neutron, Hertz on electromagnetic waves, more. Original accounts clearly annotated. 370pp. 5⅜ x 8½. 0-486-25346-5

EINSTEIN'S LEGACY, Julian Schwinger. A Nobel Laureate relates fascinating story of Einstein and development of relativity theory in well-illustrated, nontechnical volume. Subjects include meaning of time, paradoxes of space travel, gravity and its effect on light, non-Euclidean geometry and curving of space-time, impact of radio astronomy and space-age discoveries, and more. 189 b/w illustrations. xiv+250pp. 8⅜ x 9¼. 0-486-41974-6

STATISTICAL PHYSICS, Gregory H. Wannier. Classic text combines thermodynamics, statistical mechanics and kinetic theory in one unified presentation of thermal physics. Problems with solutions. Bibliography. 532pp. 5⅜ x 8½. 0-486-65401-X

Paperbound unless otherwise indicated. Available at your book dealer, online at **www.doverpublications.com**, or by writing to Dept. GI, Dover Publications, Inc., 31 East 2nd Street, Mineola, NY 11501. For current price information or for free catalogues (please indicate field of interest), write to Dover Publications or log on to **www.doverpublications.com** and see every Dover book in print. Dover publishes more than 500 books each year on science, elementary and advanced mathematics, biology, music, art, literary history, social sciences, and other areas.